GEOMETRICAL OPTICS AND OPTICAL DESIGN

OXFORD SERIES IN OPTICAL AND IMAGING SCIENCES

1. D. M. Lubman (ed)
   Lasers and Mass Spectrometry
2. D. Sarid
   Scanning Force Microscopy with Applications to Electric, Magnetic, and Atomic Forces
3. A. B. Schvartsburg
   Nonlinear Pulses in Integrated and Waveguide Optics
4. C. J. Chen
   Introduction to Scanning Tunneling Microscopy
5. S. Mukamel
   Principles of Nonlinear Optical Spectroscopy
6. J. R. Schott
   Remote Sensing: The Image Chain Approach
7. P. Mouroulis and J. Macdonald
   Geometrical Optics and Optical Design

GEOMETRICAL OPTICS AND OPTICAL DESIGN

Pantazis Mouroulis

Associate Professor
Center for Imaging Science
Rochester Institute of Technology

John Macdonald

Senior Lecturer
Physics Department
University of Reading

New York Oxford
OXFORD UNIVERSITY PRESS
1997

To my parents, Constantine and Anna Mouroulis

Oxford University Press

Oxford New York
Athens Auckland Bangkok Bogotá Bombay Buenos Aires
Calcutta Cape Town Dar es Salaam Delhi
Florence Hong Kong Istanbul Karachi
Kuala Lumpur Madras Madrid Melbourne
Mexico City Nairobi Paris Singapore
Taipei Tokyo Toronto

and associated companies in
Berlin Ibadan

Published by Oxford University Press, Inc.,
198 Madison Avenue, New York, New York, 10016

Oxford is a registered trademark of Oxford University Press

Library of Congress Cataloging-in-Publication Data
Mouroulis, Pantazis.
Geometrical optics and optical design /
Pantazis Mouroulis, John Macdonald.
p. cm. — (Oxford series on optical and imaging sciences)
Includes bibliographical references and index.
ISBN 0-19-508931-6
1. Geometrical optics.
2. Optical instruments—Design and construction.
I. Macdonald, John.
II. Title. III. Series.
QC381.M68 1996 535'.32—dc20 95-38009

2 4 6 8 9 7 5 3 1

Printed in the United States of America
on acid-free paper

Preface

This book is an attempt to make geometrical optics accessible to the common student and practicing engineer. Perhaps the recent history of geometrical optics can be seen from the point of view of textbook simplification, with an attendant increase in the number of optics practitioners. Early in the twentieth century, Conrady's *Applied Optics and Optical Design* set the standard for a complete, useful, strict, and unreadable book. But Conrady's book was perfectly appropriate for a time when lens designers had one of the most tedious jobs on earth. The mettle of the potential designer had to be tested; after surviving Conrady, the idea of spending days and nights in the company of log tables to trace one skew ray seemed perhaps not too daunting.

Computers, lasers, and all of the developments that have led to the so-called photonics revolution have caused a large increase in the number of optics practitioners. Often engineers with no formal optics training beyond that of basic physics find themselves in a position of having to deal with optics at an advanced level. Optical design software is now widely available, sometimes at an incredibly low price, and the mere availability is creating new users. Of equal importance is the fact that in the past couple of decades, the mentality, expectations, and background of the students have changed considerably. And as a rule, textbooks in geometrical optics have not kept pace with those developments.

Our aim is to provide an up-to-date treatment of the introductory aspects of geometrical optics that can be used by a variety of students at different levels, from undergraduate to introductory graduate. We also hope that those practicing engineers who find the advanced literature daunting will see this book as an easier-to-understand introduction to the topic. We do not expect this book to be the only one of its kind that the optical engineer will ever need; we simply hope that it will be the first book in the prospective engineer's long road to competence.

The book begins with a short introduction to the very basic concepts of rays, wavefronts, etc. This is not intended to compete with a first-year physics course, but to supplement the knowledge given in such a course by viewing it from a somewhat different vantage point.

The second chapter begins with exploring quantitatively the properties of plane reflective and refractive surfaces, and proceeds with a qualitative introduction to the basic properties of lenses and mirrors. It is expected that this chapter is primarily a review. Its main aim is to collect the scattered bits and pieces of the reader's optical knowledge into one coherent whole, before proceeding with the more rigorous development. The student should be expected to master such material quickly: While it is crucial to learn the alphabet, spending too much time on it comes only at the expense of further development.

The next three chapters develop the theory of first-order Gaussian optics. Chapter 3 examines image formation by a single surface and a thin lens. Chapter 4 examines lens systems and introduces some more advanced, yet fundamental, topics, such as principal planes and the optical (Lagrange) invariant. The strategy for teaching these two chapters should vary according to the level of the class. For an elementary class, Chapter 3 should be done at a deliberate pace; for this reason, it contains considerably more exercises than the previous two chapters. For a more advanced class, Chapter 3 would be primarily a review; it can be covered in the span of approximately 3 hours, by concentrating mostly on the sign convention, terminology, and notation, all of which must be very familiar before proceeding further.

Pupils and stops come in Chapter 5, followed by a section on radiometry and photometry that contains enough material to be useful in the majority of practical situations. This chapter also develops and expands the concepts of the marginal and pupil (principal) rays, which are the key to first-order system design. At this point, aided by the examples given, the reader has enough information to answer, in most cases, the following question and its variants: given an object, specify an optical system that will produce an image at a desired location, and of a specified size and irradiance, also taking into account any constraints on system size, accessibility of image/object locations, etc. However, the treatment of Gaussian optics is not really complete without expanding at least into the first two sections of Chapter 6 (telescope and microscope).

The Gaussian optics of optical instruments is a practically interminable topic; so one must be selective. In Chapter 6, we hope we have succeeded in treating a satisfactory number of topics at a sufficiently rigorous and accessible level. Inevitably, however, the section on diffractive elements offers only intuitive arguments, as we cannot develop diffraction theory in this book. The instructor can choose from among the topics of this chapter as interest dictates and time permits.

Next we approach the concept of aberrations and image quality, starting from the chromatic aberrations, which can be readily understood without new theory. However, monochromatic aberrations necessitate the development of new concepts. In the aberration chapters, we place heavy emphasis on the wavefront description. This approach simplifies and unifies the concepts, in addition to providing an immediate extension into the diffraction theory that is necessary for full evaluation of an optical system. Thus, although we do not treat the diffraction theory of aberrations, upon encountering it the reader should not find it alien.

In the aberration chapters, we have sacrificed a strict development of the theory on a couple of occasions, by offering abbreviated versions of some lengthy derivations. Including the full versions of the proofs would detract from the continuity of the text, confuse the reader, and ultimately offer very little to the literature, as enough such proofs can be found elsewhere. However, we have made an effort to include most of the important physical insights that are hidden in the proofs. In all of the aberration chapters, we have concentrated on rotationally symmetric systems, because they contain all of the fundamental concepts. The reader who wants more on aberrations is expected to resort to additional books and articles.

Chapters 8 and 9 attempt to do for aberrations what Chapter 5 did for Gaussian optics. At the end of Chapter 9, the reader has enough information to attack an image quality problem by setting up a preliminary design that does not violate any fundamental aberration rules and therefore has a chance of fulfilling the design requirements after optimization. Of course, this claim must be qualified by excluding systems with exotic aperture or field specifications that can be attacked only by experienced designers. Finally, Chapter 10 outlines the optical design process, extends the aberration discussion to higher orders, and discusses briefly topics such as off-axis pupil imagery, aspherics, and optimization.

There are six appendixes, four of which contain additional material, while the last two contain simple computer programs. The first four appendixes can be taught as desired by the instructor. Although we do not believe in the effectiveness of matrix methods as an instructional tool, we nevertheless have included them in Appendix 1 so that the reader can refer to them upon encountering a matrix-based development elsewhere. Appendix 2 on Gaussian beams is almost as long as a chapter; its importance is that it contains a treatment of Gaussian beams that is consistent with the rest of the book, rests on the concept of the optical invariant, and can be programmed into any computer using the usual paraxial ray-tracing equations of the marginal and pupil rays. In solving problems, we have found that the methods of Appendix 2 are as fast as or faster than matrix-based calculations, while the concepts are undoubtedly simpler.

In teaching optical design, our philosophy is that a proper understanding of the basics is more essential than the acquisition of fancy software. While advanced software is needed for advanced problems, we hope to have succeeded in showing that a lot of worthwhile optical design (as opposed to detailed lens design) can be done with simple means. But in any case, this is an introductory text; serious lens design begins where this book ends.

The emphasis on primary (or third-order) aberration theory that is placed in this book is not accidental. We believe that the understanding of optical system characteristics afforded by the third-order theory will become even more crucial as the practice of optical design moves increasingly towards modifying past designs with the help of expensive software. There is a danger that the student of today (and engineer of tomorrow) will be limited to a shallow knowledge of software usage, but will possess no real understanding of the underlying concepts. We therefore chose not to couple this book closely with any extensive software package, because doing so would entail one of the following two unattractive choices: doubling the size and scope of the book, or discussing several topics at a shallow level, without developing the appropriate theory.

We expect the book to be useful to students enrolled in undergraduate and graduate optics, or optics-related degree programs, and programs that have considerable optics concentration, typically physics and electrical engineering. The book contains enough material to provide the basis for at least two different courses, depending on whether the emphasis is placed on the first or on the last chapters. We also hope that the book will be of use to the increasing number of professionals who find that they need some optical design knowledge, but whose formal training did not include aberration theory and design.

In the course of writing this book, our teacher, Harold H. Hopkins, passed away. Harold, this book is a thank you note—and you are absolutely blameless for our mistakes!

Rochester, N.Y.
September 1995

P. M.

Contents

GEOMETRICAL OPTICS AND OPTICAL DESIGN

Rays and the Foundations of Geometrical Optics

1

"And I also further observed that . . . a smaller hole would make a larger *Penumbra*, which is contrary to the common Principles of Opticks: For if the Rays went in straight Lines, the bigger the Hole were, the bigger would be the *Penumbra*."

Robert Hooke, in *The Posthumous Works of Robert Hooke*, collected by Richard Waller, London, 1705. (Reprinted by Johnson Reprint Corp., New York, 1969.)

1.1. WAVES, WAVEFRONTS, AND RAYS

Light is an electromagnetic (EM) wave. The electric and magnetic field vectors obey the wave equation separately and are related through Maxwell's equations.* Traditionally, we use the electric field vector to represent the wave; so for a simple harmonic wave we can write:

$$\mathscr{E} = \mathscr{E}_0 \cos(\omega t \mp kx). \tag{1.1}$$

In this equation, $\mathscr{E}_0$ is the amplitude of the wave, $\omega = 2\pi\nu$ the angular frequency, $k = 2\pi/\lambda$ the propagation constant, and λ the wavelength. The propagation of waves is commonly described by means of *wavefronts*. A wavefront is a surface across which the phase of the wave, $\phi = \omega t \mp kx$, is constant. The condition $d\phi = 0$ then yields the speed with which the wavefronts propagate

$$v = \frac{dx}{dt} = \pm \frac{\omega}{k} \tag{1.2}$$

v is called the phase velocity. The positive sign on the right-hand side of Eq. (1.2) [or the negative sign in Eq. (1.1)] corresponds to a wave progressing

* The wave optics concepts with which we assume some familiarity can be found in several texts. At a basic level, a good example is M. Young's *Optics and Lasers*, Springer-Verlag, and at a more advanced level, R. Guenther's *Modern Optics*, Wiley.

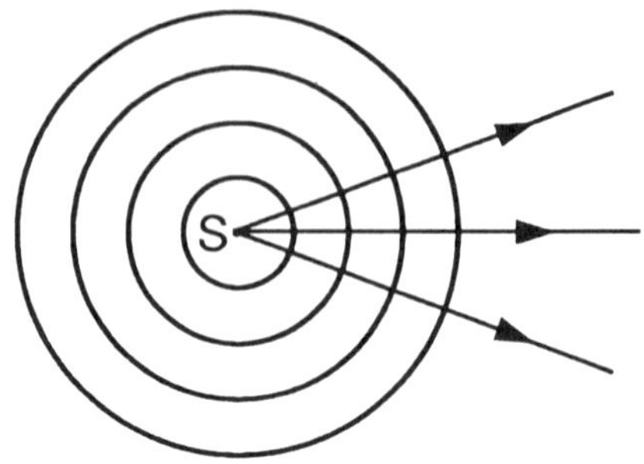

Figure 1.1. Wavefronts (plane section) and rays from a point source S. The wavefronts are concentric spheres centered on S, propagating outward with a velocity ω/k.

along the positive x axis; that is, $v > 0$, if ω and k are taken as positive. In geometrical optics, the concept of wavefront is fundamental. The *geometrical wavefronts* can be thought of as a snapshot of the physical wavefronts; in other words, the difference between the two is that geometrical wavefronts do not propagate, but are frozen in space. Figure 1.1 shows a point source, which gives rise to spherical wavefronts, centered on the source.

The direction of propagation of EM energy is given by the Poynting vector, which is the cross product of the electric and magnetic field vectors. The direction of energy flow is conveniently characterized by *rays*. For our purpose, it suffices to define the rays as lines normal to a family of wavefronts (Figure 1.1). While the geometrical wavefront has a direct physical counterpart, the same is not true of the rays; they are best thought of as a convenient geometrical abstraction.

"By the Rays of Light I understand its least Parts . . ." wrote Newton in his *Opticks*. Newton needed a means of describing the properties of light and found the concept of rays invaluable in describing geometrical optics, which in his day was most of optics. But, along with the average contemporary person, Newton incorrectly assumed that light is made of rays, as if a wide beam of light could be broken down to a bunch of very narrow beams, the rays.

Rays are not very narrow beams of light. In fact, there is really no such thing as a very narrow beam of light! A laser beam might give one the illusion of a ray. But close examination of a laser beam reveals in fact that it diverges; its cross section increases with distance. In any case, the typical laser beam more than 1 mm in diameter would not qualify as "very narrow." What if we let the beam pass through a smaller aperture, say 100 μm in diameter? Figure 1.2 gives the answer: The beam would diverge rapidly, due to *diffraction*.* It is a well-known result of physical optics (since Hooke's time at least) that the smaller the aperture, the larger the divergence.

All of this book is concerned with rays in isotropic, linear media, usually glass. In these media, our definition of rays is adequate. There are, however,

* Recent research has shown that there exist certain nonuniform beam profiles that minimize the spread of the beam over large distances; such beams are sometimes (inappropriately) called diffractionless. They have no impact on our discussion.

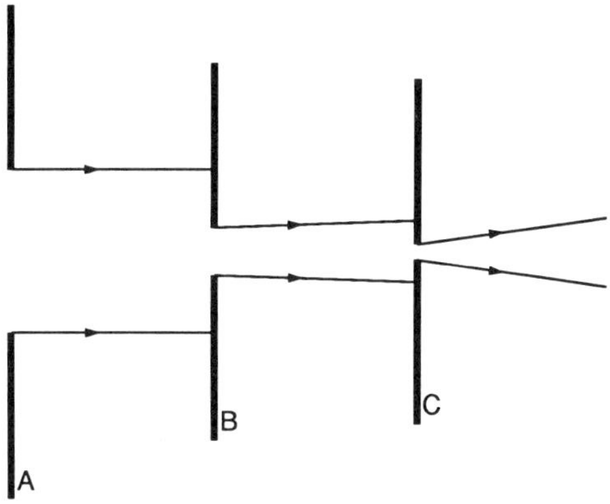

Figure 1.2. Attempt to isolate a ray through successively smaller apertures. A "parallel" beam emerges from some optical apparatus at A. Because of diffraction, the beam is already diverging after A, but the divergence is small. As the beam passes through smaller apertures, the divergence increases.

media in which the direction of propagation of EM energy is not always that of the normal to the wavefronts. Examples of such media are all crystals except those in the cubic class and stressed isotropic materials. In nonlinear media, it is even possible for light to exit the medium with a wavelength different from that with which it entered.

1.1.1. The Pinhole Camera

The operation of the pinhole camera is readily understood through the concept of rays. The pinhole camera is simply a box with a pinhole, as in Figure 1.3. The image is formed on the rear wall of the box, and the magnification is given by the ratio of the distances l and l', as indicated. The first reference to the pinhole camera was by Mo Ti in China, fifth century B.C. In later years, it became very popular for rendering landscapes. In the sixteenth and seventeenth centuries it became known as *camera obscura*, or dark room. Pinhole cameras were occasionally made large enough for a person to stand inside; artists used them to copy the contours of the image projected onto

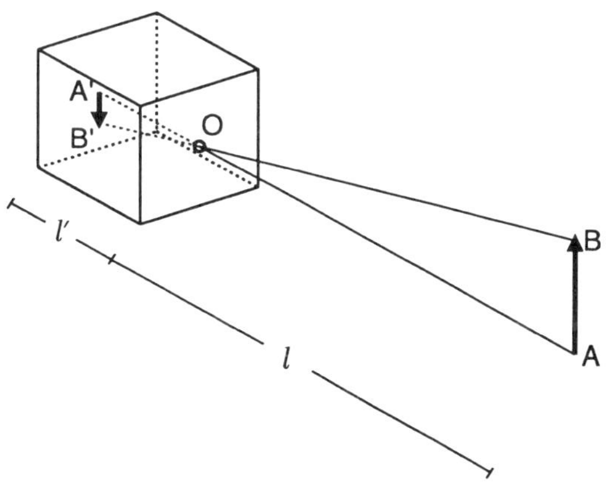

Figure 1.3. Illustration of the pinhole camera. The object, arrow AB, is imaged inverted on the screen. From the similar triangles ABO and $A'B'O$, we obtain the magnification $(A'B')/(AB)$ as l'/l.

the screen. At around that time a lens was introduced in place of the simple hole.

A fascinating demonstration of the pinhole principle in nature is described by M. Minnaert (1954). The bright elliptical patches seen on the ground when the sun's rays go through the dense foliage of a tree are in fact pinhole images of the sun, projected on a tilted plane, the ground. The pinholes are formed by small gaps between the leaves. Beeches and sycamores work best! Aristotle, in the fourth century B.C., reported seeing a crescent-shaped image of the sun during a solar eclipse, through this natural pinhole camera.

The pinhole camera is strikingly, and perhaps deceptively, simple. But it only works well for a rather narrow range of pinhole sizes. Clearly, if the opening is too large, no image is formed, because rays from a single object point may impinge anywhere within the projection of the opening on the rear wall. If the pinhole is too small, then diffraction effects become important; the rays will diverge rapidly after the pinhole, so light from a given part of the object will again illuminate a large part of rear wall. Thus the pinhole camera provides at once a demonstration of geometrical optics and its limitations.

The methods of geometrical optics are useful only when the wavelength of light can be considered as small. Geometrical optics does not work near the edges of a shadow, near the focus of a beam, and near a source. How far is "not near"? A round number to remember is 100λ. If the diameter of an area or the distance involved in an optical effect is less that about 100λ, geometrical optics is generally of little help; we must then apply the methods of physical optics.

1.2. PROPAGATION OF WAVEFRONTS, REFLECTION, REFRACTION

The speed of propagation of light inside a medium, v, is characterized by the refractive index of the medium, defined as

$$n = \frac{c}{v} \tag{1.3}$$

where c is the speed of light in vacuum. An optically dense medium is one of high refractive index; the denser the medium, the slower the speed of light in it. If we use the subscript zero to denote the quantities k and λ in vacuum, we have:

$$\omega = ck_0 = vk$$

therefore

$$\lambda = \frac{v}{c}\lambda_0 = \frac{\lambda_0}{n}$$

so that the wavelength of light is shorter inside a medium. We usually take the refractive index of air to be equal to one, but this is only an approximation. For a pressure of 1 atm, at a temperature of 14.5°C and a wavelength of 589 nm, air is found to have a refractive index of 1.0002929. For the rest of this book, we will take $n_{\text{air}} = 1$, but the reader should be aware of the level of approximation. In optical microlithography, where lenses are pushed to the limit of their performance, changes in the atmospheric conditions and the corresponding changes in the refractive index of the air must be taken into account, and preferably eliminated by keeping the system in a controlled environment.

The refractive index of a medium is a function of wavelength. This variation is called *dispersion*. The refractive index generally decreases with wavelength, except in the so-called regions of anomalous dispersion, which coincide with the absorption bands of the medium. As we will deal with essentially transparent materials, anomalous dispersion need not concern us here. The variation of n with λ makes it possible to observe the spectral colors through a prism. We will see later that the focal length of a lens depends on the refractive index of the glass from which the lens is made, and therefore also on the wavelength. This can have a profound effect on the image quality of the lens.

The propagation of wavefronts can be understood through a construction attributed to Huygens, shown in Figure 1.4. We consider each point on the current wavefront to emit a secondary spherical "wavelet," the radius of which depends on the speed of light in the medium. The new wavefront is found as the common tangent to all such wavelets. For convenience, we may take the radius of the wavelets to be λ. The procedure is shown here from an intuitive point of view, but it does have a solid mathematical foundation. It demonstrates that the propagation of EM energy can be considered as a diffraction process.

From Huygens's principle, and our definition of rays as lines normal to the wavefronts, we can obtain the laws of reflection and refraction rather easily. Figure 1.5 shows the construction for the case of reflection. The reader can

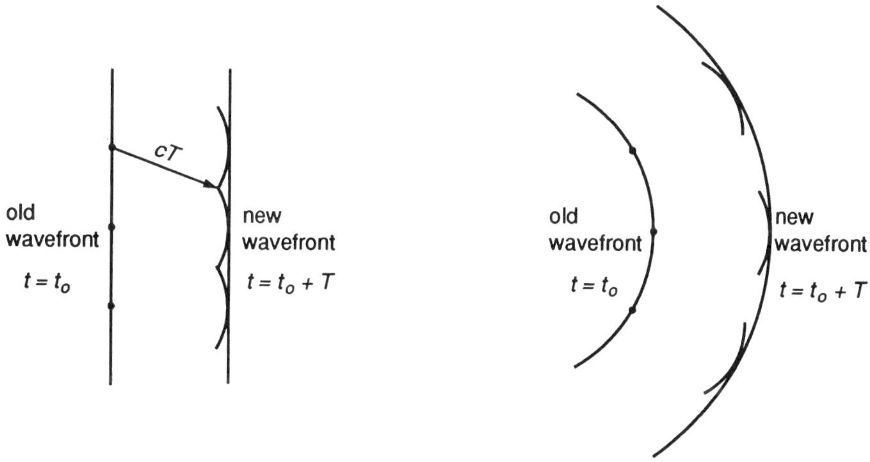

Figure 1.4. Illustration of Huygens' construction to describe the propagation of a wavefront. The radius of the wavelets is taken as $\lambda = cT$.

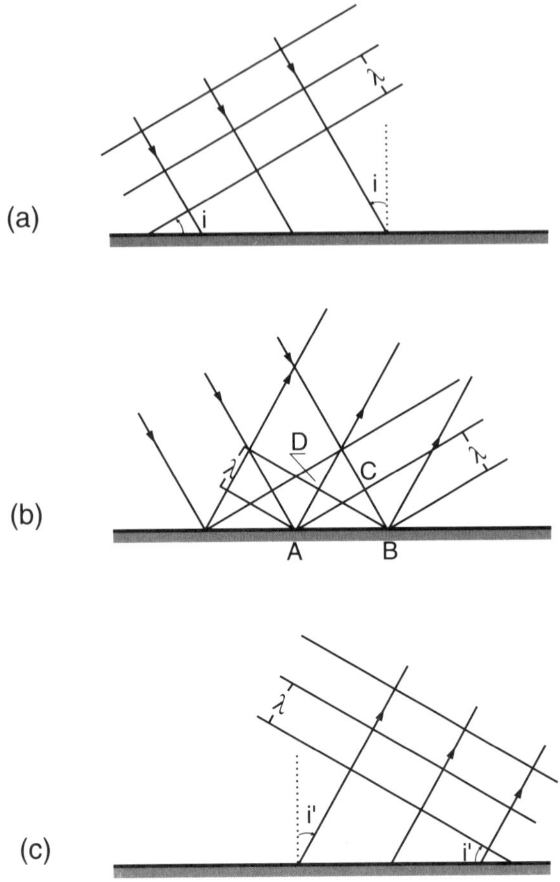

Figure 1.5. Reflection of a plane wave as described by the Huygens construction. In (b), the reflected wave goes from A to D, at the same time as the incident wave goes from C to B. Triangles ABD and ABC are congruent, giving $i = -i'$.

verify that the angles i and i' are equal. But according to the coordinate geometry convention, they are of opposite sign; so we write

$$i = -i' \tag{1.4}$$

The angles of incidence and reflection are measured from the surface normal to the ray. The incident and reflected rays will lie on the same plane with the surface normal.

Figure 1.6 shows the construction for the case of refraction. Light encounters an interface between two media, of refractive indices n_1 and n_2. Some of the incident light will be reflected, but we omit the reflected wave for clarity. Let us assume $n_2 > n_1$; so light will slow down upon entering medium 2. The wavelength will also become shorter. If we consider the successive wavefronts to be one wavelength apart, it follows that

$$n_1 \sin i_1 = n_2 \sin i_2 \tag{1.5}$$

through the sine law for triangles ABC and ABD. Equation (1.5) is the law of

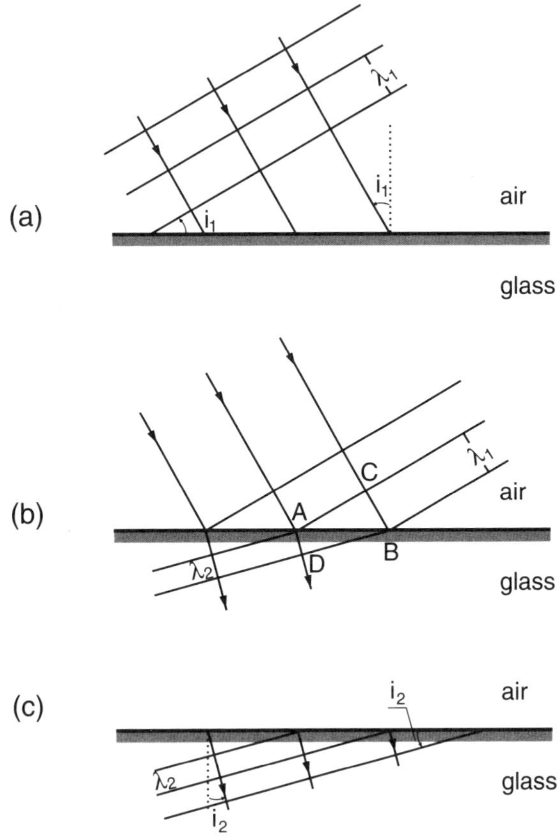

Figure 1.6. Refraction of a plane wave as described by the Huygens construction. In (b), the length $(CB) = \lambda_1$, and $(AD) = \lambda_2 < \lambda_1$ for $n_2 > n_1$. While the incident wave travels from C to B, the refracted wave travels the shorter distance (AD). We have: $\sin i_1 = (CB)/(AB)$ and $\sin i_2 = (AD)/(AB)$. Since $(CB) = \lambda_1 = v_1/v = c/n_1 v$ and $(AD) = c/n_2 v$, Snell's law follows.

refraction or Snell's law. The angles i_1 and i_2 are the angles of incidence and refraction, respectively. These angles are also measured from the surface normal. In addition, the refracted ray is along the plane defined by the incident ray and the surface normal.*

As a consequence of Snell's law, a ray will approach the normal when entering a dense medium, and will move away from the normal when entering a rear medium. Strictly speaking, refraction is not the change in angle, but the change in the speed of light, which then leads to the change in angle, as Huygens' construction shows. Refraction still occurs if light is incident normally on a surface ($i_1 = 0$), even though there is no change in angle. However, a loose terminology is sometimes applied, and one can hear that a ray incident normally on the surface is not refracted. Properly, one should say that the ray is not deviated.

* Snell's law holds in isotropic media. However, in anisotropic media we normally have two refracted rays, one of which many not obey Snell's law. It is even possible to have an infinity of allowable refracted ray directions that do not all lie on the same plane. This can occur in crystals belonging to the orthorhombic, monoclinic, and triclinic classes. These exceptions are beyond our current scope, but the reader should be aware that our development is restricted to isotropic media..

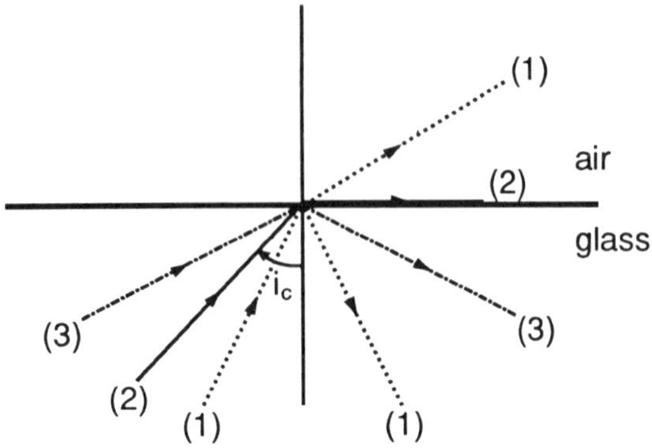

Figure 1.7. Illustration of the critical angle i_c and total internal reflection. For ray 1, part of the energy is reflected and part transmitted. For ray 3, all the energy is reflected.

When light enters a rare medium at increasingly larger angles of incidence, there will be an angle of incidence for which the corresponding angle of refraction is 90° (Figure 1.7). For larger angles of incidence, light will no longer be refracted, but instead it will undergo total internal reflection (TIR). The word *total* implies that all of the incident energy is reflected and none transmitted. If the angle of refraction is 90°, the corresponding angle of incidence is called the *critical angle* (i_c). Applying Snell's law, and assuming the rare medium to be air, we have

$$i_c = \sin^{-1}\left(\frac{1}{n}\right)$$

If the rare medium is not air, the value of the critical angle is given by $n_1 \sin i_c = n_2$. Some applications of total internal reflection can be found in reflecting prisms, discussed later, and in fiber optics.

1.3. FERMAT'S PRINCIPLE

The idea that light travels along straight lines has been prevalent through the ages. The most obvious optical phenomenon, shadows, is explained through rectilinear propagation, though not fully. Plane geometry, which was well developed in antiquity, says that the straight line is the shortest path between two points. This was probably why Hero of Alexandria asserted that light must travel along the shortest possible path between two points. Hero was indeed able to describe the phenomenon of reflection using this idea.

Later the idea was recast in terms of time of travel. Pierre de Fermat (1601–65) was able to derive the law of reflection through the principle that light travels along the path of minimum time, which is not necessarily identical with the shortest path, as the speed of light varies from one medium to another. For light, the time of travel is proportional to the *optical path*, rather than the geometrical path. Thus, if light travels between points A and B inside a medium of refractive index n, the time of travel t is

$$t = \frac{AB}{v} = \frac{n(AB)}{c} = \frac{[AB]}{c}$$

where

$$[AB] \equiv n(AB) \tag{1.6}$$

or, the optical path $[AB]$, denoted by square brackets, is the product of the geometrical path (AB) and the refractive index n.

Today we know that Fermat's original formulation is incomplete. The principle that governs the propagation of light is still known as Fermat's principle, but it states that *the time of travel is stationary*, rather than minimum. We will call a quantity stationary with respect to a variable when its first derivative is zero. If no further conditions are imposed on the second derivative, we do not know whether we have a minimum, maximum, or point of inflection. This indeed means that the time that light takes to travel from point A to point B may be maximum. Millions of perfectly respectable rays take maximum paths every day! For example, consider a concave reflector as in Figure 1.8. Let us say that light is to travel from point A to point B after reflection at Q. Let us also draw an ellipse, with foci A and B, going through Q as shown. As the radius of curvature of the mirror is arbitrary, it is clearly possible to have a mirror that lies inside the ellipse. For this case, taking a point P in the vicinity of Q, and applying a well-known property of the ellipse, we have

$$(AQ) + (QB) = (AQ') + (Q'B) > (AP) + (PB)$$

Since AQB represents the actual ray path, for which the law of reflection holds, it follows that the path AQB is a maximum.

It is more useful to restate Fermat's principle in terms of the optical path, rather than time of travel. Thus, using the symbol P for the optical path, Fermat's principle is written as

$$\frac{\partial P}{\partial q_i} = 0 \tag{1.7}$$

where q_i stands for the variables specifying the optical path, such as the (x, y, z) coordinates of points along the ray, or the (L, M, N) direction cosines of the ray.

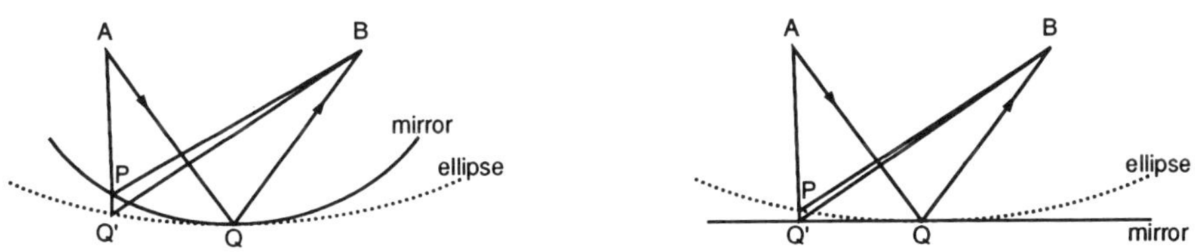

Figure 1.8. Illustration of Fermat's principle of the stationarity of the optical path. The optical path $[AQB]$ is maximum in (a) and minimum in (b).

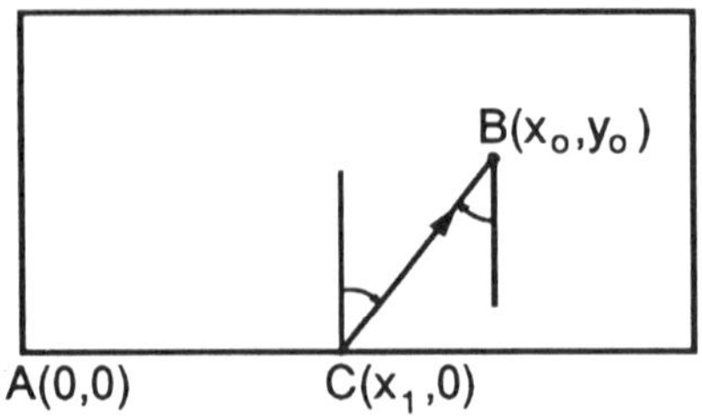

Figure 1.9. The lifeguard problem. The lifeguard, standing at A, must get to point B in the shortest possible time.

Using Fermat's principle, Eq. (1.7), we can derive the laws of reflection and refraction. The derivation is shown in several texts, for example, Welford (1986) or Möller (1988), and is left as an exercise for the student. The procecure shown in the following example should be followed.

Example 1.1. The lifeguard problem. Richard Feynman, in his famous *Lectures on Physics*, used a variation of the following example to illustrate Fermat's principle. In Figure 1.9, a lifeguard is standing at one end of a swimming pool. At the deep end of the pool, point B with coordinates (x_0, y_0), someone is drowning. The lifeguard can run with velocity v_1 and swim with velocity v_2 $(v_1 > v_2)$. Determine the point C $(x_1, 0)$ at which the lifeguard should dive to get to the drowning person as soon as possible.

There are two ways of solving this problem. The first way, solid but unimaginative, involves writing an expression for the time t it takes the lifeguard to reach B, $t = t(v_1, v_2, x_1, x_0, y_0)$, and then taking $\partial t/\partial x_1 = 0$. The time that the lifeguard spends running is $t_1 = x_1/v_1$. The time that she spends swimming is $t_2 = (BC)/v_2$, where $(BC) = [(x_0 - x_1)^2 + y_0^2]^{1/2}$. The total time is $t = t_1 + t_2$. Upon substituting and taking the derivative $\partial t/\partial x_1 = 0$, we obtain an expression for x_1 as a function of the known coordinates and velocities. In other words, this is an application of Fermat's principle.

The second way is to realize that this is essentially a critical angle problem. The lifeguard's path is the same as the path that would have been traversed by a ray of light, if light had velocities v_1 and v_2 in the two media. The lifeguard's path is governed by the same principle as the optical path.* Therefore, we can write:

$$\sin i_c = \frac{v_2}{v_1} \qquad \text{and} \qquad (B'C) = x_0 - x_1 = y_0 \tan i_c$$

By eliminating i_c from these two equations, x_1 can be determined.

Fermat's principle is the foundation of geometrical optics as it governs the behavior of rays. However, it is not a fundamental principle. The fundamental equations that govern the behavior of light are Maxwell's equations, and

* Well, not exactly. Clearly the lifeguard's path must always be minimum, not just stationary. But in this case, the maximum path is ill-defined; so only the minimum condition is meaningful.

Figure 1.10. Alternative form of Fermat's principle.

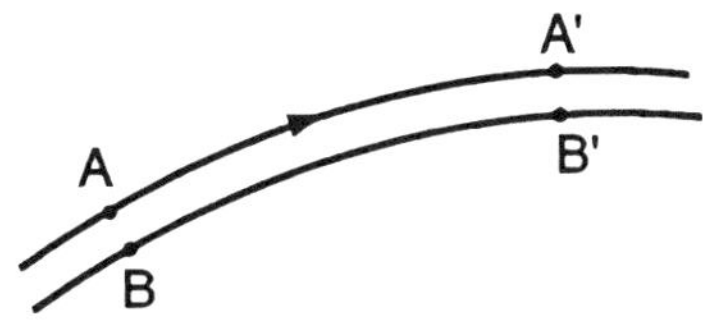

Fermat's principle can be derived from them (see, for example, Born and Wolf, 1975). In the same sense, Fermat's principle can be restated in terms of the phase of the wave, rather than the path of rays. So from a physical viewpoint, it is the phase $\phi = \omega t \pm kx$ that is stationary. Clearly, though, the two statements are mathematically identical, since at any given time t the phase difference is simply $2\pi/\lambda_0$ times the optical path difference.

We give here an alternative formulation of Fermat's principle that will be useful in proving some theorems later. Referring to Figure 1.10, let $A \cdots A'$ be a ray, specified by coordinates (x, y, z) and direction cosines (L, M, N), and let $B \cdots B'$ be a nearby path, not necessarily a ray, with coordinates $(x + \delta x, y + \delta y, z + \delta z)$ and direction cosines $(L + \delta L, M + \delta M, N + \delta N)$. Fermat's principle can now be stated as follows: The difference in optical path length between the ray and the nearby path is at least of the second degree in the variables specifying the nearby path with respect to the ray, that is,

$$\delta P \equiv [A \cdots A'] - [B \cdots B'] = a\,\delta x^2 + b\,\delta y^2 + c\,\delta z^2 + d\,\delta L^2 + \cdots + g\,\delta x\,\delta y + \cdots + O(3) = O(2) \tag{1.8}$$

where $O(m)$ is shorthand for a sum of terms of order m or higher in the differences.

At the limit, as $[B \cdots B']$ approaches $[A \cdots A']$, all of the differences become differentials and any first derivative of P becomes zero; that is, we obtain Eq. (1.7).

By defining the rays as normals to the wavefronts, our development of geometrical optics owes its validity to the physical concepts of the wavefront and the Poynting vector. It is possible, however, to develop an equally valid description of geometrical optics based on the stationarity of the optical path, that is, Fermat's principle. In this development, which has been of great historical importance, the rays are the primary concept and the wavefronts are derived from the rays. It then has to be proven that there exists a surface that is normal to all rays. This is clearly not true for an arbitrary family of lines. A family of lines that can be intercepted at right angles by a single smooth surface is called a *normal congruence*. Rays emanating from a point source are clearly a normal congruence; the theorem of Malus and Dupin (Born and Wolf, 1975) states that they will remain a normal congruence after any number of refractions and reflections from smooth surfaces.

1.4. IRRADIANCE AND THE INVERSE-SQUARE LAW

We have already identified the direction of propagation of EM energy with the rays. But rays can also be used to calculate the distribution of optical power incident on a given surface. To show this, consider a point source emitting total

Figure 1.11. The irradiance is proportional to the ray density.

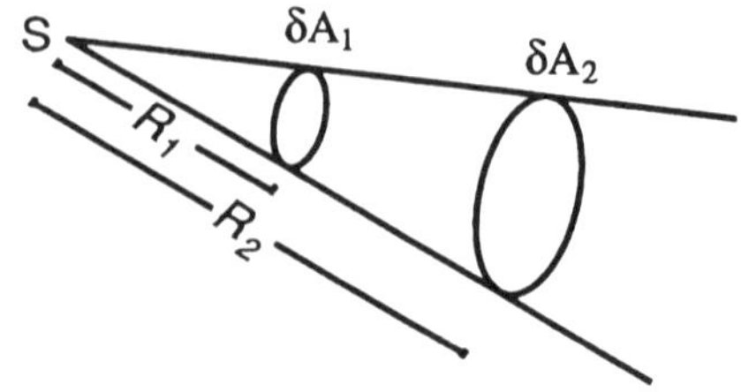

power Φ. The irradiance is defined as the power incident on a surface per unit area. For a sphere of radius r, the total area is $4\pi r^2$; so the irradiance will be given by

$$E = \frac{\Phi}{4\pi r^2} \tag{1.9}$$

Equation (1.9) is one way of expressing the famous inverse-square law. In the words of Robert Hooke, "Light . . . doth decrease in Duplicate Proportion to its Distance of Propagation from the Luminous Body."

Instead of considering the entire sphere, we may simply take the power confined in a small cone, as in Figure 1.11; the area intercepted by the cone increases as the square of the distance from the source (center of the sphere). We now define the ray density ρ as the number of rays per unit area:

$$\rho = \frac{\delta N}{\delta A} \tag{1.10}$$

where δN is the number of rays inside the small cone (this number is a constant, because of rectilinear propagation). For the area elements δA_1 and δA_2, the ray densities are $\rho_1 = \delta N/\delta A_1$ and $\rho_2 = \delta N/\delta A_2$. Therefore,

$$\frac{\rho_2}{\rho_1} = \frac{\delta A_1}{\delta A_2} = \frac{R_1^2}{R_2^2} \tag{1.11}$$

Thus, like the irradiance, the ray density obeys the inverse square law; so we may set $\rho \propto E$. This result can be extended to an arbitrary ray configuration, that is, for any number of reflections or refractions. We avoid here a strict proof of this theorem; the interested reader may consult Born and Wolf (1975), Section 3.1.2.

The preceding discussion illustrates the link between radiometry and geometrical optics. It pays at this point also to develop the link between radiometry and electromagnetic theory, which is ultimately our starting point. The irradiance associated with a light wave is defined as the time-averaged value of the magnitude of the Poynting vector, $\mathscr{P}$. For nonmagnetic media, we have the following results from basic physics:

$$\mathscr{H} = n\mathscr{E}\left(\frac{\varepsilon_0}{\mu_0}\right)^{1/2} \tag{1.12}$$

and

$$|\mathscr{P}| \equiv \mathscr{P} = \mathscr{E}\mathscr{H} = n\mathscr{E}^2\left(\frac{\varepsilon_0}{\mu_0}\right)^{1/2} \tag{1.13}$$

where $\mathscr{E}$ and $\mathscr{H}$ are the instantaneous amplitudes of the electric and magnetic fields, respectively, and n the refractive index of the medium. The irradiance E is

$$E = \frac{1}{\tau}\int_{-\tau/2}^{\tau/2} \mathscr{P}\, dt \tag{1.14}$$

where τ the integrating (averaging) time of the photodetector, assumed large compared to the period of the light wave. Substituting from Eq. (1.13) into Eq. (1.14),

$$E = n\left(\frac{\varepsilon_0}{\mu_0}\right)^{1/2} \frac{1}{\tau}\int_{-\tau/2}^{\tau/2} \mathscr{E}^2\, dt \tag{1.15}$$

Substituting for $\mathscr{E}$ for a simple harmonic wave [Eq. (1.1)] and noting that the average value of $\cos^2$ is $\frac{1}{2}$, we have for the irradiance:

$$E = \tfrac{1}{2}n\mathscr{E}_0^2\left(\frac{\varepsilon_0}{\mu_0}\right)^{1/2} \tag{1.16}$$

This last equation links radiometry and electromagnetic theory by relating the measured irradiance to the electric field amplitude.

1.5. THE BASIC POSTULATES OF GEOMETRICAL OPTICS

We can now summarize the set of propositions on which our development is based:

1. Rays are normal to the wavefronts and vice versa.
2. The rays satisfy the laws of reflection and refraction.
3. The optical path is stationary with respect to the variables that specify it.
4. The optical path lengths along any rays between two wavefronts are equal.
5. The irradiance at any point is proportional to the ray density at that point.

Proposition 4 is a direct consequence of the definitions of wavefronts and rays. We can consider a point and its image through an ideal optical system to be wavefronts of zero area at the limit; thus we may also state that the optical path difference between a point and its image along any two rays is zero.

The following corollaries of Fermat's principle must be added here:

1. In a uniform medium, light travels along straight lines.
2. The optical paths are reversible; that is, if light can travel from point A to A', then it can also travel from A' to A via the same path.
3. The optical path difference between two neighboring rays is zero.

Regarding the last corollary, notice that the stationarity of the optical path expressed through Eq. (1.8) relates the difference in optical path between a ray and a nearby path that is not a ray. If, however, the nearby path is also a ray, then the difference in optical path must be zero, otherwise the stationary value would be ill-defined.

Finally, we should remember that the ray description will be expected to hold away from a focus and away from the edges of shadows.

PROBLEMS

1. Calculate the critical angle for a water–air interface ($n_{\text{water}} = 1.33$) and for a glass–air interface with $n_{\text{glass}} = 1.82$.

2. The following version of the so-called Lorentz–Lorenz formula can be used to find the variation of the refractive index of a gas with temperature or pressure:

$$A \cong \frac{RT}{p}\frac{n^2 - 1}{3}$$

   where R the molar gas constant (8.31441 J/mol K) and $A \cong 4.6 \times 10^{-6}$, the so-called molar refractivity, which can be taken as approximately independent of pressure. Using this formula, determine the change in the refractive index of the air between a rainy day (low pressure) and a clear day (high pressure). Use typical values for atmospheric pressure and assume a temperature of 14.5°C.

3. A man is 1.80 m tall, and he can just see his full-length image in a vertical plane mirror 3 m away. His eyes are 1.7 m from floor level. Determine the vertical length of the mirror, and the height of the bottom of the mirror from the floor.

4. Figure 1.12 shows a meridional section of a cylindrical glass rod and of a *multimode step-index optical fiber*. (A meridional section is one that contains

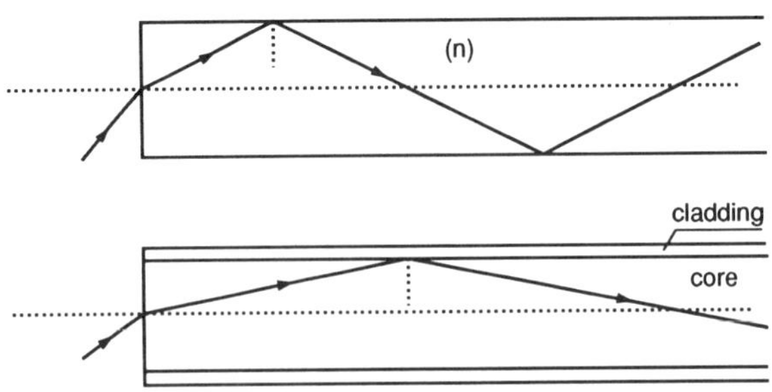

Figure 1.12

the axis.) If a ray is incident at the glass–air or core-cladding interface at an angle greater than the critical angle, it will be confined in the rod/fiber. Determine the critical angle inside the glass rod for $n = 1.52$, and inside the fiber for $n_{\text{core}} = 1.51$ and $n_{\text{clad}} = 1.49$. Notice that the critical angle condition also determines the maximum angle of incidence for the ray entering the fiber, if it is to be confined in the fiber core. Calculate this angle. Then attempt the same calculation for the rod. What do you conclude?

5. The refractive index of a glass is given by the following empirical relation (known as the Cauchy formula):

$$n = c_1 + \frac{c_2}{\lambda^2}$$

where $c_1 = 1.5$, $c_2 = 3 \times 10^4$ (nm^2). Determine the critical angle for the glass–air interface at $\lambda_1 = 650$ nm and $\lambda_2 = 450$ nm.

6. By applying the principle of stationarity of the optical path, derive Snell's law at an interface between two media of refractive indices n_1 and n_2. You may assume that the refracted ray is restricted to the plane of incidence from symmetry considerations.

7. For the right-angle prism shown in Figure 1.13, determine the maximum angle of incidence θ for the ray to be totally internally reflected at the hypotenuse face. Take the refractive index of the glass to be 1.52. First show that a ray with $\theta = 0$ will be totally internally reflected. (Notice that this effect places a limit on the convergence or divergence angle of the incident beam, if the prism is to be used as a reflector.)

Figure 1.13

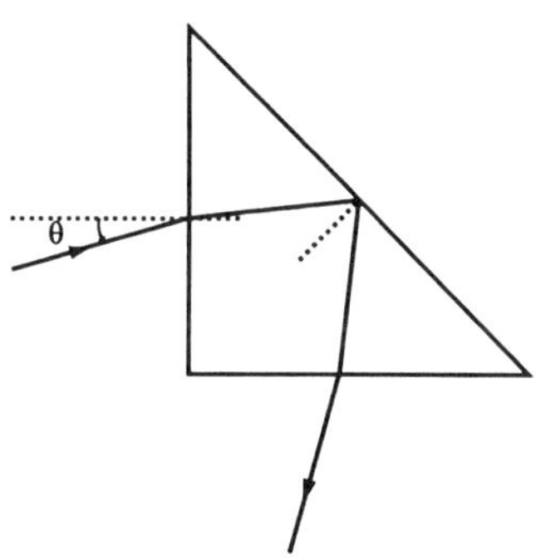

8. Calculate the magnitude of the electric field produced by a 100-W light bulb at a distance of 2 m.

Review of Elementary Ray Optics

2

"But marvellouse are the conclusions that may be performed by glasses concave and convex of circulare and parabolicall fourmes..."

Leonard Digges, *A Geometrical Practise, named Pantometria,* London, 1571.

The optical components that concern us in this book belong to one of the following three categories:

1. Reflective (mirrors, prisms);
2. Refractive (lenses, prisms);
3. Diffractive (gratings, holographic and binary optical elements).

A plane mirror may be used simply to redirect the radiation; a lens or a curved mirror is used to focus the radiation or form an image. A diffraction grating is used to disperse the radiation, that is, separate it into its component wavelengths. A holographic optical element is used to redirect and possibly focus the radiation; it can be thought of as a hybrid of grating and lens and can be used in either capacity. Diffractive components have some special characteristics and are discussed in Chapter 6.

2.1. PLANE SURFACES

We can derive some important properties of plane mirrors and refracting surfaces using only the laws of reflection and refraction and plane geometry. We give these properties in the form of solved problems or theorems. The treatment is meant as a review and is not exhaustive. Some further aspects of reflection relative to reflecting prisms are discussed in Chapter 6.

2.1.1. Reflecting Surfaces

Problem 2.1. For a plane mirror, derive the law of reflection from Fermat's principle.

Solution. In Figure 2.1, we assume that light will travel from point A to point B through a point C on the mirror surface, and we shall prove that the point C is such that $i_1 = -i_2$. Define a convenient coordinate system with its origin at A and with the x axis along the perpendicular to the mirror (AE). If the mirror is in air, the optical path $P = (AC) + (CB)$, where $(AC) = y/\sin i_1$ and $(CB) = -(y_0 - y)/\sin i_2$. The negative sign in the last equation is due to the fact that i_2 is a negative angle. Fermat's principle can be written as $\partial P/\partial y = 0$. Upon making the substitutions, we obtain $1/\sin i_1 + 1/\sin i_2 = 0$, from which $i_1 = -i_2$. Q.E.D.

Problem 2.2. A beam of light is incident on a plane mirror. The angle of incidence is θ. If the mirror is then tilted by an angle ϕ, show that the reflected ray forms an angle 2ϕ with its original direction.

Solution. From the law of reflection, we see that the total ray deviation (that is, the angle between the incident and the emergent ray) is twice the angle of incidence. Therefore, before tilting the mirror, the reflected ray forms an angle 2θ with the incident ray. After tilting, the new angle of incidence will be $\phi + \theta$, and the new reflected ray will be at $2(\phi + \theta)$ from the incident. The difference between the two positions of the reflected ray is then $2(\phi + \theta) - 2\theta = 2\phi$.

Problem 2.3. Consider two plane mirrors forming an angle ϕ, as in Figure 2.2. Show that the ray deviation is independent of the angle of incidence if the ray escapes after two reflections.

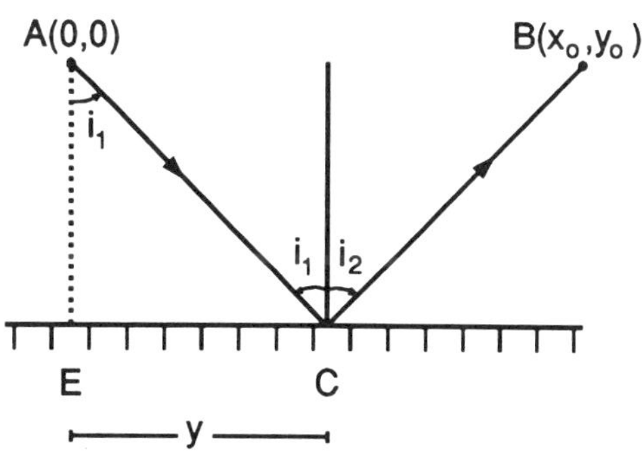

Figure 2.1. The law of reflection from Fermat's principle.

Figure 2.2. $\delta = 2\phi$ for any angle of incidence i_1 such that the ray escapes after two reflections.

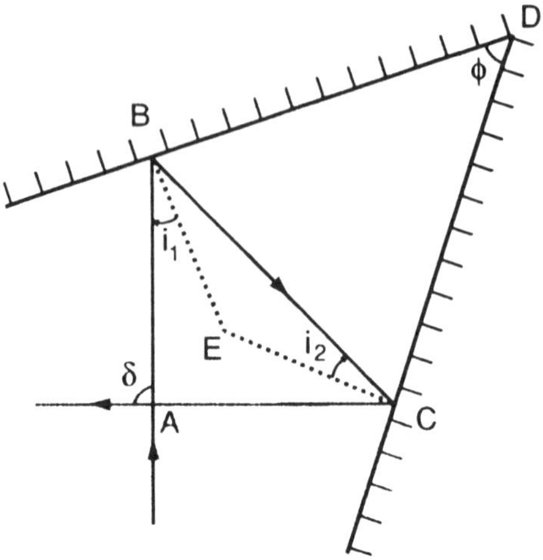

Solution. The ray deviation is δ. From triangle (ABC) and using the law of reflection, we have:

$$\delta = 2i_1 + 2i_2 \tag{2.1}$$

From quadrangle ($BECD$), we obtain the angle $B\hat{E}C = \pi - \phi$. From triangle (BEC), we have $i_1 + i_2 + (\pi - \phi) = \pi$, or $i_1 + i_2 = \phi$, and therefore, from Eq. (2.1), $\delta = 2\phi$. (The same result is just as easily obtained using the sum of the angles of quadrangle $ABCD$.) Such a two-mirror system is sometimes called a "constant-deviation system."

As an important special case, notice that if $\phi = 45°$, then $\delta = 90°$. This result is often used in the design of reflecting prisms, examined in Chapter 6.

2.1.2. Refracting Surfaces

Problem 2.4. Consider a block of transparent material of refractive index n and thickness d, as in Figure 2.3. Show that the refractive index is given by the thickness d divided by the apparent thickness x.

Figure 2.3. Thickness and apparent depth.

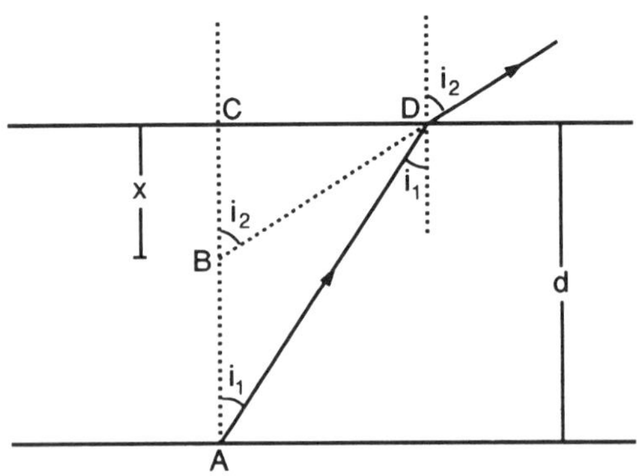

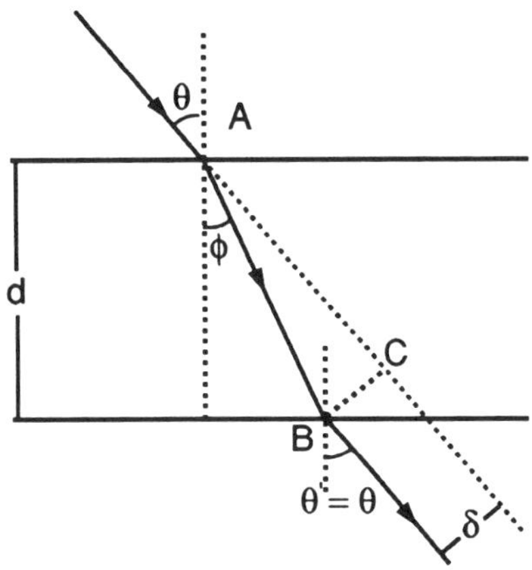

Figure 2.4. Ray displacement by a plane parallel block of glass.

Solution. Consider a ray from the bottom of the block that is refracted at the interface. The ray appears to have come from point B; so the apparent thickness is $(BC) = x$. From triangle (ACD), we have $d = (CD)/\tan i_1$, and from triangle (BCD) we have $x = (CD)/\tan i_2$. Upon dividing these two equations, we obtain $d/x = \tan i_2/\tan i_1 \cong \sin i_2/\sin i_1$ for small angles. Snell's law gives $n \sin i_1 = \sin i_2$ (assuming the second medium to be air); therefore, $d/x = n$.

Q.E.D.

Notice that if we had a means of measuring the apparent thickness (e.g., a microscope), we could use this method to obtain an estimate for the refractive index of the block.

Problem 2.5. Determine the ray displacement $\delta = \delta(n, d, \theta)$ for a ray that goes through a glass block of thickness d, as in Figure 2.4. Show first that the emergent ray is parallel to the incident ray.

Solution. Successive application of Snell's law at the two interfaces gives: $\sin\theta = n \sin\phi = \sin\theta'$; therefore $\theta = \theta'$. From triangle (ABC), we have $\delta = (AB)\sin(\theta - \phi) = (d/\cos\phi)\sin(\theta - \phi)$. Upon expansion of the $\sin(\theta - \phi)$ and substitution of ϕ from Snell's law, we obtain

$$\delta = d \sin\theta\left(1 - \frac{\cos\theta}{\sqrt{n^2 - \sin^2\theta}}\right) \tag{2.2}$$

Notice that as $n \to 1$, $\delta \to 0$, and also as θ or $d \to 0$, $\delta \to 0$, as expected.

Problem 2.6: Thin prism deviation. For a thin refracting prism with apex angle ϕ (Figure 2.5), determine the ray deviation in terms of the refractive index and the apex angle. Assume that the angle of incidence is small.

Figure 2.5. Refraction by a thin prism.

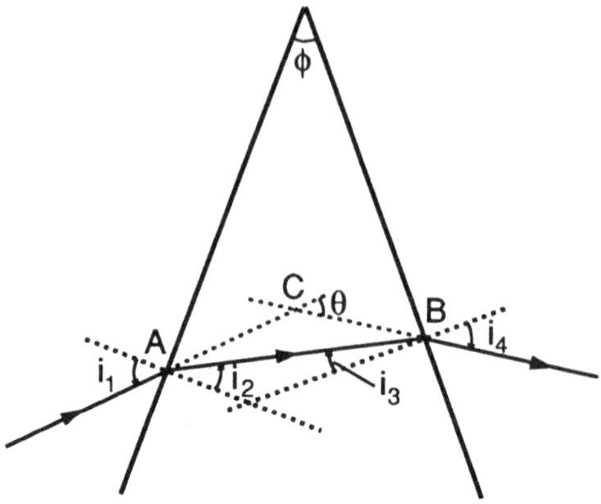

Solution. From triangle (ABC), we obtain $\theta = (i_1 - i_2) + (i_4 - i_3)$. Snell's law gives $\sin i_1 = n \sin i_2$. For small angles, we can set $i_1 = ni_2$, and similarly, $i_4 = ni_3$. Substituting these values, we obtain $\theta = (n - 1)(i_2 + i_3)$. By following the same reasoning as in Problem 4, Section 2.1.1, we can show that $i_2 + i_3 = \phi$; substitution into the expression for θ yields the desired result

$$\theta = (n - 1)\phi \tag{2.3}$$

Problem 2.7. Consider a parallel beam of light incident on a prism as in Figure 2.6. Determine the width of the emerging beam cross section as a function of ϕ, n, and w_1. (For a simplified case, take the beam to be normally incident on the first face.)

Solution. The figure gives $w_1 = (AB) \cos \phi$, and $w_2 = (AB) \cos \phi'$. Therefore, $w_2/w_1 = \cos \phi'/\cos \phi$, or, using also Snell's law,

$$w_2 = \frac{w_1(\sqrt{1 - n^2 \sin^2 \phi})}{\cos \phi} \tag{2.4}$$

Figure 2.6. Change in the beam diameter upon passing through a prism. Notice that the cross section of the emergent beam is not circular, because w_1 is not affected in the direction perpendicular to the paper.

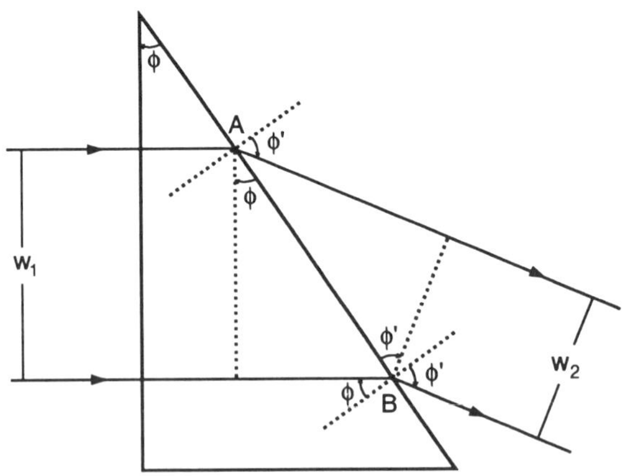

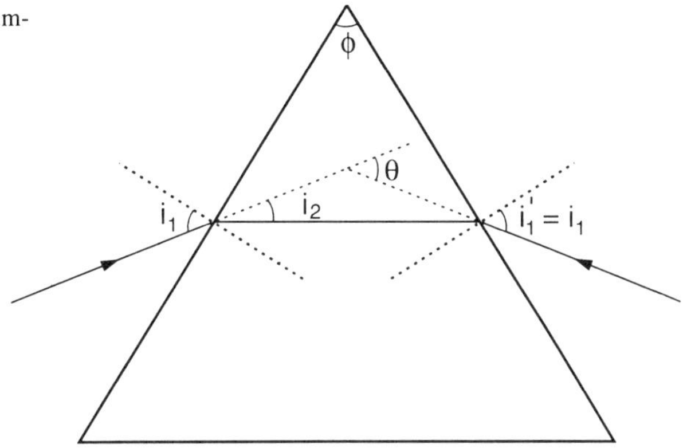

Figure 2.7. Thick-prism minimum-deviation condition.

Notice that the prism can be used to change the beam width: A similar device is used in the Cinemascope (wide-screen projection) system.

Problem 2.8: Minimum-deviation condition. For a thick prism, the minimum ray deviation is found to be when the angle of incidence and the angle of emergence are equal, as in Figure 2.7. Show that under these conditions

$$n = \frac{\sin(\frac{1}{2}(\theta + \phi))}{\sin(\phi/2)} \tag{2.5}$$

Solution. From the figure, and following a reasoning similar to Problem 2.4, the reader can justify the following relations: $\phi = 2i_2$ and $\theta = 2i_1 - \phi$, or $i_2 = \phi/2$ and $i_1 = (\theta + \phi)/2$. Snell's law gives $\sin i_1 = n \sin i_2$. Upon substituting for i_1 and i_2, Eq. (2.5) follows.

2.2. CURVED SURFACES: FOCUSING*

2.2.1. Focusing in the paraxial region

Focusing is the bending of rays toward a common point of intersection. Legend has it that Archimedes, living in the Greek colony of Syracuse around 214 B.C., focused the sun's rays with large mirrors in order to burn the invading Roman fleet. In so doing, Archimedes formed an image of the sun on the ships' sails (if indeed it is true that he performed that feat). Focusing and image formation go together. The exception to this rule is the pinhole camera, discussed in Chapter 1. There is no mechanism for focusing the pinhole camera; every object appears "in focus," irrespective of its distance from the pinhole.

* This section is a qualitative review of the basic properties of lenses and curved mirrors. The formulae that justify them are developed in subsequent chapters. If your knowledge of simple graphical raytracing is getting a bit rusty, you probably have a lot to gain from reviewing this section carefully. The importance of the simple graphical ray-tracing methods reviewed here is that they allow quick visualization of the action of an optical component.

However, the images produced with pinhole cameras are not sharp, which is like saying that all objects are out of focus! This was observed long ago, and around the sixteenth century A.D., when lenses became more easily available, the pinhole was replaced with a lens. This allowed focusing, and resulted in a much sharper and brighter image.

Let us consider a parallel bundle of rays. Such a ray configuration can arise from a point source at infinity, such as a star. Let this parallel bundle of rays encounter an optical component, such as a lens or a mirror (Figure 2.8). The rays will somehow be redirected by this component, and we distinguish the following three possibilities.

1. The rays can remain parallel (Figure 2.8a). We then say that the component has *zero power.*
2. The rays can converge so that they pass through a common point after the component. The component is then said to have *positive power* (Figure 2.8b).
3. The rays can diverge so that they appear to come from a point located before or inside the component. We then say that the component has *negative power* (Figure 2.8c).

Let us now examine some of the difficulties that we encounter with these simplified definitions. To begin with, the above are not the only three possibilities; clearly, the rays may not all meet at a common point. But unless this condition is fulfilled (at least approximately), no image is formed. Sunlight reflected from a rough surface does not form a recognizable image of the sun.

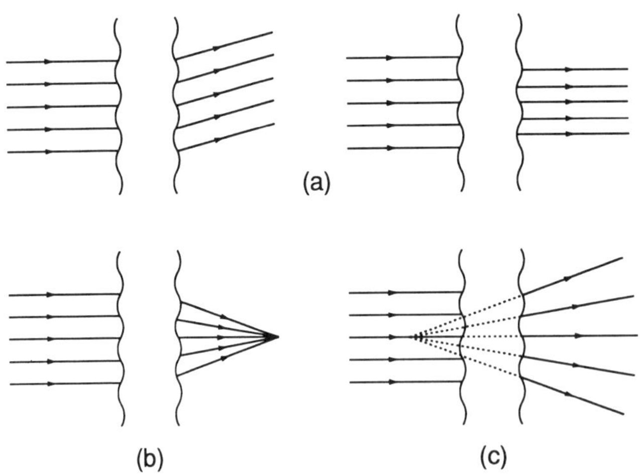

Figure 2.8. (a) A zero-power component. (b) A positive component. (c) A negative component. (*Note*: A component here implies a single thin lens or mirror. See also Figures 2.10 and 2.11.)

Figure 2.9. Focusing of a parallel beam by a concave spherical mirror. Rays infinitesimally close to the axis pass through $\mathscr{F}'$. Rays farther away from the axis will intercept it at different points after reflection.

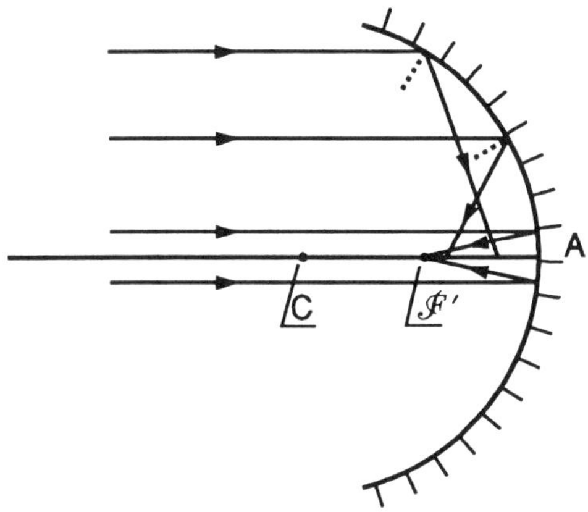

So we restrict our attention to those three possibilities because we are primarily interested in image-forming systems.

The second difficulty lies again in the statement that all rays meet at a common point. Figure 2.9 shows that this statement is generally only true in the *paraxial region*, that is, infinitesimally close to the system axis. The concave mirror shown in Figure 2.9 is spherical; it is possible to design an *aspheric* mirror that would concentrate all rays at the same point (ignoring diffraction limitations). For a parallel input beam, the necessary shape is that of a parabola.* Aspherics will be examined briefly later, but there are good reasons for concentrating on spherical surfaces. A practical reason is that spherical surfaces are much easier to make than aspherics; so most optical systems utilize only spherical surfaces. A pedagogical reason is that spherical surfaces are easier to study. But the most important reason is that all smooth surfaces of revolution are indistinguishable from the sphere very near the axis.

This last fact can be seen from the equation describing a smooth surface of revolution around the z axis, with the apex at the origin,

$$z = \tfrac{1}{2}c(x^2 + y^2) + O(4) \tag{2.6}$$

where c is the surface curvature and $O(4)$ a quantity of degree four or higher in x and y. Near the axis, that is, for small x and y, we can neglect higher orders, and all smooth surfaces of revolution become identical. This applies to the refracting or reflecting surfaces as well as the wavefront shape. So the wavefront shape near the axis can always be taken as spherical, and the normals to the wavefront, the rays, will then all meet at a common point, the center of curvature of the wavefront. Thus we have two equivalent definitions of perfect image formation, namely, that the wavefronts are spherical or that the rays all meet at a common point. Departure from these conditions constitutes *aberrations*, to be discussed later.

* "... a Parabolical Section, that is of all Glasses the most burning," wrote Giovanni Battista della Porta in his *Magia Naturalis* some time in the sixteenth century A.D.

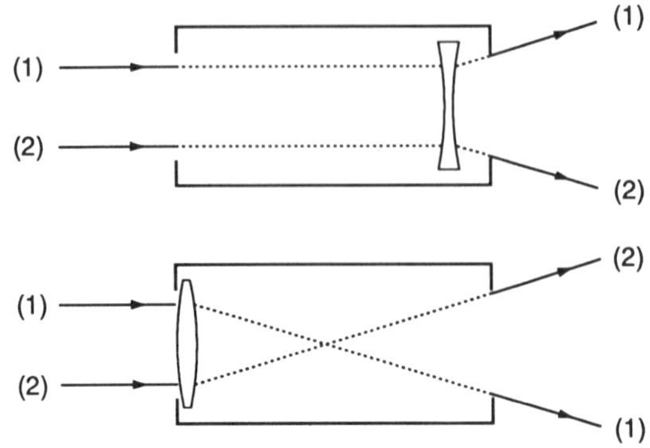

Figure 2.10. Two "blackbox" systems with apparently identical output but different function. The two outputs are not identical upon closer examination.

The third and final difficulty is suggested by Figure 2.10. If we treat the optical system as a "blackbox," is there a difference between the two systems shown? At the output of both boxes, we obtain a similarly diverging beam, yet by peeking inside the boxes we can see that they act differently. Of course we suspect that there is a very important difference between the two cases, but is it discernible at the blackbox output? After some pondering, we may come up with the answer: If we were to trace a single ray, such as the top one, we would find that in case (1) it exits at the top, but in case (2) it exits at the bottom. So box (2) reverses the order of the rays. From this we can conclude that it is really a positive system, even though it appears to produce a diverging beam.

We have every reason to be pleased with our solution to this dilemma, until we encounter the system shown in Figure 2.11. In this case, the system inside the two blackboxes is the same, but the blackbox outputs are different. Do they have the same power? Is the power positive or negative? Based on our simple definitions we would have to answer that case (1) has negative power and case (2) has positive power. Yet it does not appear satisfactory that the same system should have different power, depending on where one looks. What if we were to trace the top ray only? In case (1), it exits above the axis, which would be expected from a negative system; so this case seems clear. In case (2), the top

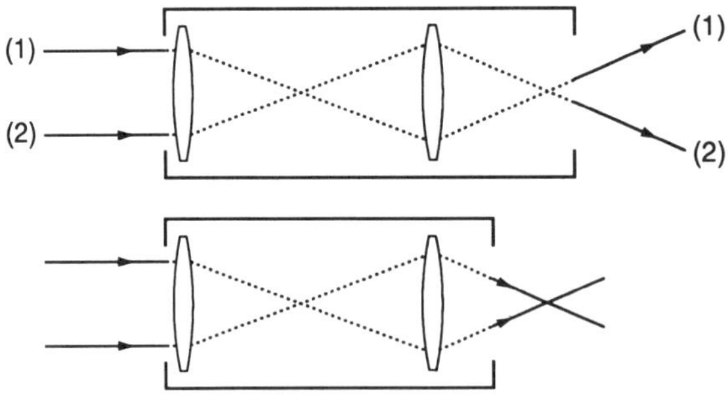

Figure 2.11. Two "blackbox" systems with apparently different output but identical internal construction.

ray exits below the axis and bends upward, which is not what we expect from a positive system. The answer is that the system has negative power, but full justification of this answer can only be given in Chapter 4. For now, the conclusion is that of the three simple definitions used above, only the definition of a zero-power system is robust; those of the negative and positive power are not, based on what we have said so far. The reason for using these definitions at this stage is because they apply to the most commonly encountered optical components, thin (or not-too-thick) lenses, and mirrors.

2.2.2. Graphical Ray Tracing for Thin Lenses

To illustrate the preceding concepts, consider a thin lens. A thin lens is a theoretical construction: It comprises two refracting surfaces that are taken as spherical near the axis, separated by an infinitesimally small distance. Clearly, such a device is physically impossible, but it is of great theoretical importance, and it provides a good approximation to the behavior of real lenses in many cases. In the diagrams that follow, we draw the lens as having a finite thickness in order to indicate its power, but refraction is assumed to take place along a single plane, normal to the axis.

Let the thin lens be in air, and let a parallel bundle of rays be incident from the left, as in Figure 2.12. In this and the following figures, unless otherwise stated, the rays are assumed to be infinitesimally close to the axis, although, obviously, they cannot be drawn that way. The convergence angle of the rays relates to the power of the lens. The more the convergence, the more the power, and for a negative system, the more the divergence, the less (algebraically) the power. We use the symbol K for power. The quantity $1/K$ is called the *equivalent focal length* F. In Figure 2.12, F is the distance between the lens and the point at which an incoming ray parallel to the axis intersects the axis after refraction, $\mathscr{F}'$. (We will re-examine these definitions more quantitatively later.)

The point $\mathscr{F}'$ is called the *secondary*, or *image-space, focal point.* The name implies that there is another focal point, and indeed that is obviously the case, as we can see by considering a ray incident on the lens from the right; the ray will cross the axis at $\mathscr{F}$. The point $\mathscr{F}$ is then called the *primary*, or *object-space, focal point.* Curved mirrors also have a focal length, defined exactly as above, but there is obviously only one focal point, $\mathscr{F}'$.

The terms *object space* and *image space* refer to the fact that the points

Figure 2.12. A thin lens in air.

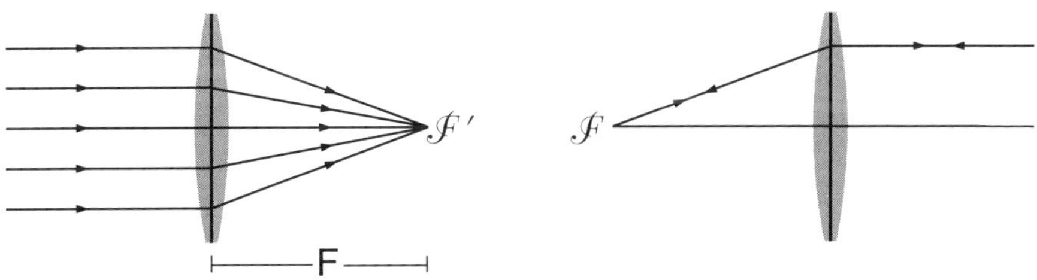

$\mathscr{F}$, $\mathscr{F}'$ are associated with the object and image, respectively. The object and image spaces should not be thought of as separable physical spaces; when we say that a quantity is in object space, we mean that it is object related. In our example, $\mathscr{F}'$ is in the image space because a ray coming from the left reaches it after refraction, that is, after imaging. Similarly, $\mathscr{F}$ is in the object space because a ray may reach it before refraction.

There are some simple rules that allow us to determine graphically (approximately) the location and size of the image, in the paraxial approximation. For thin lenses, these are:

1. Ray parallel to the axis crosses the axis at $\mathscr{F}'$ after refraction.
2. Ray crossing the axis at $\mathscr{F}$ will emerge parallel to the axis after refraction.
3. Ray through the middle of the lens emerges undeviated.

Let us now use these rules, before proving them, to determine the location and size of the image for a given object. To do so, we need two rays from the top of the object (any two of the three above). Their intersection after refraction determines the top of the image. The construction for a positive lens is shown in Figure 2.13. It is most instructive to use a positive lens in order to observe the image in reality, rather than merely looking at the diagrams. Try the following exercise: Take a positive lens and put it in contact with an object, such as this book page. Notice that near the middle of the lens the image appears to be of equal size as the object. As you increase the distance between

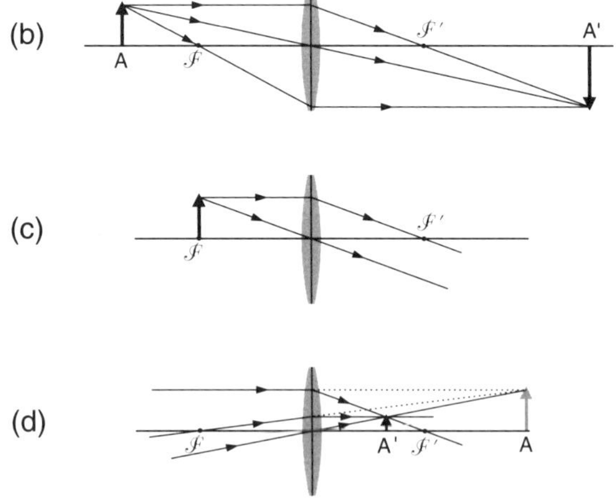

Figure 2.13. Image formation by a positive lens. (a), (b), (c) Real object; (d) virtual object. (a) Virtual, erect, magnified image; (b) real, inverted image; (c) image at infinity; (d) real, erect, minified image. Notice that (a) and (d) are essentially the same if the direction of the rays is reversed.

lens and paper, the image becomes larger (Figure 2.13a). There is a point where the image reaches a maximum size; then it becomes increasingly blurred. If you keep increasing the distance between lens and paper, and assuming you maintain an adequate distance between the lens and your eye, you will eventually see an inverted image (Figure 2.13b).

What happens in the region where we do not see a clear image? Most people tend to guess that the image is at infinity, as in Figure 2.13c. But this is not at all true. There is nothing wrong with an image at infinity! If you can see the stars, then you can also see an image at infinity (that is, if you are not short-sighted). The region where there is no recognizable image corresponds to the region where a real image is formed behind your eye.*

The image in Figure 2.13b is real, and the image in Figure 2.13a is *virtual*. You can think of a virtual image as one that is formed by producing the rays backwards, that is, the image formed by a diverging bundle of rays. A virtual image cannot be formed on a screen, since the rays (and the EM energy) are not directed towards it but away from it. For light going from left to right, a virtual image is to the left of the lens. When the image is at infinity, the distinction between real and virtual loses its meaning.

It is also possible to have a virtual object. For light going from left to right, a virtual object would be to the right of the lens. This implies that there is another optical system to the left of our lens, forming a real image at the same position as the virtual object, and that we are interrupting the path of the rays with our lens, before that image is formed. For a real object, rays diverge from it. For a virtual object, rays are directed (converge) towards it, but are of course intercepted by the lens before they meet (Figure 2.13d). Virtual objects are encountered very frequently, in multielement systems.

It is instructive to consider these examples from the convergence/divergence viewpoint. In Figure 2.13a, we have a highly diverging beam. A positive lens will decrease the divergence, but does not have enough power to make the beam converge; so the beam remains divergent and a virtual image is formed. In Figure 2.13b, we have a less diverging beam, and in this case the lens has enough power to produce eventually a converging beam and a real object. In Figure 2.13d, the incident beam is converging; the positive lens makes it converge even faster.

Let us now consider a thin negative lens. If you have one handy, notice that the image of any object looks upright and smaller when you look through the lens (Figure 2.14a). That is true for the image of any real object, but when the object is virtual, it is possible to have a virtual or a real image (Figure 2.14b,c). Finally, when the object is at $\mathscr{F}$, the image is at infinity; this statement applies to both positive and negative lenses, for light coming from left to right. These figures can be interpreted from the convergence viewpoint, just as before.

---

* To be precise, the image need not be formed behind your eye, but behind the *near point*, which is the closest point of clear vision, located typically 25 cm in front of your eye.

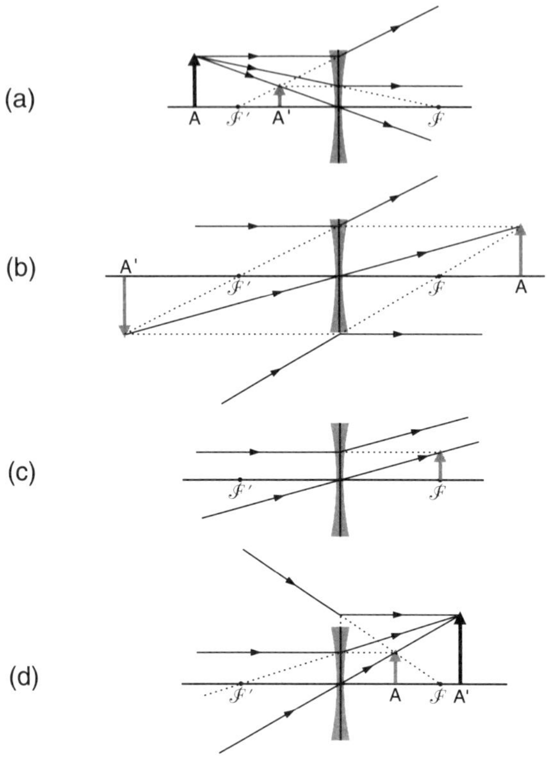

Figure 2.14. Image formation by a negative lens. (a) Real object; (b), (c), (d) virtual object. (a) Virtual, erect, minified image; (b) virtual, inverted image when A is after $\mathscr{F}$; (c) image at infinity; (d) real, erect, magnified image for A between the lens and $\mathscr{F}$. (a) and (d) are essentially the same but for the reversal of the direction of the rays.

We see that a positive lens can form both virtual and real images, and the same is true of a negative lens. But it pays to memorize the following cases, which apply for real object only:

1. A positive lens with the object to the left of $\mathscr{F}$ produces a real, inverted image.
2. A positive lens with the (real) object to the right of $\mathscr{F}$ produces a virtual, erect, and magnified image.
3. A negative lens with a real object produces a virtual, erect, and minified image.

We distinguish six different types of thin lenses, depending on the relative surface curvatures (Figure 2.15). The planoconvex, biconvex, and positive meniscus types are positive lenses, while the planoconcave, biconcave, and negative meniscus are negative lenses. When the two curvatures of the biconvex or biconcave types are equal, they are called *equiconvex* and *equiconcave*, respectively. Thus a positive lens is thicker in the middle, and a negative lens is thicker at the edges (it is shown later that this may not be true for very thick lenses).

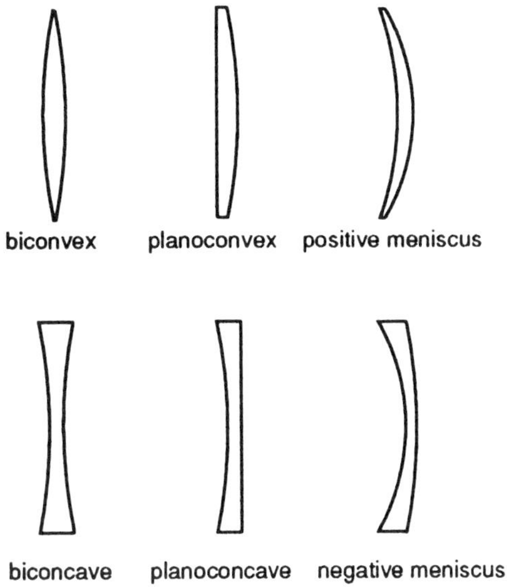

Figure 2.15. Different types of thin lenses.

In the paraxial approximation, the shape of the lens is of no importance; the only relevant parameter is the focal length. More accurate theory, examined in later chapters, shows that the shape of the lens is very important in determining the quality of the image.

2.2.3. Graphical Ray Tracing for Mirrors

The rules for graphical ray tracing with mirrors are:

1. Ray parallel to the axis crosses the axis at $\mathscr{F}'$ after reflection.
2. Ray crossing at $\mathscr{F}'$ will emerge parallel to the axis after reflection.
3. Ray crossing the axis at the center of curvature will be reflected back on itself.

The point $\mathscr{F}'$ for spherical mirrors is located halfway between the apex and the center of curvature. Figure 2.16 shows the different cases for a concave mirror. Notice that a concave mirror produces a virtual and erect, or real and inverted, image, in a manner very similar to a positive lens. The concave mirror has positive power, because it produces a converging beam from an incident parallel beam. The inevitable difference between mirrors and lenses is that a virtual image must be formed to the right of the mirror, if the light is incident from the left. But the nature of the virtual image is unchanged, as it is still formed by diverging ray bundles.

The convex mirror (Figure 2.17) has negative power, and like the negative lens produces a virtual and minified image given a real object. We can then summarize these simple cases to remember:

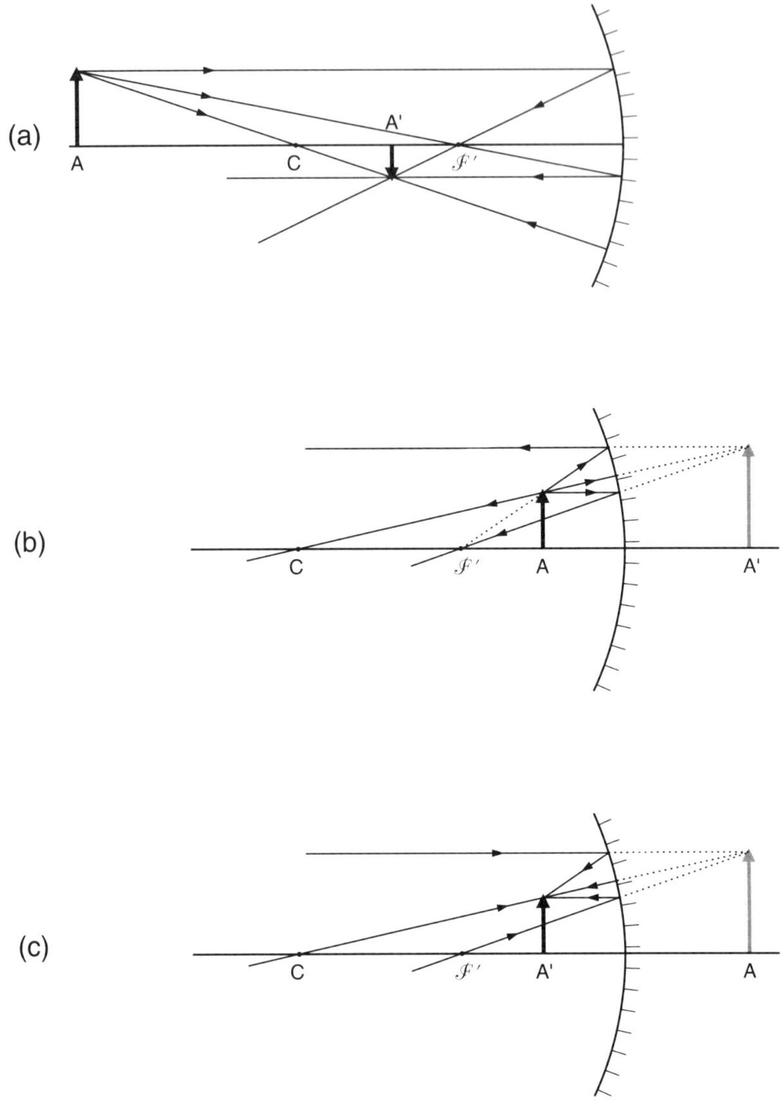

Figure 2.16. Image formation by a concave mirror. (a), (b) Real object; (c) virtual object. Notice that (b) and (c) are the same, but for the reversal of the direction of the rays.

1. A concave mirror produces a real, inverted image if the object is to the left of $\mathscr{F}'$.
2. A concave mirror produces a virtual, erect, and magnified image if the (real) object is to the right of $\mathscr{F}'$.
3. A convex mirror produces a virtual, erect, and minified image for any real object.

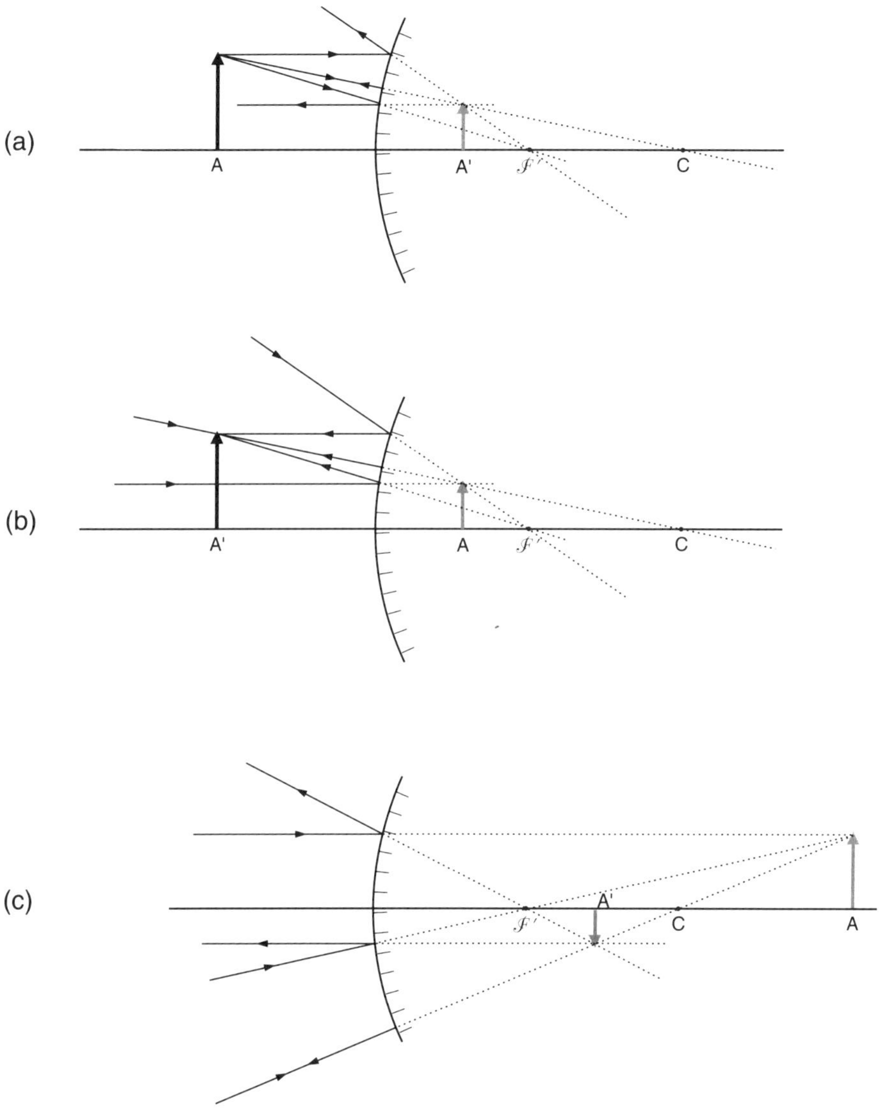

Figure 2.17. Image formation by a convex mirror. (a) Real object; (b), (c) virtual object. Again, (a) and (b) differ only in the direction of the rays.

Having now examined these graphical ray-tracing rules, we should perhaps emphasize that the only correct way of determining the image is algebraically, through the equations developed in the next chapters. But graphical ray tracing is helpful in visualizing the effect of a lens quickly. Also, at least at the beginning, *it is a good idea to make a diagram of the system and a couple of rays as a means of verifying your algebraic solution.*

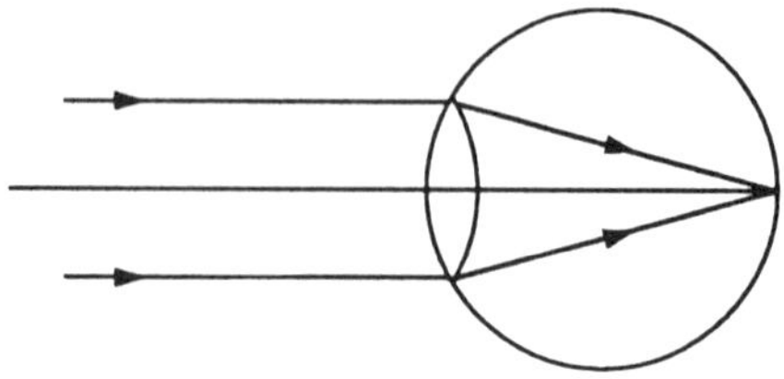

Figure 2.18. A schematic emmetropic eye. The cornea and the crystalline lens have been combined into one thin lens for simplicity. The image of a distant object is formed on the retina.

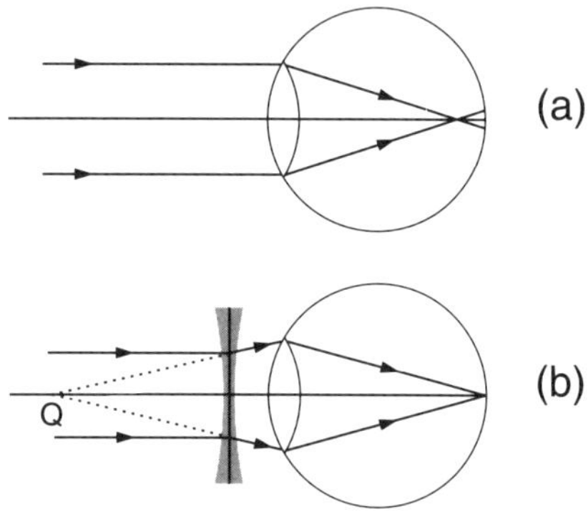

Figure 2.19. (a) A schematic myopic eye. (b) With spectacle correction.

Example 2.1: Spectacle correction. The eye has two refracting elements: the external surface (cornea), and the crystalline lens, located inside the eyeball. The image of an external scene is formed on a photoreceptor mosaic, called the retina, that covers the inside rear surface of the eyeball. Since the image location is fixed, the eye must change its power in order to bring into clear focus objects near and far. This is called *accommodation*, and it is accomplished by changing the power of the crystalline lens. The normal or *emmetropic* eye can focus an object located anywhere between the near point and infinity (Figure 2.18). The short-sighted or *myopic* (young) eye has a near point located closer than that of the normal eye, and cannot focus an object an infinity (Figure 2.19). Thus we can say that the myopic eye has too much power (or too long an eyeball). Therefore, a negative lens is required for correction. Most people with defective vision are myopic; you can tell by the fact that their eyes look smaller behind their glasses, as you would expect with a negative lens.

The opposite of myopia is *hypermetropia* or *hyperopia*. Here the near point is farther away than that of a normal eye.* The eye appears to have too little power, or to be too short. To correct for this condition, a positive lens is required (Figure 2.20). The eyes of hypermetropic people look larger behind their glasses, since their positive lenses act as magnifiers.

It can be argued that hypermetropic people can see behind their head! In

* This is not to be confused with *presbyopia* or "old-age vision," which also results in a receding near point due to hardening of the crystalline lens and subsequent loss of accommodation.

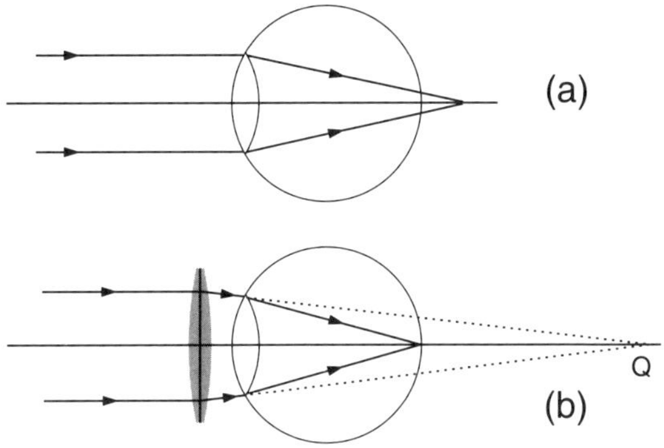

Figure 2.20. (a) A schematic hypermetropic eye. (b) With spectacle correction.

Figure 2.20, the positive lens forms an image at point Q, which is located behind the eye. The image formed by the spectacle lens becomes the object for the eye, which then forms a final image on the retina. Thus the object that the eye is seeing is located behind it. The critical difference is, of course, that it is a virtual object. So hypermetropic people can see behind their head, but only if the object is virtual!

To practice these concepts, try thinking of the spectacle example from the convergence viewpoint. Try guessing whether people are myopic or hypermetropic by looking at the apparent size of their eyes seen through their glasses; after a while you may discover that you can even tell approximately the power of the spectacle lenses. Also, you should experiment with any simple lenses and mirrors you can find, and then try to explain what you see using the simple graphical ray-tracing rules of this chapter.

☑ PROBLEMS

1. A thin prism is said to have a power of one *prism-dioptre* if it produces a ray deviation of 1/100 rad. For a refractive index of 1.52, what is the apex angle of a prism of power equal to one prism-dioptre?

2. Over the visible region, the refractive index of a glass is given by the formula $n = c_1 + c_2/\lambda^2$, where c_1, c_2 are constants, with $c_2 = 3 \times 10^4\ \text{nm}^2$. White light passes through a thin prism of 5° apex angle made from this glass. Determine the difference in the deviation angle between the extreme red (700 nm), the green (550 nm), and the violet (400 nm) rays.

3. A parallel beam of light has an elliptical cross section, with semiaxis ratio 2:1. Design a prism system that can transform the cross section of this beam to circular.

4. A converging beam of light, with convergence angle $\pm 10^\circ$, encounters an air–glass interface. Determine the angle of convergence of the beam in the glass ($n = 1.5$).

5. Through a simple graphical ray trace, determine the approximate image location and magnification, in the following cases: (a) concave mirror, real object at C; (b) concave mirror, real object between C and $\mathscr{F}'$; (c) convex mirror, virtual object between $\mathscr{F}'$ and C; (d) negative lens, real object at $\mathscr{F}'$; (e) negative lens, virtual object at $\mathscr{F}$; (f) positive lens, virtual object at $\mathscr{F}'$; (g) convex mirror, image at infinity.

6. A positive thin lens of 10 cm focal length is followed by a negative thin lens of -5 cm focal length. The distance between the two lenses is 5 cm. Show that this is a zero-power system, using the graphical ray-tracing rules of this chapter.

7. A positive thin lens of power $K = 0.1\ \text{cm}^{-1}$ is followed by a convex mirror of radius $r = 5$ cm, located 5 cm behind the lens. What is the power of the system? (Solve this graphically.)

8. Repeat Problem 6, but with a negative lens, $K = -0.1\ \text{cm}^{-1}$, and a concave mirror, $r = -15$ cm, located 5 cm behind the lens.

9. A parallel bundle of rays is incident on a refracting surface, as in Figure 2.21. On one side of the surface is air, and on the other glass, $n = 1.5$, as noted. Determine if the surface has positive or negative power, using simple, qualitative arguments derived from Snell's law.

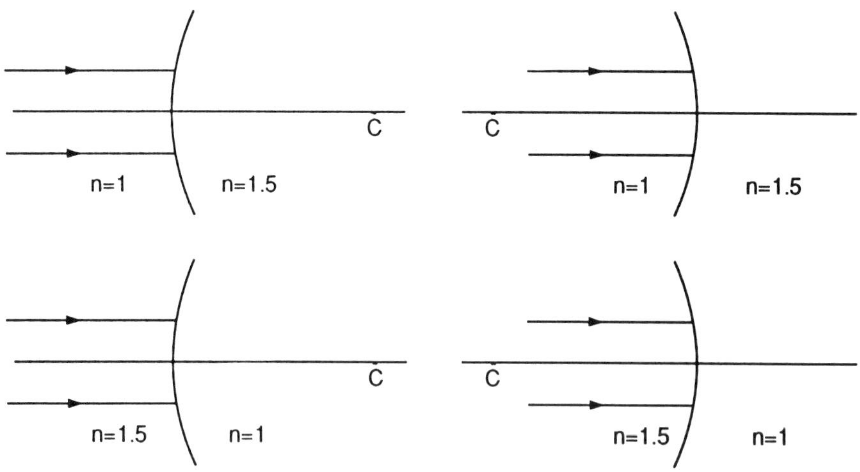

Figure 2.21

10. You are looking into a large aquarium at Sea World through its plane glass wall. Suddenly, you are startled by a beluga whale, which swims right past you. (a) Does the beluga appear closer to the aquarium wall than it actually is, farther away, or the same? (Show ray paths.) (b) How large is its image compared to its actual size? (c) Does the aspect ratio of the beluga (length to height) appear larger than it actually is, smaller, or the same?

Imagery by a Single Surface and a Thin Lens 3

"... I present just one, single lens. Nothing more than a minute magnifying glass, typically not much larger than the head of a pin, yet, capable of producing an image that is only a little lower down the scale of magnifying power than a modern instrument..."

from **Brian J. Ford,** *Single Lens: The Story of the Simple Microscope*, Harper & Row, New York, 1985.

Given a lens or mirror and an object, the following questions are pertinent:

1. Where is the image?
2. How large is it?
3. How bright is it?
4. What is the image quality?

These problems are perhaps more commonly encountered in their inverse form: Specify an optical element that will produce a given image. The simple paraxial theory of this chapter addresses the first two of the above questions, the location and size of the image. After the chapter on radiometry, we will be able to answer question (3), while question (4) is addressed in the chapters on aberrations.

3.1. THE SIGN CONVENTION

Consider a spherical surface separating two media of different refractive indices, n, n', as in Figure 3.1. We must apply a sign convention to the various lengths and angles shown in the figure. Let a ray from the axial object point A be refracted at the surface and intersect the axis after refraction at A'. The following convention applies:

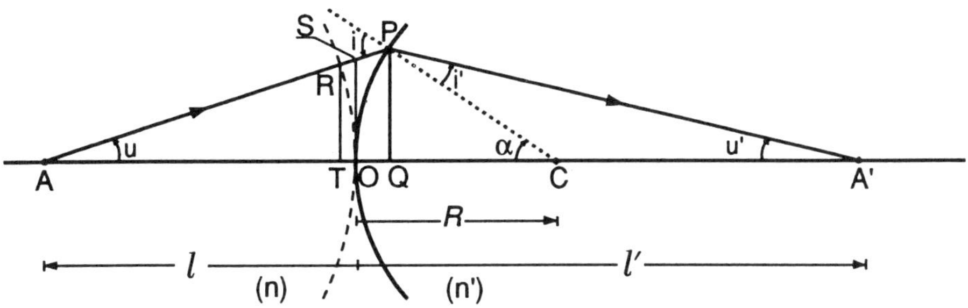

Figure 3.1. The sign convention and imagery by a single surface. Point P is taken very close to O, so that $O \to Q$. The distances l, l', etc. are indicated with a single arrow, corresponding to their algebraic sign.

1. Unless otherwise stated, light travels from left to right.
2. The point of intersection between the surface and the axis is taken as the local origin of coordinates, O.
3. All distances are measured from O; distances to the top and right of O are taken as positive.
4. Angles are measured from the axis to the ray and are positive if counterclockwise. Similarly, the angles of incidence and refraction are measured from the surface normal, and are positive if counterclockwise.*
5. For light going from right to left, the refractive index is taken as negative.

The last rule may appear somewhat mystifying for the moment; it will be explained fully in Section 3.4 on mirrors. Notice that in the diagram all distances and angles are marked by a single arrow, to indicate whether they are positive or negative. This is a good practice to adopt.

While there are some alternative sign conventions, this one has some very important advantages. It agrees fully with Cartesian coordinate geometry (and therefore with other branches of physics), which makes it easy to remember. Also, it is almost universally employed in lens design, which is the logical endpoint of geometrical optics. Other sign conventions may appear convenient for the most simple problems but become impractical for more advanced problems.

3.2. THE PARAXIAL APPROXIMATION

In the paraxial approximation, we restrict our attention to rays that are infinitesimally close to the axis. In Figure 3.1, point P is infinitesimally close to O; so that all of the angles shown, u, u', i, i', and α are very small. Let θ stand

* The word *angle* here implies the smaller of the two angles that the ray forms with the axis or surface normal.

for any of the above angles. Then we can write the power-series expansion for $\sin\theta$ or $\tan\theta$ as follows:

$$\sin\theta = \tan\theta = \theta + O(2)$$

where $O(2)$ denotes a quantity of at least second degree in the small variable θ. The paraxial region is defined as the region where $O(2)$ is negligible.

It is instructive at this point to discuss in a little more detail the paraxial intersection height between the surface and the ray. Referring to Figure 3.1, one might be tempted to define the intersection height as the segment (QP). However, there are two more possibilities, at least equally valid is principle: (1) the segment (OS), found through the intersection between the ray and the tangent to the surface at the apex, and (2) the segment (TR), found as the intersection between the wavefront (OR) and the ray. These choices are equivalent paraxially since they differ from each other only by $O(2)$. To show this, consider the following relations, noting that $(OA) = (RA) = l$.

$$(OS) = |l \tan u| = |lu| + O(2)$$

and

$$(TR) = |l \sin u| = |lu| + O(2)$$

from which $(OS) = (TR)$ paraxially. Also, $(QP)/(OS) = (AQ)/(AO)$, and $(AQ) = (AO) + (OQ)$. The segment $(OQ) \equiv z$ can be found from Eq. (2.6) as $(QP)^2/2R + O(2)$. Since $(QP) = |R \tan\alpha| = |R\alpha|$ paraxially, it follows that $(OQ) = R\alpha^2/2$; that is, (OQ) is of the second degree in the small angle α, and therefore $(AO) = (AQ)$ and $(QP) = (OS)$ paraxially.

Thus all three choices are equivalent paraxially. However, we will find later (Ch. 5) that we can extend the paraxial equations to encompass finite quantities. In that case, the choice of the paraxial ray height becomes important. Anticipating that discussion, we note that there is no advantage in using (QP) to represent the paraxial ray height, but that the segments (OS) and (TR) each provide geometrically consistent equations that can be extended beyond the paraxial region.

3.3. IMAGERY BY A SINGLE SURFACE

3.3.1. The Conjugate Equation

We now wish to derive an equation that relates the object and image locations, given the characteristics of the interface, n, n' and the curvature $c \equiv 1/R$. There are two avenues: through direct application of Fermat's principle, and through Snell's law. Since the latter is a corollary of the former, the two approaches are equivalent. We chose to apply Snell's law because it is simpler and it gives us the opportunity to develop some additional useful concepts.

Again referring to Figure 3.1, Snell's law gives $n \sin i = n' \sin i'$. Paraxially, this becomes

$$ni = n'i' \tag{3.1}$$

Geometry gives: $i = |\alpha| + |u|$ and $i' = |\alpha| - |u'|$. From our sign convention, $\alpha < 0$ and $u' < 0$; so $|\alpha| = -\alpha$ and $|u'| = -u'$, while $|u| = u$. Substituting, we obtain

$$\begin{aligned} i &= u - \alpha \\ i' &= u' - \alpha \end{aligned} \tag{3.2}$$

Equations (3.2) should be regarded as defining the sign of i and i'. They can be applied in all cases, without resorting to the intermediate step of using the absolute values. This step was only used here to show that the definitions make geometrical sense. The reader can readily see that definitions (3.2) are equivalent to the second part of rule 4 of the sign convention.

Substitution of Eq. (3.2) into Eq. (3.1) gives

$$n(u - \alpha) = n'(u' - \alpha) \tag{3.3}$$

We also have $\tan u \cong u = -h/l$, and similarly, $u' = -h/l'$, and $\alpha = -h/R$. (Notice that the minus sign on the right-hand side of these equations is necessary according to the sign convention: For example, $u > 0$, $h > 0$, $l < 0$; so we cannot simply write $u = h/l$.) Substituting these values into Eq. (3.3), we obtain

$$n\left(\frac{h}{l} - \frac{h}{R}\right) = n'\left(\frac{h}{l'} - \frac{h}{R}\right)$$

which, upon cancelling h and rearranging terms becomes

$$\frac{n'}{l'} - \frac{n}{l} = c(n' - n) \tag{3.4}$$

where $c = 1/R$. We will call this the *conjugate equation*, as it relates the object and image locations (object and image are called conjugates). Notice that Eq. (3.4) does not contain the height h; that is, l' is independent of h. Therefore, *in the paraxial approximation, all rays from A will end up at a common point of intersection A'*. So paraxial image formation is perfect, in the sense that all rays meet at a common point, or equivalently, the wavefronts are spherical.

We strongly advise the reader to adopt the following, very helpful notation: any image-space quantity is primed, and denoted by the same symbol as the corresponding (unprimed) object-space quantity. This will be particularly important when solving problems that involve more than two surfaces or optical elements.

3.3.2. Power and Focal Lengths of a Surface

From Eq. (3.4), the product

$$K = c(n' - n) \tag{3.5}$$

is called the power of the surface. If K is positive, that is, if c and $(n' - n)$ have the same sign, then the surface is positive, or converging. If K is negative, that is, if c and $(n' - n)$ have different signs, then the surface is negative or diverging. This can also be predicted graphically (Problem 2.5). Thus, we can rewrite Eq. (3.4) as

$$\frac{n'}{l'} - \frac{n}{l} = K \tag{3.6}$$

Dimensionally, the power is inverse length and therefore can be measured in any length unit. In optics, the unit m^{-1} is given the special name *dioptre*, denoted by Δ or D. The dioptre is also used to denote the curvature of a wavefront $(1/l)$, or a surface c.

In Chapter 2 we encountered the concept of focal length in a qualitative way. Equation (3.4) or (3.6) allows us to define it quantitatively. For a ray coming from the axial object point at infinity ($l = \infty$), we have $n/l = 0$. In this case, the image distance l' is by definition the focal length f':

$$f' = \frac{n'}{c(n' - n)} = \frac{n'}{K} \tag{3.7}$$

This is the image-space, or second focal length f'. The object-space, or first focal length, is obtained by setting $l' = \infty$; therefore,

$$f = -\frac{n}{c(n' - n)} = -\frac{n}{K} \tag{3.8}$$

The corresponding geometrical construction is shown in Figure 3.2. As Eqs. (3.7) and (3.8) show, the first and second focal points, $\mathscr{F}$ and $\mathscr{F}'$, are not equidistant from O. To justify the negative sign in Eq. (3.8) and the positive sign in eq. (3.7), refer to Problem 2.9, and relate the location of the points $\mathscr{F}, \mathscr{F}'$ to the sign of the power K (n and n' are taken as positive for light coming from the left).

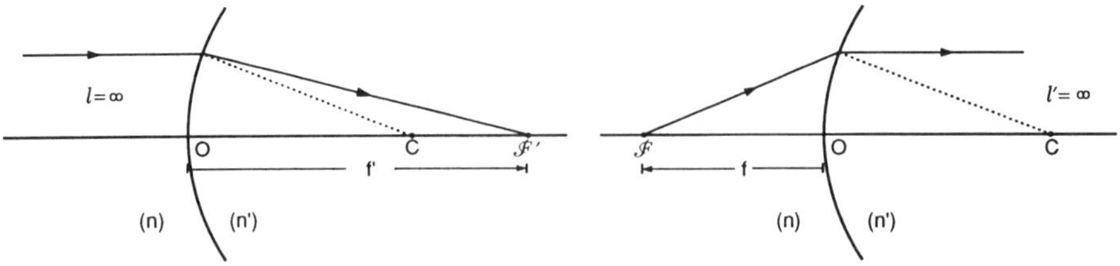

Figure 3.2. Focal lengths of a single surface.

3.4. MIRRORS

There is a trick that allows us to treat reflective surfaces as if they were refractive, so we do not need to remember additional formulae. Recall the law of reflection: $i = -i'$. This can be written as $\sin i = \sin(-i') = -\sin i'$. Rewriting this as $(1) \sin i = (-1) \sin i'$ and comparing it with Snell's law, $n \sin i = n' \sin i'$, we can see that the effect of the mirror can be described as a refraction, where the refractive index becomes negative, in the "new" medium n'. This is the justification for rule 5 of the sign convention, that the refractive index becomes negative for light traveling from right to left.

The sign reversal of n upon reflection is more than merely an interesting trick. It has physical justification in the fact that the sign of the term kx in the phase of a harmonic light wave, $\phi = \omega t \mp kx$, determines the direction of propagation of the wave, as discussed in Section 1.1. For a wave traveling from left to write, we have $v = \omega/k > 0$, while for a wave in the opposite direction, $v > 0$; so we must conclude that k changed sign. Recalling that $k = nk_0$, we see that the sign change can be assigned to the refractive index n. Thus a negative refractive index corresponds to the physical situation of negative wave velocity.

We may now derive the conjugate formula for mirrors. We apply the formula for the single surface (3.4), but with $n' = -n$, to obtain

$$\frac{1}{l'} + \frac{1}{l} = 2c = \frac{2}{R} \tag{3.9}$$

The power of the mirror is given from Eq. (3.5), again for $n' = -n$ as

$$K = -2nc \tag{3.10}$$

where n is the object-space refractive index. The corresponding focal length f' is given from Eq. (3.7) as

$$f' = \frac{1}{2c} = \frac{R}{2} \tag{3.11}$$

This is independent of n, as it should be, since the focusing power of a mirror is not affected if it is immersed in a liquid of refractive index $n \neq 1$.

In air, we have $n = 1$, $n' = -1$; so Eq. (3.10) becomes $K = -2c$. But for light going from right to left before reflection, we have $n = -1$, and $n' = 1$, so now the power $K = 2c$. Far from being a confusing disadvantage, this consequence of the sign convection means that a concave (positive, converging) mirror will always have positive power, and a convex (negative, diverging) mirror will always have negative power, independent of the direction of propagation of the light. Also notice that Eqs. (3.9) and (3.11) do not depend on the direction of propagation.

A plane mirror has no power ($c = 0$). Normally, a plane mirror is used inside an optical system in order to fold the beam. Such a mirror is not included in our sign convention, which is restricted to systems that have a symmetry of revolution around a common axis. When a folding plane mirror is encountered, it should be ignored, as it does not affect the calculation of image distance and size.

Should one memorize the mirror formulae? At the beginning, the answer is no. It is more rewarding to memorize the general equations for a single surface (3.4) and (3.5), and then apply the sign convention in order to arrive at the mirror relations.

3.5. IMAGERY BY A THIN LENS

3.5.1. Thin Lens Conjugate Equation

We will assume for simplicity that the lens is in air; the case with $n,n' \neq 1$ is treated in Section 3.5.4. The thin lens comprises two refracting surfaces separated by a negligible distance; so the problem is solved simply by applying the single surface conjugate equation twice. The reader should observe the use of a consistent notation, which makes it easy to obtain a solution without reference to a diagram. The essence of the notation is to use primes for image-space quantities and to use an appropriate subscript to indicate the surface with which a quantity is associated. Suppose then that we have an object located a distance l_1 away from the first surface. The object-space refractive index for the first surface is $n_1 = 1$, and the image-space index $n_1' = \mu$, where μ the refractive index of the glass from which the lens is made. (The unusual symbol μ for the refractive index is used here because we need to reserve n for the object-space refractive index, as shown later.) Let the first surface have a curvature c_1, so from Eq. (3.4)

$$\frac{n_1'}{l_1'} - \frac{n_1}{l_1} = c_1(n_1' - n_1) \rightarrow \frac{\mu}{l_1'} - \frac{1}{l_1} = c_1(\mu - 1) \tag{3.12}$$

using the substitutions noted above. For the second surface, we have $n_2 = \mu$ and $n_2' = 1$; so

$$\frac{n_2'}{l_2'} - \frac{n_2}{l_2} = c_2(n_2' - n_2) \rightarrow \frac{1}{l_2'} - \frac{\mu}{l_2} = c_2(1 - \mu) \tag{3.13}$$

The image produced by the first surface becomes the object for the second surface. Since the separation of the two surfaces is negligible, the image distance for the first surface is the same as the object distance for the second surface (i.e., $l_1' = l_2$). Thus by adding Eqs. (3.12) and (3.13), we obtain

$$\frac{1}{l_2'} - \frac{1}{l_1} = (\mu - 1)(c_1 - c_2) \tag{3.14}$$

We can now simplify the notation by dropping the subscripts from the left-hand side of Eq. (3.14). This is because the distances l_1 and l_2' may be

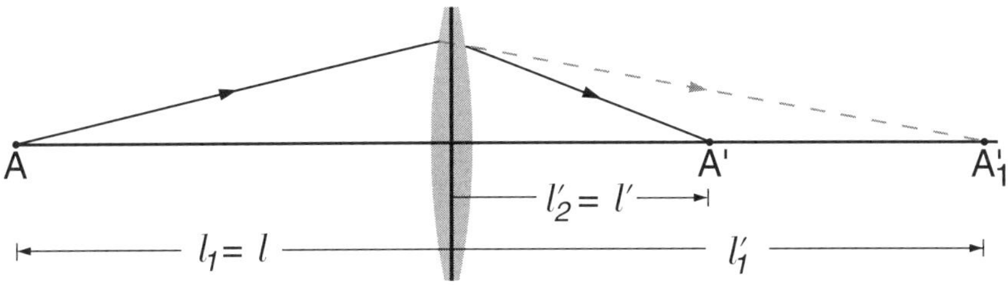

Figure 3.3. Refraction by a thin lens. The thickness of the lens is considered infinitesimal; so all refraction is assumed to take place along a single plane perpendicular to the axis at O.

measured from either surface 1 or surface 2, as the apexes of the two surfaces are considered coincident. Thus we finally obtain

$$\frac{1}{l'} - \frac{1}{l} = (\mu - 1)(c_1 - c_2) \tag{3.15}$$

This is the conjugate equation for a thin lens in air.* We can now refer to Figure 3.3, which shows the object location at A, the thin lens at O, the final image at A', and the intermediate image produced by the first surface at A'_1. The lens is drawn as having a finite thickness for clarity, but the thin lens approximation means that all refraction is assumed to take place at a single plane that is normal to the axis at O.

3.5.2. Power of a Thin Lens in Air

From Eq. (3.15), the product

$$K = (\mu - 1)(c_1 - c_2) \tag{3.16}$$

is called the power of the thin lens. As with a single surface, if $K > 0$, then the lens is positive or converging, and if $K < 0$, the lens is negative or diverging. For practice, the reader should verify that even for a nonsymmetric lens, such as a positive or negative meniscus, K does not depend on the orientation of the lens (see Problem 3.8).

Let us now rewrite Eq. (3.15) as

$$\frac{1}{l'} - \frac{1}{l} = K \tag{3.17}$$

An interesting physical insight is given by this equation. Referrring to Figure 3.4, notice that $1/l$ is the curvature of the incident wavefront at the lens, while $1/l'$ is the curvature of the emergent wavefront. If we rewrite Eq. (3.17) as $1/l' = 1/l + K$, we can see that *the power of the lens is the algebraic increment in curvature*

* Equation (3.15) is sometimes called the *lensmaker's equation*, but no lensmaker has been known to produce a lens with zero thickness! In Chapter 4, where we discuss the effects of lens thickness, we will derive an equation that involves all of the constructional parameters of a lens.

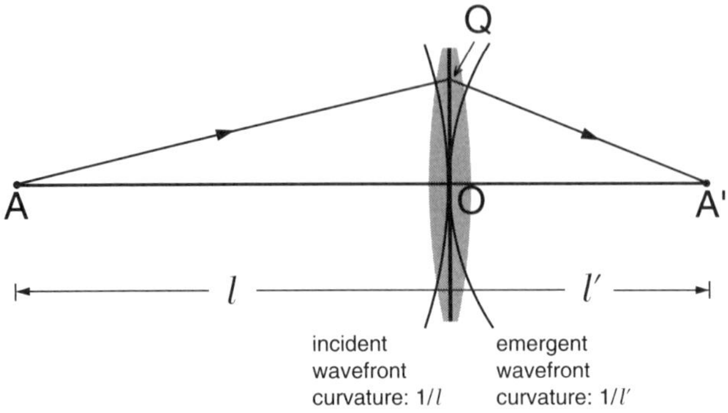

Figure 3.4. Thin-lens refraction seen from the viewpoint of wavefront curvature change.

added to the incident wavefront. This suggests a slightly different way of writing Eq. (3.17), which is often used in optometry (e.g., Keating, 1988). The term *vergence* is used to indicate the curvature of the incident and emergent wavefronts. Thus setting $V \equiv 1/l$, $V' \equiv 1/l'$, Eq. (3.17) becomes $V' = V + K$, showing that the emergent vergence is the sum of the incident one and the lens power. When writing the conjugate equation in this form, V, V', and K are all measured in dioptres.

We may also examine image formation by a thin lens from the viewpoint of Fermat's principle. Recall the basic postulates of geometrical optics (Section 1.5): The optical path length from the object A to the image A' is constant, along any ray. Consider now the ray AQA' in Figure 3.4, and also the ray AOA' along the axis. Clearly, the geometrical path AQA' is larger than AOA'. For the optical paths (or equivalently, the times of travel) to be equal, the ray AOA' must spend more time in the glass, since glass has the effect of slowing down the light by a factor n. It follows that the lens must be thicker in the middle. It is possible to derive the conjugate formula directly from Fermet's principle, through such considerations. As an exercise, the student may attempt to justify qualitatively why a negative lens, forming a virtual image, must be thinner in the middle.

3.5.3. Focal Lengths of a Thin Lens

Just as we did for a single surface, we can define the first and second focal lengths of a thin lens. From Eq. (3.17), for an object at infinity ($1/l = 0$), we obtain $l'_\infty \equiv f' = 1/K$, while for $1/l = 0$, we have $l_\infty \equiv f = -1/K$. Therefore, for a thin lens in air, the first and second focal lengths are opposite,

$$f' = -f \tag{3.18}$$

f and f' must have different signs since they are measured on opposite sides of the lens. This holds whether the lens is positive or negative. However, it is desirable to characterize a positive lens by a positive focal length, and a negative lens by a negative focal length. Thus we define the *equivalent focal length* (EFL) F as

$$F = \frac{1}{K} \tag{3.19}$$

When we speak of, for example, a lens of focal length $+5$ cm, we mean a positive lens for which $F = \pm 5$ cm.

3.5.4. This Lens Refraction: The General Case

The thin lens may not be in air. By now the reader should be able to handle this case by following the same procedure used in deriving Eq. (3.15). However, we treat this as part of the theory in order to highlight some important results.

There are two possible cases, as the media on either side of the lens may or may not be the same. Let n,n' be the object- and image-space refractive indices, respectively. Then by application of the conjugate equation for a single surface, we obtain:

$$\frac{\mu}{l'_1} - \frac{n}{l_1} = c_1(\mu - n)$$

$$\frac{n'}{l'_2} - \frac{\mu}{l'_1} = c_2(n' - \mu)$$

By adding, and setting $l_1 \equiv l$, $l'_2 \equiv l'$, we obtain:

$$\frac{n'}{l'} - \frac{n}{l} = c_1(\mu - n) + c_2(n' - \mu) \tag{3.20}$$

The right-hand side of Eq. (3.20) is defined as the power of the thin lens, K. For $l = \infty$ (and $l' = f'$), we find the second focal length from Eq. (3.20) as

$$f' = \frac{n'}{c_1(\mu - n) + c_2(n' - \mu)} = \frac{n'}{K}$$

while for $l' = \infty$, $l = f$, we have

$$f = -\frac{n}{c_1(\mu - n) + c_2(n' - \mu)} = -\frac{n}{K}$$

The equivalent focal length is still defined as $1/K$; so we have

$$f' = n'F \qquad f = -nF \tag{3.21}$$

for a thin lens not in air. In the special case $n = n'$, we obtain from Eq. (3.20)

$$\frac{1}{l'} - \frac{1}{l} = \left(\frac{\mu}{n} - 1\right)(c_1 - c_2) \tag{3.22}$$

which shows that as $n \to \mu$, $K \to 0$, as expected.

3.5.5. Many Surfaces in Contact

From Eq. (3.20), we notice that the power of the thin lens equals the sum of the powers of the individual surfaces; that is, $K_1 = c_1(\mu - n)$, $K_2 = c_2(n' - \mu)$, and

$$K = K_1 + K_2$$

This result can be immediately extended to any number of surfaces, and therefore, any number of thin lenses in contact

$$K = K_1 + K_2 + K_3 + \cdots \tag{3.23}$$

Example 3.1. The Mangin mirror (Figure 3.5) is a mirror with an additional refractive surface, which helps improve the image quality. What is the power of the mirror, if assumed thin? Notice that light encounters three surfaces, the refractive surface being counted twice. Thus we can write

$$K = K_1 + K_2 + K_3 = c_1(n - 1) + c_2(-n - n) + c_1(-1 + n)$$

or

$$K = 2c_1(n - 1) - 2nc_2$$

Notice the sign reversal for the refractive indices after reflection in this derivation.

In practice lenses have finite thickness; so the approximation provided by Eq. (3.23) becomes worse as the number of surfaces (or lenses) increases. The shape of the lens is also important, as shown in Chapter 4. Equation (3.23) should be applied with great caution, as the real power of the system can be very different from the prediction based on the thin-lenses-in-contact assumption.

The expression "thin lenses in contact" is often a source of confusion for the tyro, who may think that the contacting surface radii must match. Within the context of thin lens theory, two surfaces are "in contact" when we can ignore their separation. In practice this can only mean that the two surfaces are as close as they can get, given the limitations imposed by different radii. Thus the contact may be at the edges (a circle) or at the apexes (a point). In both cases,

Figure 3.5. The Mangin mirror.

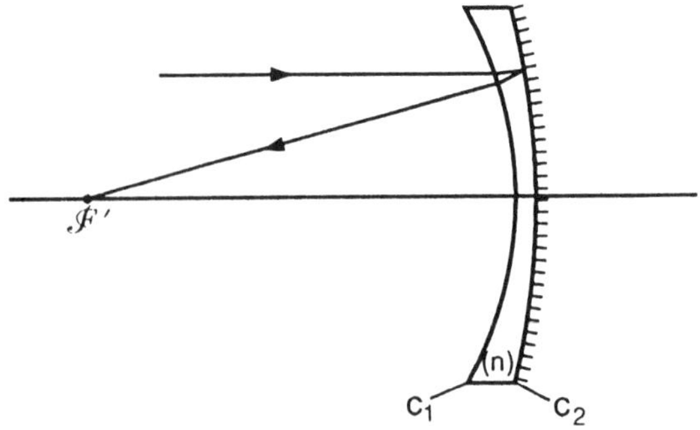

the area of contact is zero, which means that no energy is transferred from one lens to the other without going through air, unless of course the contacting surface curvatures are identical.

Such differences are encountered whenever one tries to apply the thin lens concept to real lenses. The reader should keep in mind that the thin lens is a purely theoretical and physically unrealizable concept. Its value lies in the fact that it simplifies many preliminary computations of an optical system.

3.5.6. Throw

The object-to-image distance is called *throw*, and given by

$$T = l' - l \tag{3.24}$$

the minus sign being necessary because of our sign convention. When the object and/or image is virtual, the throw can be anywhere between zero (object at the lens) and infinity (object or image at focal plane). For an important special case, let us restrict both object and image to be real. (This of course means that the lens must be positive.) Consider the lens to be in air for simplicity. We then have $1/l' - 1/l = 1/F$, and by eliminating l' from Eq. (3.24),

$$T = -\frac{l^2}{l + F}$$

For T to be minimum, we require $\partial T/\partial l = 0$. The derivative is easily shown to be

$$\frac{\partial T}{\partial l} = -\frac{l^2 + 2lF}{(l + F)^2}$$

which becomes zero for $l = 0$ and for $l = -2F$. The first case is the trivial case of the object and image coinciding at the lens. The second case is illustrated in Figure 3.6. Substitution of $l = -2F$ into the conjugate formula gives $l' = 2F$; so the minimum throw for real object and image is $T = 4F$. In this case, the magnification $M = l'/l = -1$ (see Section 3.6).

Even the casual user of optics is well advised to remember this rule: If object and image must be real, the minimum distance between them is $4F$. Inversely, if T is given, the focal length of the lens has a maximum value of $T/4$.

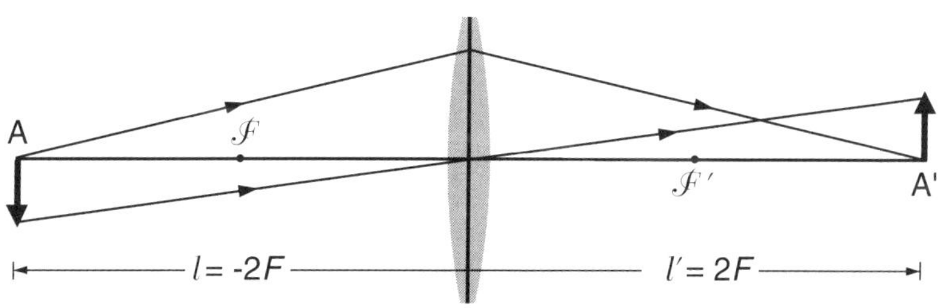

Figure 3.6. Minimum-throw condition for real conjugates. The magnification $M = -1$.

3.6. IMAGERY OF AN EXTENDED OBJECT. MAGNIFICATION

For a single refracting surface (Figure 3.7) suppose that the axial object and image points are A, A', and let η and η' be the extended object and image heights, respectively. The image is taken to be on a plane normal to the axis at A', called the *paraxial image plane.* Consider a ray from the edge of the object that passes through the apex. The angles i and i' are then the angles of incidence and refraction, for which the paraxial form of Snell's law gives $ni = n'i'$. But from the figure, $i = \eta/l$, $i' = \eta'/l'$; so by substitution, we have $n\eta/l = n'\eta'/l'$. Therefore, the magnification M is

$$M = \frac{\eta'}{\eta} = \frac{nl'}{n'l} \tag{3.25}$$

If $M > 0$, the image is erect (η, η', of the same sign), and if $M < 0$ the image is inverted (η, η' of the opposite sign).

It is instructive to point out in some detail the assumptions that are contained in the derivation of Eq. (3.25). Notice first that the location of the axial image point A' is established exactly, without any approximations, from Eq. (3.4) [Of course the paraxial approximation was used in deriving Eq. (3.4), but for a point on axis the paraxial formulae are exact.] We then demand that the rest of the image be located on the plane that is normal to the axis at A'. From what we have said so far, this is not at all guaranteed; so we must show that Eq. (3.25) holds for all paraxial rays. In Figure 3.8, take a ray from B that intersects the surface at X. Let B'_x be the intersection between this ray and the normal to the axis at A'. From the figure,

$$u'_x = \frac{\eta'_x}{l' - l'_x} + O(3)$$

$$u_x = \frac{\eta}{l - l_x} + O(3)$$

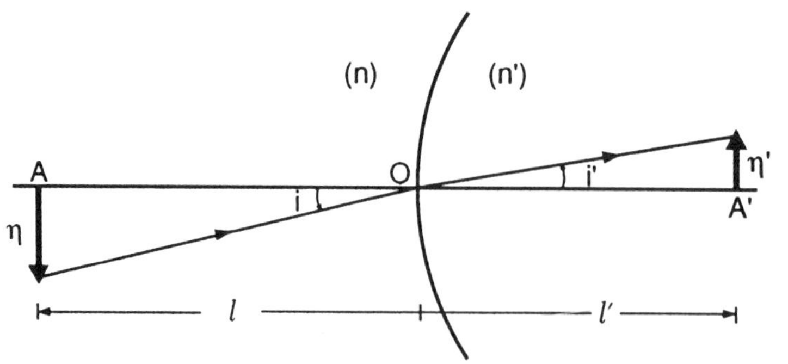

Figure 3.7. Imagery of an extended object by a single surface.

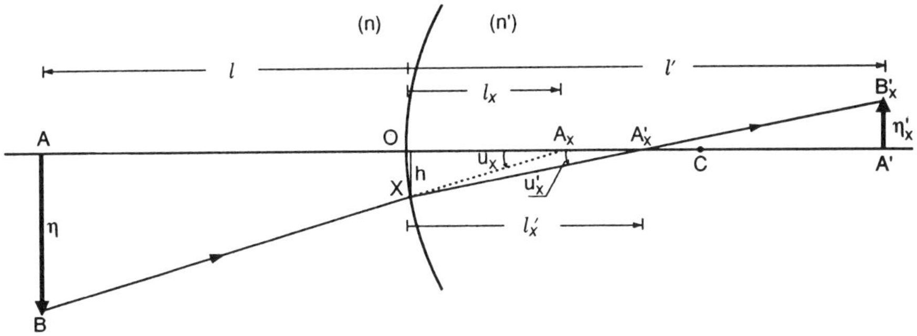

Figure 3.8. Proof that all paraxial rays from B cross the image plane at $B'_x \equiv B'$.

Therefore,

$$\frac{\eta'_x}{\eta} = \frac{u'_x}{u_x}\frac{l' - l'_x}{l - l_x} + O(3) \tag{3.26}$$

Now we apply the conjugate equation (3.4) twice, for the axial conjugates (A, A') and (A_x, A'_x), to obtain: $n'/l' - n/l = c(n' - n)$, and $n'l'_x - n/l_x = c(n' - n)$. By subtracting the last two equations, and after some rearrangement of terms, we have

$$\frac{l' - l'_x}{l - l_x} = \frac{nl'}{n'l}\frac{l'_x}{l_x}$$

Substitution of this last equation into Eq. (3.26) gives

$$\frac{\eta'_x}{\eta} = \frac{nl'}{n'l}\frac{u'_x l'_x}{u_x l_x} + O(3) \tag{3.27}$$

We also have $u_x = -h/l_x + O(3)$, and $u'_x = -h/l'_x + O(3)$. These two equations can be rewritten as $u'_x l'_x = -h + O(3)$, and $u_x l_x = -h + O(3)$. Therefore, the ratio

$$\frac{u'_x l'_x}{u_x l_x} = 1 + O(2)$$

so Eq. (3.27) becomes

$$\frac{\eta'_x}{\eta} = \frac{nl'}{n'l} + O(2)$$

In the paraxial approximation, $O(2)$ is neglected, and we obtain Eq. (3.25). Since point X is arbitrary, we have proven that all paraxial rays from B will intersect the plane that is normal to the axis at A', at a common point, whose height is given by $\eta nl'/n'l$.

In determining the magnification formula for a thin lens, we have once again the opportunity to appreciate the convenience of a consistent notation. Thus let the object be of height η_1, at a distance l_1. The first surface produces an image of height η'_1, at a distance l'_1. Since this image becomes the object for the second surface, the object height $\eta_2 = \eta'_1$, and the object distance $l_2 = l'_1$, for a thin lens. The total magnification will be given by the ratio of the final image height to the object height,

$$M = \frac{\eta'_2}{\eta_1} = \frac{\eta'_2}{\eta_2}\frac{\eta'_1}{\eta_1} = M_2 M_1 \tag{3.28}$$

where M_1 and M_2 are the magnifications produced by the individual surfaces. (Notice that the formula $M = M_1 M_2$ does not depend on the assumption that the two surfaces are in contact!) Therefore, using Eq. (3.25),

$$M = \frac{\eta_2 l'_2}{n'_2 l_2}\frac{n_1 l'_1}{n'_1 l_1} = \frac{n_1 l'_2}{n'_2 l_1}$$

since $n'_1 = n_2\ (=\mu)$ and $l'_1 = l_2$. We can now simplify the notation by dropping the subscripts. Let the object distance be l, the image distance l', and let the object- and image-space media have refractive indices n and n', respectively. The thin lens magnification is then given by

$$M = \frac{nl'}{n'l} \tag{3.29}$$

which is the same as for a single surface. Notice that the refractive index of the lens μ does not appear in the formula. If we have the same refractive index on both sides of the lens, then $M = l'/l$.

For a mirror, $n' = -n$; therefore,

$$M = -\frac{l'}{l} \quad \text{(mirror)} \tag{3.30}$$

3.7. THE ONE-COMPONENT DESIGN PROBLEM

A number of imaging situations can be handled with only a single component. For example, we may want to form a real image of a distant object on photographic film; a single positive lens can perform this task, in the sense of providing an image at the desired distance and with the desired magnification.*

The general one-component problem, with the lens in air for simplicity, is summarized as follows. We have three equations, $1/l' - 1/l = 1/F$, $M = l'/l$, and $T = l' - l$. These equations contain five unknowns, l, l', T, M, F. Given any two of these, the other three can be found.

* In the context of this discussion, "one component" means a single thin lens. In practice, however, it means any grouping of lenses whose action can be reduced to that of a single thin lens, for the purposes of calculating the image distance and magnification. For an everyday example, notice that a typical camera lens comprises more than one lens, yet it is given a focal length $\mathscr{F}$, just like a single thin lens. A detailed discussion of how this reduction is made can be found in the next chapter; for now, it suffices to note that it can be made.

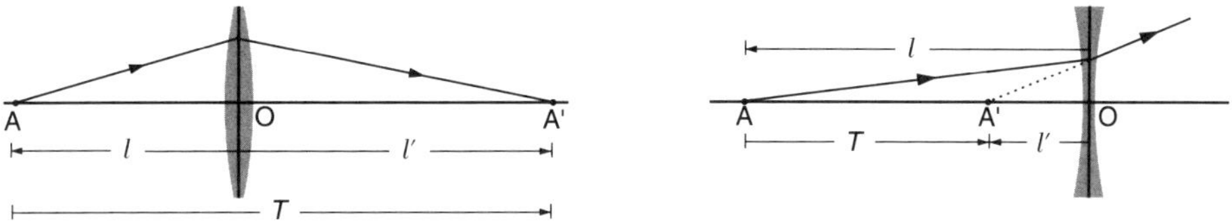

Figure 3.9. The segment $(AA') \equiv T$ is divided internally or externally in the ratio 1:M by the lens.

Often, the throw and magnification are given. In this case, using the above three equations, the student can show that

$$F = -\frac{TM}{(M-1)^2} \tag{3.31}$$

Figure 3.9 shows the geometry. The lens divides the distance T internally (positive lens) or externally (negative lens) in the ratio 1:M.

In the one-component problem, the following equations are also found very useful:

$$l' = (1 - M)F \qquad \text{and} \qquad l = (1/M - 1)F \tag{3.32}$$

proof of which is left as an exercise for the student.

Example 3.2. Consider a projection lens, such as found in a slide projector. Suppose that the slide-to-screen distance T is 10 m, the slide semidiagonal (maximum object height) is 2 cm, and the required image size (semidiagonal) is 1.6 m. From Chapter 2, we recall that when both object and image are real, the lens must be positive, and further, that the image is inverted. Therefore, we must set $M = -160/2 = -80$. With these numbers, we obtain $F \cong 12.2$ cm. If we had not realized that we should set $M < 0$, we would have obtained $\mathscr{F} < 0$, which should warn us that something is wrong, as a negative lens cannot produce a real image from a real object. (Remember: Until you are confident, it is a good idea to make a simple ray diagram to verify your algebraic solution.)

Example 3.3. Suppose that we are asked to specify a mirror that must produce an inverted real image, three times larger than the (real) object, and with the image 1 m away from the object. With these data, if we set $T = l' - l = 1$ m and $M = -l'/l = -3$, we obtain $l = 0.5$ m, $l' = 1.5$ m. Further, by applying $1/l' + 1/l = 2c$, we get $c = 1.333\ \text{m}^{-1}$. Unfortunately, there is a problem with this solution: We have obtained a convex mirror (positive c), and both object and image distances are positive, that is, behind the mirror, for light coming from the left. Therefore, object and image are virtual (draw a diagram to verify!). So we have obtained a consistent solution, but not the one we want.

To solve this problem, we must realize that if both object and image are to be real, they must lie to the left of the mirror; so both l' and l must be negative. From the magnification formula, we obtain $M = -l'/l = -3 \rightarrow l' = 3l$; so $|l'| > |l|$. This tells us that the image will be farther away from the mirror than the object; therefore, the throw $T = l' - l < 0$. Therefore, setting $T = -1$ m, we obtain $l = -0.5$ m, $l' = -1.5$ m, and $c = -1.333\ \text{m}^{-1}$. A ray diagram will convince us that we now have the right solution.*

While some optical problems have a one-component solution, many do not. For example, mechanical constraints may make it impossible to place the lens at the required position. Or, since $F \propto T$, the required T might make F impossibly small; for $T = 0$, only the trivial solution $l = l' = 0$ is possible with a single lens. In Chapter 4, we examine the two-component solution, which removes these limitations.

3.8. OTHER TYPES OF MAGNIFICATION

3.8.1. The Angular Size of an Object

We should all be familiar with the common magnifying glass. How do we go about specifying such a lens? We may begin with the constraints, of which we can immediately see three: (1) $M > 1$, if it is to be a magnifier; (2) the object is real; and (3) the image must be located at least 25 cm in front of the observer, if it is to be clearly visible. From these, and recalling the graphical ray-tracing results of Chapter 2, we conclude immediately that the lens must be positive, and the object must be placed between the first focal point and the lens. A casual glance at the formula $M = l'/l$ would seem to suggest that the farther away the image (the larger l'), the greater the magnification, which should be advantageous. At the limit, if we set $l' = \infty$, then we obtain $M = \infty$, which sounds too good to be true. Indeed, there are two obvious difficulties with this: (1) As experience suggests, the image does not look infinitely large, and (2) the conjugate equation degenerates to $F = -l$; so no conclusion may be drawn about F. What are we missing?

The answer is that when either the object or the image distance approaches infinity, the concept of the linear magnification, η'/η, loses its meaning. If the image is at infinity, it must also have an infinite linear size, if it is to subtend a finite angular size (that is, if it is to be visible). Similarly, however small the angular size may be, assuming it remains finite, it corresponds to an infinite linear size. Thus it is not the linear size of the image η', that is of importance in this case, but the angular size, β'. Figure 3.10 shows the object at the focal plane; β' is conveniently determined by drawing the undeviated ray through the middle of the lens. (All other paraxial rays from B will emerge parallel to

---

*Of course, we could have avoided all this discussion by wording the problem so as to give $T = -1$ m at the beginning. But you should be aware that your boss or client will not take pains to pose the problem in a way consistent with your sign convention!

Figure 3.10. When the image is at infinity, its angular size is $\beta' = \eta/\mathscr{F}$.

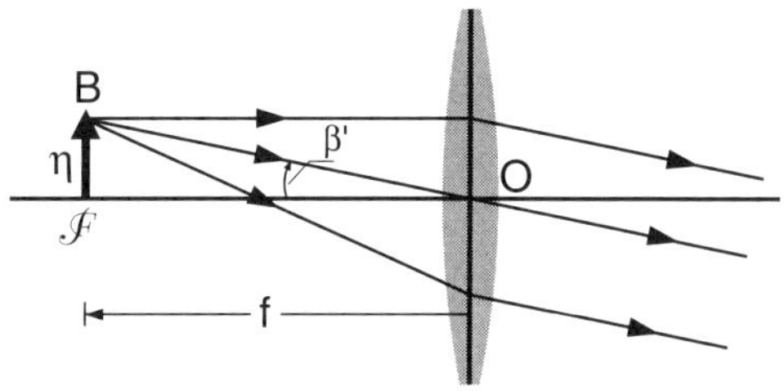

BO.) In this case, then, the angular image size is given by $\beta' = \beta = \eta/f = -\eta/F$. This shows that the smaller F, the larger β'; so we conclude that a lens with as short a focal length as possible is needed, if we want maximum magnification. Therefore, by introducing the concept of angular size, we have resolved both difficulties mentioned in the previous paragraph.

Since in the paraxial approximation all angles are infinitesimal. the discussion of a finite angular object size is attempting a rather fine distinction. Once again, we are anticipating the discussion of Chapter 5, that the paraxial equations can be used with finite quantities within a reasonable approximation. Alternatively, we may simply consider that our discussion is limited to the region where $\sin\theta = \theta$ to within a desired degree of accuracy.

If we are using finite quantities and we are seeking a high degree of accuracy, then there are three choices for representing the angular object size: β, $\sin\beta$, and $\tan\beta$. Clearly, all three are equivalent paraxially, but when their differences become important, it is shown in Chapter 5 that the optimum choice for the angular object size is $\tan\beta$.

3.8.2. Visual Magnification

If an object is to be viewed through a magnifier, what is an appropriate definition of magnification? Here, rather than the ratio of image to object size, it is more meaningful to determine the gain in apparent size obtained through the use of the lens. Thus the apparent image size obtained with the lens must be compared to the maximum object size as seen with no lens. Naturally, the closer the object, the larger it looks and the more detail can be resolved, until it reaches the minimum distance of clear vision, called the *near point*. This is located at about 25 cm from the eyes of people in their thirties with normal vision. Therefore, we define the visual magnification as

$$M_{\text{vis}} = \frac{\beta'}{\beta_{25}} \tag{3.33}$$

where β' is the angular size of the image looking through the lens, and β_{25} the angular size of the object, when located 25 cm away from the eye. From Figure 3.11, $\beta_{25} = \eta/(-25\text{ cm})$.

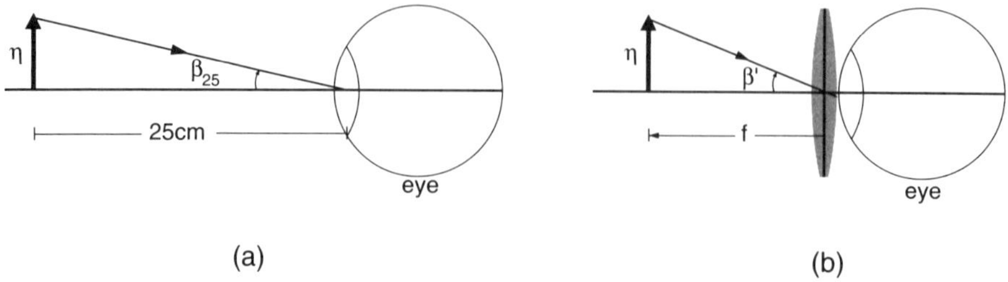

Figure 3.11. To define a meaningful visual magnification, we compare the angles β_{25} and β', for (a) unaided versus (b) aided vision. All lenses are assumed thin and in contact.

If the image formed by the lens is at infinity, we obtain

$$M_{\text{vis}} = -\frac{\eta/F}{-\eta/25\text{ cm}} = \frac{25\text{ cm}}{F} \tag{3.34}$$

showing again that a small F is required for large magnification. However, the maximum visual magnification is not obtained with the image at infinity, but with the image at the near point. Assuming the eye to be in contact with the lens (Figure 3.12), we have $l' = -25$ cm, and the angular image size is $\beta' = \eta'/l' = \eta/l$. The conjugate equation gives $1/l = 1/l' - 1/F$. By substituting into Eq. (3.33), we obtain

$$M_{\text{vis}} = \frac{25\text{ cm}}{F} + 1 \tag{3.35}$$

This is the maximum value of visual magnification that can be achieved with a simple magnifier.

Before closing this section, we note that it is possible to define an angular magnification to be the angular image size divided by the angular object size. However, this concept is most useful when both the object and the image are at infinity, a condition that cannot be satisfied with a single lens. We therefore postpone the discussion of angular magnification until the next chapter.

Figure 3.12. Maximum visual magnification is obtained with the image at the near point ($l' = -25$ cm).

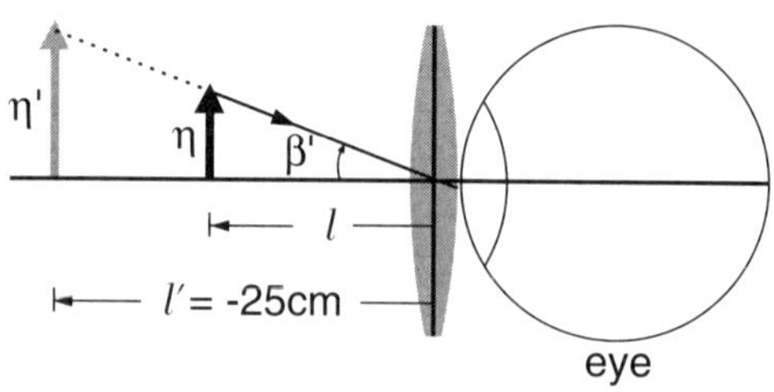

3.8.3. Longitudinal Magnification. Imagery of a Volume

In certain applications, a lens is required to image a volume rather than a plane. We must then consider the imaging properties of the lens not only across, but also along the axis. Suppose for the moment that the object is an axial element of length dl. Differentiation of the conjugate formula (3.17) gives $dl'/l'^2 - dl/l^2 = 0$, from which we obtain the longitudinal magnification

$$M_{\text{long}} = \frac{dl'}{dl} = \left(\frac{l'}{l}\right)^2 = M^2 \tag{3.36}$$

where M stands for the *transverse* magnification, η'/η. Note that when we simply say "magnification," the word *transverse* is normally implied. Thus we see that the longitudinal magnification is the square of the transverse one. In practice, we are usually interested in a finite axial segment δl, rather than an infinitesimal quantity dl. From the derivation of Eq. (3.36), we see that M_{long} varies with l, so Eq. (3.36) is only an approximate formula for small δl. We derive here an exact formula that applies for finite δl.

Consider a cube of side length a, as in Figure 3.13. What does its image look like? Take first $ABCD$ as the object plane, and determine its image $A'B'C'D'$. Let the transverse magnification be M_0 between these two planes. Plane $EFGH$ is located a small distance δl away from $ABCD$. Its image, $E'F'G'H'$, is located at a distance $(K'L')$ away from $A'B'C'D'$, where $(K'L')$ is the image of the axial segment (KL). On the basis of Eq. (3.36), we expect that $(A'B') = M_0 a$, while $(K'L') = M_0^2 a$. But if the size

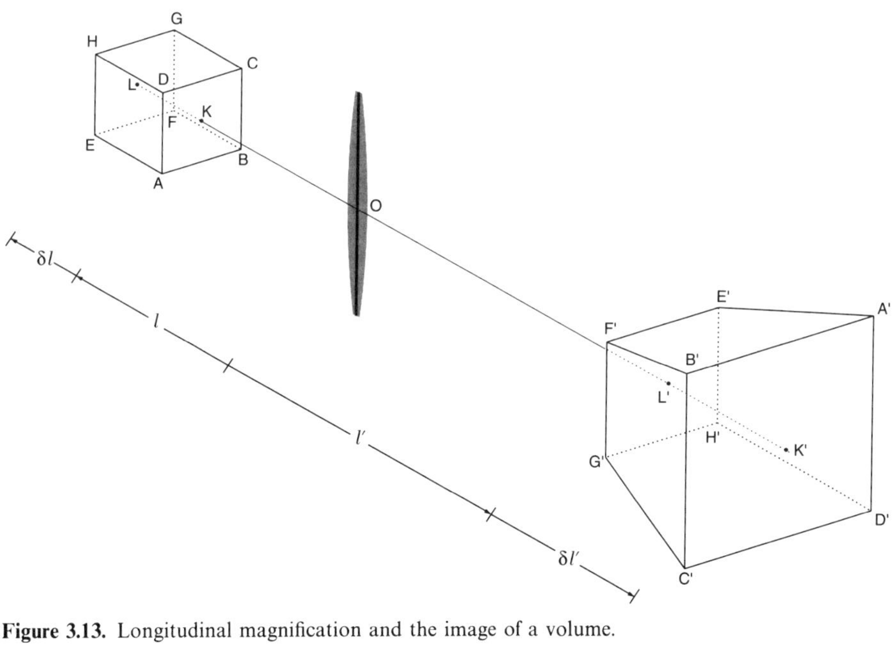

Figure 3.13. Longitudinal magnification and the image of a volume.

a is small enough for Eq. (3.36) to hold, then $EFGH$ would be imaged with a magnification that is approximately equal to M_0. This would imply that the image of a cube is a parallelepiped, but unfortunately, this conclusion is wrong. In reality, the image of a cube is a truncated pyramid, as Figure 3.13 shows. To prove this, let $(OK) = l$, $(OK') = l'$. and $(OL) = l + \delta l$, $(OL') = l' + \delta l'$. The distances δl and $\delta l'$ are considered as finite. Further, let the magnification between K and K' be M_0, and that between L and L' be M. We then have the following four equations:

$$\frac{1}{l' + \delta l'} - \frac{1}{l + \delta l} = \frac{1}{F}$$

$$\frac{1}{l'} - \frac{1}{l} = \frac{1}{F}$$

$$M = \frac{l' + \delta l'}{l + \delta l}$$

$$M_0 = \frac{l'}{l}$$

By subtracting the first two, and after a little algebra, we obtain

$$\frac{\delta l'}{\delta l} = \frac{l'}{l}\frac{l' + \delta l'}{l + \delta l}$$

which by virtue of the two magnification formulae becomes

$$M_{\text{long}} = \frac{\delta l'}{\delta l} = M_0 M \tag{3.37}$$

This equation reduces to Eq. (3.36) when δl and $\delta l'$ are very small, that is, $M_0 \cong M$. However, since in general $M_0 \neq M$, we have justified the truncated pyramidal shape. The condition $M_0 = M$ is satisfied exactly in the case of telescopic systems (discussed later), even when δl is finite. Thus for telescopic systems, the image of a cube is a parallelepiped.

PROBLEMS

1. For Figure 3.14 (see page 59), give the sign of all the quantities indicated. Also verify that the relation $u = -h/l$ applies in this case, too.

2. For Figure 3.15 (see page 59), give the sign of all the quantities indicated. (Note: light travels from right to left.)

3. A spherical mirror forms an erect image that is one-third the size of the object. What is the mirror curvature? If the object is 100 cm from the vertex, determine the image distance l'.

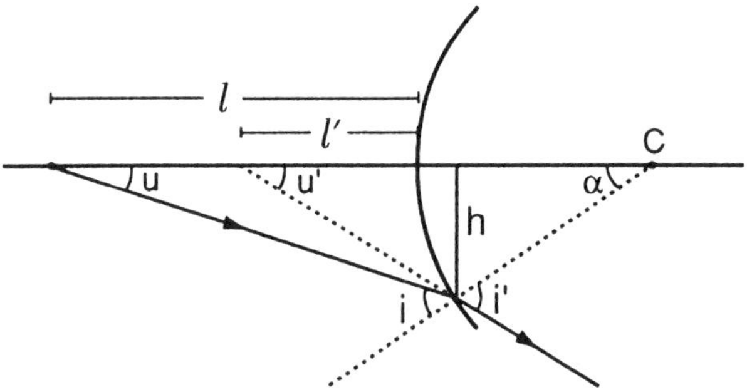

Figure 3.14

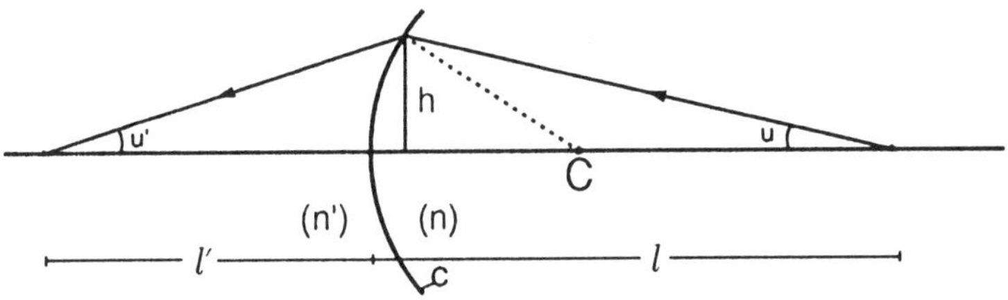

Figure 3.15

4. A thin lens of 5 cm focal length forms an image of a real object located 15 cm away. What is the magnification?

5. We must produce an upright, magnified image of an object located 3 cm away from a convenient lens holder. What is the focal length of the lens, if the required magnification is $4\times$?

6. (a) Calculate the position and magnification of the image of a small object located in air, 300 mm to the left of a convex glass surface ($n = 1.5$) of radius of curvature 100 mm. Consider refraction at this single surface only, (b) A virtual object is located 300 mm to the right of the same surface (again, light is incident from the left). Where is the image and what is its magnification?

7. The dioptre (D) is a unit of lens or mirror power, equal to $1\ \text{m}^{-1}$. (Note: the same name is also used as a unit of curvature $c = 1/R$.) A planoconvex lens is known to have a power of $+3$D. The radius of the curved face is measured with a spherometer, and found to be equal to 22.97″ cm. What is the index of refraction of the glass?

8. A negative meniscus lens has surface radii of $+15$ and $+10$ cm, and it is made of glass with $n = 1.6$. (a) Sketch the lens; (b) find its focal length and

power; (c) verify that the power is the same if the lens is turned around, so that both radii become negative (sketch the lens again!).

9. An equiconcave lens ($|R_1| = |R_2| = 20$ cm, $n = 1.5$) is placed in contact with a positive meniscus ($R_1 = 15$ cm. $R_2 = 25$ cm, $n = 1.6$). Determine the power of the combination, treating both lenses as thin and neglecting their separation. Sketch the system.

10. A thin positive lens of power K casts a real image on a screen, m times larger than the object. Determine the lens-to-screen distance.

11. For a thin lens in air, show that $l' = (1 - M)\mathscr{F}$, and $l = (1/M - 1)\mathscr{F}$, in the usual notation.

12. For a given system, we have $T = 200$ mm, and we require a real image. We have two lenses at our disposal, $F_1 = 40$ mm and $F_2 = 60$ mm. Which one would you use and why? (Jusify your answer by determining l or l' in each case.)

13. A single thin lens ($c_1 = 0.2\ \text{cm}^{-1}$ $c_2 = -1\ \text{cm}^{-1}$, $n = 1.5$) has water as the object-space medium ($n_w = 1.33$). The object is 3 cm away from the lens. Describe what you would see if your eye were (a) very close to the lens. (b) 50 cm away from the lens, and (c) 10^7 cm away from the lens.

14. Use the data of Problem 3.13, but with $c_1 = 0.1\ \text{cm}^{-1}$ and $c_2 = -0.2\ \text{cm}^{-1}$. Determine the location of the image and the magnification. Sketch the system and trace a ray from the axial object point.

15. You are an optics consultant, and your first customer is a myopic swimmer who would like her swimming goggles to correct her vision, just like her eyeglasses (which have a power of -3D). The catch: The goggles must work in water as well as in air. Can you specify a lens that would do the job?

16. The visual magnification of a simple magnifier is given as (a) $5\times$, (b) $250\times$. What is its focal length? (assume image at infinity). [Note: Case (b) might seem extreme, but was in fact achieved by one of the most famous microscopists of all time, Anton van Leeuwenhoek, in the seventeeth century. But the thin-lens approximation gives only a very rough idea of the behavior of Leeuwenhoek's lenses.)

17. For a given throw T, there are two possible lens positions, symmetrical about the middle of the segment AA' (see Figure 3.9). If these two positions are separated by d, show that $F = (T^2 - d^2)/4T$. (Notice that this equation can provide a method of measuring the focal length of a lens through T and d.)

18. We are observing visually through a negative thin lens two identical rulers, one in the left half of the lens and one in the right half. The two rulers are located at different distances from the lens, l_1 and l_2 (assumed known), and are therefore seen at different magnifications. If m divisions of the left ruler appear equal in size to n divisions of the right ruler. determine the focal length of the negative lens. (Try doing this as an experiment!)

19. You have a positive lens of $F = 50$ mm. The object is a tilted line segment 10 mm long. The middle of the segment is located 70 mm away from the lens, on the axis. The angle between the segment and the axis is 45°. Draw a diagram showing the approximate form of the image (justify your answer with calculations).

Gaussian Optics 4

"First I prepared a tube of lead, at the ends of which I fitted two glass lenses, both plane on one side while on the other side one was spherically convex and the other concave. Then placing my eye near the concave lens I perceived objects satisfactorily large and near, for they appeared three times closer and nine times larger than when seen with the naked eye alone."

Galileo Galilei, *The Starry Messenger,* A.D. 1610 (translated by Stillman Drake)

In this chapter we examine systems that comprise two or more surfaces that are not in contact. We consider first the passage of rays through each surface in succession, as the obvious means of determining the image. Then we show that even complex optical systems can be reduced to a set of only three parameters (the equivalent focal length and the location of a pair of conjugate planes), for the purpose of determining the image paraxially. These two approaches are complementary; both are necessary in order to understand and evaluate multilens systems. The ideas on which this chapter is based were first developed by Gauss; so the general process and methods bear his name.

4.1. THE PARAXIAL HEIGHT AND ANGLE VARIABLES

In the derivation of the single-surface conjugate formula of the previous chapter, we encountered briefly the quantities u and h, which we took to represent the angle between a paraxial ray and the axis, and the intersection height of the ray at the surface, respectively. When dealing with multielement systems, it is more common to use the variables u and h instead of the conjugate distance l. Recall that these three variables are related through $u = -h/l$; so there is no fundamental difference between the two approaches. But there are several advantages in using u and h, and we will point these out as we encounter them. Let us first rederive the singe-surface conjugate equation in terms of the new variables. Substitute $\alpha = -h/R$ into Eq. (3.3) to obtain (see also Figure 3.1):

$$n(u + h/R) = n'(u' + h/R)$$

After rearranging terms, and using $1/R \equiv c$, we obtain

$$n'u' - nu = -hc(n' - n) \tag{4.1}$$

or

$$\boxed{n'u' - nu = -hK} \tag{4.2}$$

This is the new form of the paraxial refraction equation. We can immediately see an advantage that is important when performing manual calculations: The previous variables l,l' appeared in the denominator, whereas the new form of the conjugate equation contains only products. This simplifies the algebra considerably, when dealing with more than one or two lenses.

In deriving Eq. (4.1), we have assumed that the angles u and u' are infinitesimal, that is, that the paraxial approximation holds. However, in practice, we can only insert finite values for u and u'. Thus a question arises about the validity of inserting finite values into a paraxial formula (notice that this was not a problem previously because l and l' are normally finite). The answer is that Eq. (4.1) provides an exact description of paraxial rays and a first-order approximation to the behavior of real rays (that is, rays that form a finite angle with the axis or are at a finite distance away). In Chapter 5, we explain in more detail that the validity of this approximation increases if u is taken to represent not an angle, but the sine or tangent of an angle. This is the reason why we call u the *paraxial angle variable*, rather than the paraxial angle.

Let us now derive the conjugate equation for a thin lens. Using Eq. (3.20), and substituting $u = -h/l$, $u' = -h/l'$, we obtain:

$$-\frac{n'u'}{h} + \frac{nu}{h} = c_1(\mu - n) + c_2(n' - \mu)$$

The right-hand side represents the sum of the powers of the two surfaces, $K_1 + K_2 = K$, where K the power of the thin lens in the general case, with different media on either side. Therefore, the thin-lens conjugate equation becomes

$$n'u' - nu = -hK \qquad \text{(thin lens)}$$

which is formally the same as Eq. (4.2) but with the understanding that K represents a different quantity. From this equation, we can see that if $h = 0$ (ray through the middle of the lens), and if $n' = n$, then $u' = u$; that is, the ray emerges undeviated.

For both a single surface and a thin lens, we have seen that the magnification is given by $M = nl'/n'l$. Upon substituting l and l' in terms of u, u', we immediately obtain:

$$\boxed{M = \frac{nu}{n'u'}} \tag{4.3}$$

4.2. PARAXIAL RAY TRACING FOR SYSTEMS OF MANY SURFACES

4.2.1. Notation

During the thin lens discussion of the previous chapter, we mentioned the advantages of a proper notation. The consistency of the notation becomes paramount when dealing with a system of several surfaces. It is no exaggeration to say that the entire problem of ray tracing through many surfaces can be reduced to one of devising an appropriate notation.

In Figure 4.1 we have a system of k surfaces, generally not in contact. For all surfaces, all image-space quantities are primed and denoted by the same symbol as the corresponding object-space quantities. Every quantity, whether it refers to the system (such as c, n, d) or to the ray (such as u, h) is given a subscript to denote the surface number with which it is associated. Since the image space for surface i becomes the object space for surface $i + 1$, there is some redundancy in the notation, which helps to avoid confusion and ambiguity. For example, the image height for surface i is the same as the object height for surface $i + 1$, $\eta_i' = \eta_{i+1}$. Similarly, $u_i' = u_{i+1}$, and $n_i' = n_{i+1}$.

4.2.2. Magnification

There is a very convenient formula for the magnification produced by a system of many surfaces, which is made possible by the introduction of the new variables, h and u. Referring to Figure 4.1, the magnification is:

$$M = \frac{\eta_k'}{\eta_1} = \frac{\eta_1'}{\eta_1}\frac{\eta_2'}{\eta_2} \cdots \frac{\eta_{k-1}'}{\eta_{k-1}}\frac{\eta_k'}{\eta_k} \tag{4.4}$$

since $\eta_1' = \eta_2$, $\eta_2' = \eta_3, \ldots$. Using Eq. (3.25), we then have

$$M = \frac{n_1}{n_k'}\frac{l_1'}{l_1}\frac{l_2'}{l_2} \cdots \frac{l_k'}{l_k} \tag{4.5}$$

since intermediate refractive indices cancel. Apart from being cumbersome, this formula can create computational problems when any intermediate object or image is either at infinity or at the surface. An alternative formula can be

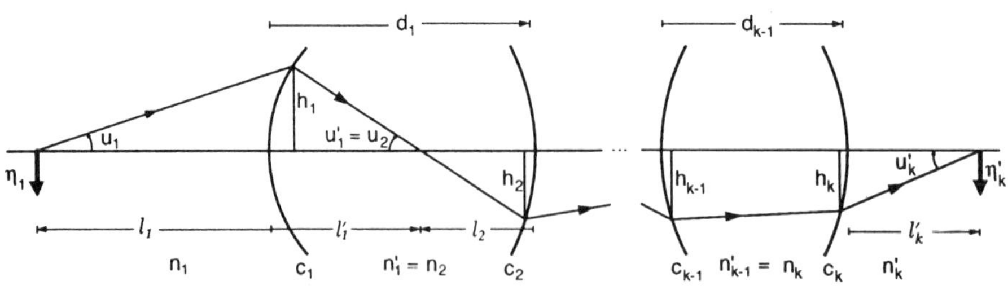

Figure 4.1. A consistent notation for paraxial ray tracing through many noncontacting surfaces.

determined in terms of the paraxial angle variables: From Eqs. (4.4) and (4.3) we get

$$M = M_1 M_2 \cdots M_k = \frac{n_1 u_1}{n_1' u_1'} \frac{n_2 u_2}{n_2' u_2'} \cdots \frac{n_k u_k}{n_k' u_k'}$$

Since $u_i' = u_{i+1}$ and $n_i' = n_{i+1}$, all intermediate angles and refractive indices cancel, and we have

$$\boxed{M = \frac{n_1 u_1}{n_k' u_k'}} \tag{4.6}$$

A comparison of Eqs. (4.5) and (4.6) shows the considerable simplification achieved through the use of the paraxial angle variable. Also, because Eq. (4.6) is independent of all intermediate object and image locations, it remains valid even when one or more of them are zero or infinite.

4.2.3. Paraxial Ray Tracing through Many Surfaces

The general problem is stated as follows: Given an optical system comprising any number of surfaces with known curvatures, separations, and refractive indices, determine the image of an arbitrary object. The problem is easy to solve after one has established a consistent notation and sign convention. (In fact, it requires so little intelligence that it can be done by computer or pocket calculator.) There are two steps that are repeated for each surface: refraction through the surface, and transfer to the next surface. Thus we have, for the ith surface (refer also to Figure 4.1) for refraction and transfer, respectively:

$$\frac{n_i'}{l_i'} - \frac{n_i}{l_i} = c_i(n_i' - n_i) \tag{4.7}$$

$$l_{i+1} = l_i' - d_i \tag{4.8}$$

Using the additional obvious relation $n_i' = n_{i+1}$, these two equations can be easily programmed. The negative sign on the right-hand side of Eq. (4.8) is necessary because of the sign convention: The surface separation d_i is measured from the ith to the $i + 1$ surface, and is taken as positive if it is from left to right, as usual.

More often, we use the formulae that contain the paraxial angle variable u, instead of Eqs. (4.7) and (4.8). These are, for refraction and transfer, respectively:

$$\boxed{n_i' u_i' - n_i u_i = -h_i c_i (n_i' - n_i)} \tag{4.9}$$

$$\boxed{h_{i+1} = h_i + d_i u_i'} \tag{4.10}$$

where $n_i' = n_{i+1}$ and $u_i' = u_{i+1}$. The transfer formula is derived through simple geometry, as in Figure 4.2. But we can also derive Eq. (4.10) without any reference to a diagram, by substituting $u_{i+1} = u_i' = -h_{i+1}/l_{i+1}$ and $u_i' = -h_i/l_i$ into Eq. (4.8).

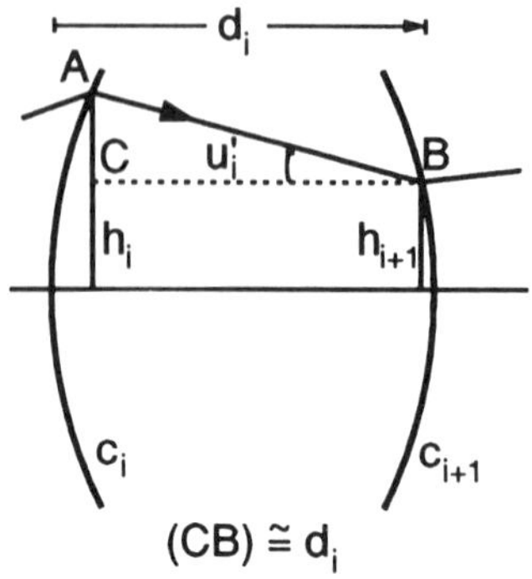

Figure 4.2. Derivation of the transfer formula: From triangle ABC, $(AC) = h_i - h_{i+1} = -d_i u_i'$ with $h_i - h_{i+1} > 0$, $d_i > 0$, $u_1' < 0$, as shown.

To begin a ray trace, we need h and u at the first surface, and then we begin by applying the refraction equation. Sometimes the object itself is taken as the first surface, in which case we begin by applying the transfer equation between object and first surface.

Since both the location of the image and the magnification can be found using a single ray, the reader may well wonder if anything can be gained by tracing any more rays paraxially. In Chapter 5, we show that all Gaussian properties of the system are established by tracing two rays; the second ray gives us information about the size of the individual lenses. The need for this information has been ignored so far, but it is easy to imagine its relevance in a practical situation.

The repetitive nature of paraxial ray tracing lends itself to different schemes that attempt to make it appear easier or more systematic. The success and value of these schemes is largely a matter of individual taste. A manual scheme for tabulating the results of the two preceding equations usually comes under the name "*y-u*" or "*ynu*" trace, where the symbol y is used in place of h for the paraxial height variable. It is also possible to put the refraction and transfer equations in matrix form, which allows one to characterize the entire system with a single matrix. We develop this approach briefly in Appendix 1 because it is relatively popular, especially in the field of optical resonator design. It should be realized, however, that in most cases the amount of computation is not reduced by the matrix methods. Also, there are some mathematical facts peculiar to the approach that must be memorized, but do not add to the physical understanding obtained so far.

Example 4.1. Consider two positive thin lenses in air, of focal lengths F_1 and F_2, separated by a distance $d = F_1 + F_2$ (Figure 4.3a). Determine the image location and magnification for an object located at the primary focal point $\mathscr{F}_1$.

Take any ray from $\mathscr{F}_1$ that intersects the first lens at a height h_1. The ray will emerge parallel to the axis after refraction ($u_2 = u_1' = u_1 - h_1/F_1 = 0$). After the second refraction, it will cross the axis at $\mathscr{F}'_2$, the secondary focal point of lens 2. The transfer equation gives $h_2 = h_1$; so the angle $u_2' = u_2 - h_2/F_2 = -h_1/F_2 = -u_1 F_1/F_2$. The magnification is $M = u_1/u_2' = -F_2/F_1$.

Note that by using u and h we did not have to deal with the intermediate

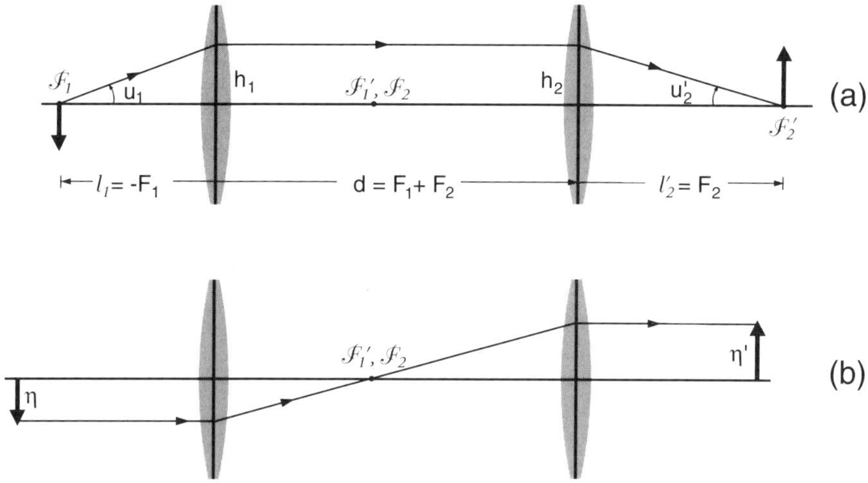

Figure 4.3. A system of two lenses separated by the sum of their focal lengths.

object and image distances that are infinite. In this case, it has turned out that the magnification $M = -F_2/F_1 = l'_2/l_1$, but remember that the intermediate distances do not normally cancel.

The system of Figure 4.3 has a very interesting property: The linear magnification is always equal to $-F_2/F_1$, independent of the location of the object. This can be shown easily by taking a ray from the top of any object of height η that is parallel to the axis in the object space. This ray will cross the axis at the common focal point of the two lenses, and will emerge parallel to the axis after the second lens, at a height η' (Figure 4.3b), which is the image height (since the ray originated from the top of the object). Similar triangles give $\eta'/\eta = -F_2/F_1$. For practice, the reader is advised to attempt solving this problem using Eqs. (4.7), (4.8), and (4.5), or the alternative set, (4.9), (4.10), and (4.6).

Example 4.2: Backward ray tracing. Recall our sign convention: Light travels from left to right unless otherwise stated, and if light travels from right to left, then the refractive index is taken as negative. Normally, there is no need to ray trace an entire system from right to left, because it is just as easy to reverse it and trace it from the left, or to reverse the roles of object and image. But in systems that include reflecting surfaces, ray tracing from right to left is unavoidable. As we did with mirrors, we can still apply the same equations for refraction and transfer, except that in addition to changing the sign of n, *we must also change the sign of the axial separation d between surfaces when light travels from right to left*. The reason for this is illustrated in Figure 4.4, which shows a schematic of a so-called Cassegrain mirror, used in telescopes. Surface 1 is the one that the ray encounters first. The separation d between surfaces 1

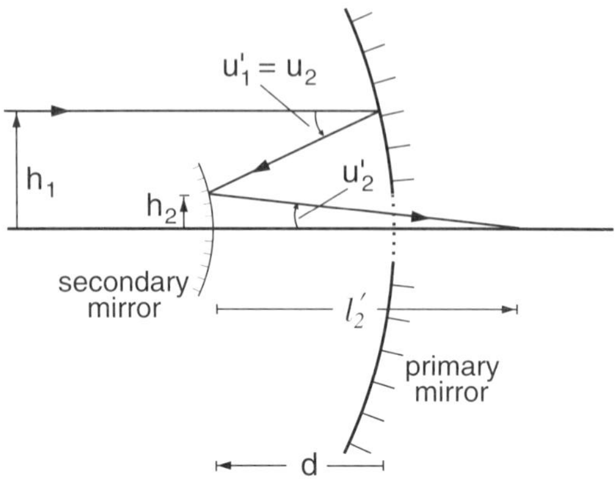

Figure 4.4. The Cassegrain mirror.

and 2 is always measured from surface 1 to surface 2, but in this case d is to the left of surface 1; so it is negative.

For a numerical example, take $R_1 = -400$ mm, $R_2 = -100$ mm, $n_1 = n'_2 = 1$, $n_2 = n'_1 = -1$, and $d = -160$ mm, and let us determine the image location for an object at infinity. All we need to do is trace a ray from the axial object point through the system. In this case, the ray will be parallel to the axis ($u_1 = 0$), and we can assume any arbitrary height h_1. The refraction formula (4.9) gives

$$n'_1 u'_1 = -h_1 c_1 (n'_1 - n_1) \rightarrow u'_1 = -2h_1 c_1$$

The transfer formula (4.10) gives

$$h_2 = h_1 + du'_1 = h_1 - 2dh_1 c_1 = h_1(1 - 2dc_1)$$

Finally reflection at the second surface gives

$$\begin{aligned} u'_2 &= -u'_1 - h_2 c_2 (n'_2 - n_2) = -u'_1 - 2h_2 c_2 = 2h_1 c_1 - 2c_2 h_1 (1 - 2dc_1) \\ &= 2h_1(c_1 - c_2 + 2dc_1 c_2) \end{aligned}$$

The image distance from the second mirror is

$$l'_2 = -h_2/u'_2 = -\frac{1 - 2dc_1}{2(c_1 - c_2 + 2dc_1 c_2)}$$

which is independent of h_1, as expected. Substituting the values given, we obtain $l'_2 = 200$ mm. Of course, this example was only used to illustrate the method: If all that is required is the image location, then it may be simpler to apply the equations involving l and l'.

4.3. THE OPTICAL INVARIANT

By writing the magnification $M = \eta'/\eta$, we obtain from Eq. (4.3): $nu\eta = n'u'\eta'$. The same result is obtained through Eq. (4.6) if we omit the subscripts. Thus the product $nu\eta$ is invariant upon imaging through a single surface, a thin lens,

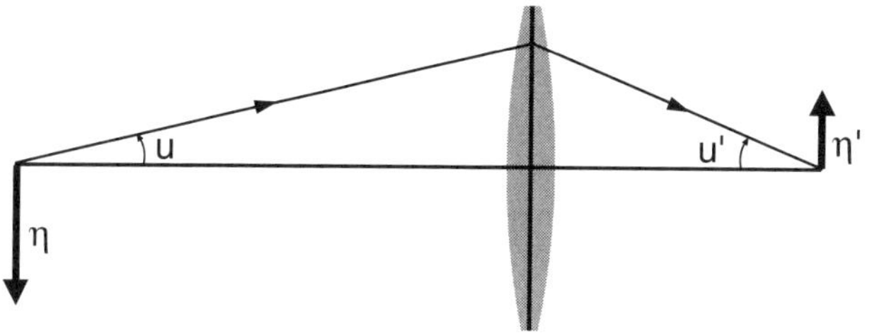

Figure 4.5. For $n = n'$, the product $u\eta = u\eta'$. This result applies for a single thin lens as well as a general system.

or any general multisurface system. This invariant is called the optical, Lagrange, or Helmholtz invariant. We will use the name optical invariant, and denote it by the symbol H. Thus

$$\boxed{H = nu\eta = n'u'\eta'} \tag{4.11}$$

At first sight this might not look like anything more than a trivial restatement of the magnification formula. But, as we show in detail in Chapter 5, the optical invariant determines both the maximum amount of light that a lens system can accept and the maximum number of resolvable spots in the image (i.e., the information capacity of a system); so we can see that it is indeed of fundamental importance.

Figure 4.5 shows graphically the implications of the magnification formula or the optical invariant. If the image is smaller than the object as shown, then the convergence angle, as represented by u', must be proportionately larger than u and vice versa. This is an important physical intuition to develop when thinking about the action of a lens system.

If one of the conjugates (e.g., the object) is infinite, then $\eta \to \infty$ and $u \to 0$, so Eq. (4.11) becomes indeterminate. In this case, we can determine the optical invariant as follows. In Figure 4.6a, consider a general optical system, for which we need only know the position of a pair of conjugate planes, O and O'. Let the object height at O be η_0, and the image height at O' be η'_0. Then take a ray that is parallel to the axis and crosses the object plane at a height η_0; in the image space, this ray will pass through the image-space focal point $\mathscr{F}'$ and cross the image plane at a height η'_0. Also, take a ray passing through O, forming an angle u_0 with the axis. In the image space, this ray will cross the axis at O', and the focal plane at a height η'. The optical invariant, applied for the conjugates O and O', gives:

$$nu_0\eta_0 = n'u'_0\eta'_0$$

From the figure, $u'_0 = \eta'/z'_0$, if z'_0 is measured from O'. Also, $u'_\infty = -\eta'_0/z'_0$. By eliminating u'_0 and z'_0, we have

$$nu_0\eta_0 = -n'u'_\infty\eta'$$

Ray 1 can be thought of as coming from the axial object point at infinity, and ray 2 as coming from the edge of the object at infinity. Therefore, we can substitute η_0

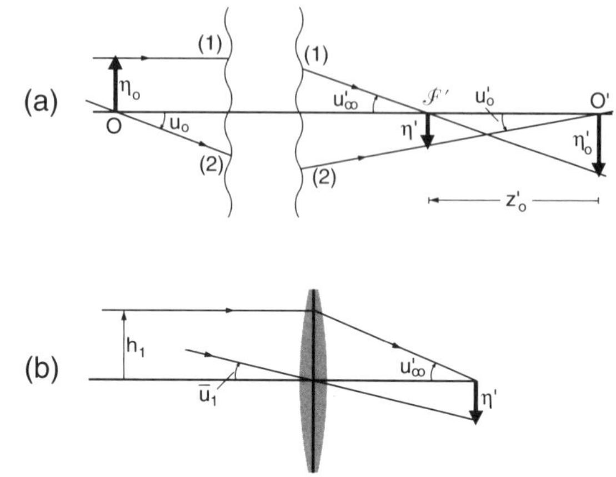

Figure 4.6. (a) Optical invariant for a general system with the object at infinity. (b) A simplified version of (a).

with h_1, the paraxial height variable at the first surface, and u_0 with $\bar{u}_1$, representing the angular object size (we use the bar to indicate that this ray comes from the edge of the object rather than the axial object point). Therefore, when the object is at infinity, the optical invariant can be written as

$$H = n'u'_{\infty}\eta' = -nh_1\bar{u}_1 \tag{4.12}$$

Having derived Eq. (4.12), we can dispose of the auxiliary planes O and O', and we can simplify the diagram even further by considering a single thin lens, as in Figure 4.6b.

4.4. PRINCIPAL PLANES

In this section we show how a complex, multielement system can be reduced to a set of only three parameters, for the purpose of determining the image location and magnification for any object. We first summarize the relevant theory in a qualitative manner (Section 4.4.1), and then provide the necessary proofs in Section 4.4.2.

4.4.1. Definition and Properties of the Principal Planes

It is desirable to extend the definition of the focal length to encompass thick lenses, and also compound lenses, comprising more than one elements. But that immediately creates a difficulty. In the case of a thin lens, the focal length is the distance between the lens and the focal point; but for a thick or compound lens, where should it be measured from? (e.g., the first or the last surface?).

We can expect that the focal point, and consequently the focal length, will be established through a process similar to that followed for thin lenses, which was to trace a ray parallel to the axis, and determine its intersection with the axis after refraction. Consider then a thick lens as in Figure 4.7, and an incoming ray, parallel to the axis. By extending the initial and final ray segments until

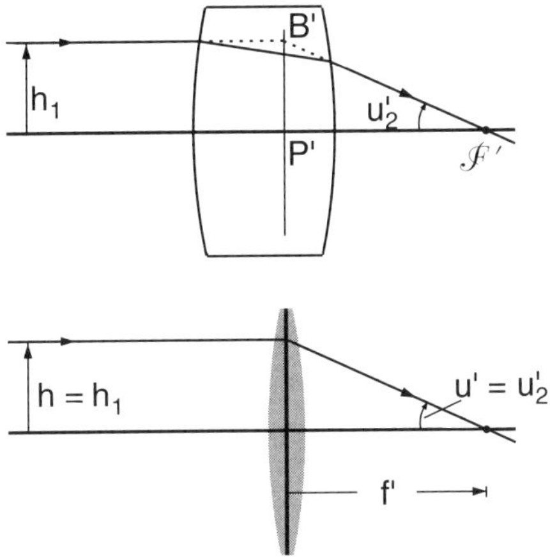

Figure 4.7. A thick lens and an equivalent thin lens.

they intersect, we locate the plane P' shown. From the point of view of an observer in the image space, it looks as though all refraction took place at point B'. So a thin lens located at P' and with focal length equal to $P'\mathscr{F}'$ would have the same effect as the thick lens. Therefore, Figure 4.7 suggests the answer to our question: The focal length of the thick lens is equal to $(P'\mathscr{F}')$.

It is shown in the following that it is possible to characterize a thick or compound lens by the location of two planes and a quantity called the equivalent (or effective) focal length (EFL) or its inverse, the power K. One of the two planes mentioned is the plane normal to the axis at P', and is determined as in Figure 4.7, by tracing a ray parallel to the axis and determining the intersection between the initial and final segments. The other plane is found by an identical construction, if we take a ray parallel to the axis, incident from the right. These two planes are called *principal planes*. P' is called the secondary or image-space principal point, and the plane normal to the axis at P' is the image-space principal plane. Similarly, P is called the primary or object-space principal point, with the corresponding primary principal plane normal to the axis at P (Figure 4.8). The principal planes are conjugates; that is, one is the image of the other through the system; also, the magnification between them is unity. Thus we have the following definition: *Principal planes are the conjugates of unit magnification.*

Because of the unit magnification, any ray crossing the first principal plane at a height h will emerge from the other principal plane at the same height. The angles that the ray forms with the axis will in general be different (Figure 4.9). Although we have drawn P to the left of P' for clarity, that need not be the case; P and P' can, in principle, be located anywhere inside or outside the system.

The principal planes provide a tremendous simplification: We can treat a complex system as a thin lens, provided the conjugate distances are measured from the principal planes. For example, if l and l' are the object and image

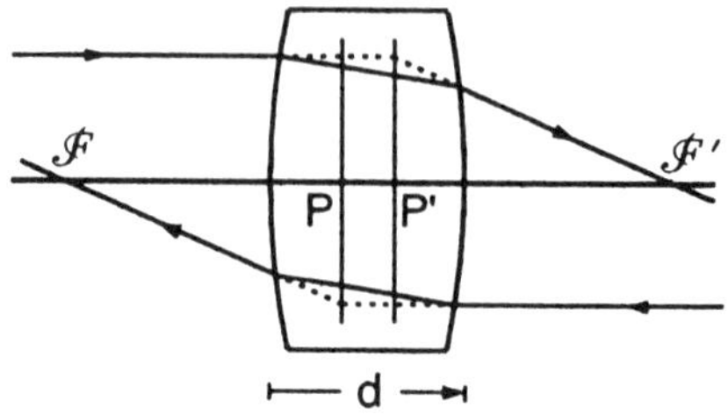

Figure 4.8. Location of the principal planes.

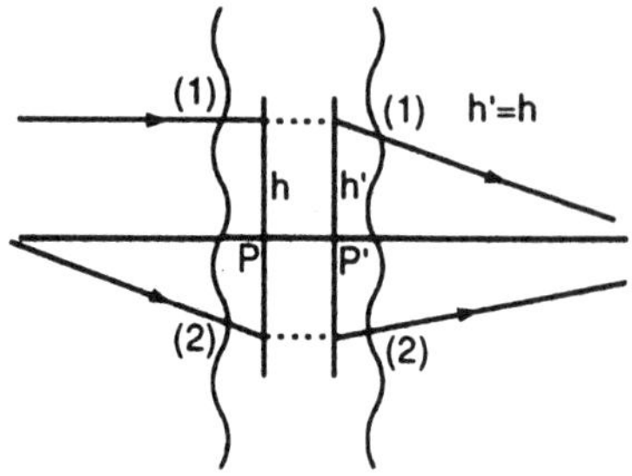

Figure 4.9. Unity magnification between the principal planes.

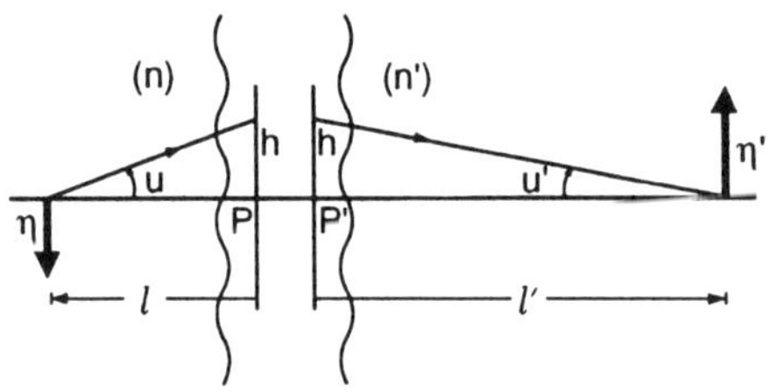

Figure 4.10. If the conjugate distances are measured from the principal planes, the thin-lens formulae apply: $n'/l' - n/l = K$, $M = nl'/n'l$. Also note that $u = -h/l$ and $u' = -h/l'$; so the formulae $n'u' - nu = -hK$ and $M = nu/n'u'$ also apply.

distances from the primary and secondary principal planes, respectively, then the thin-lens equation $n'/l' - n/l = K$ holds, and so does the magnification formula, $M = nl'/n'l$ (Figure 4.10). This is the basic principle (and the beauty) of Gaussian optics: *All thin-lens formulae transfer straight across to complex optical systems in their Gaussian-equivalent form* (i.e., when the complex systems are described in terms of their principal planes).

There is, however, one inevitable difference between a thin lens and a multisurface system: For the latter, the separation between principal planes cannot be ignored. This results in an image shift equal to (PP') when compared with the effect of a thin lens of the same focal length. This is illustrated in Figure 4.11. But apart from that difference in location, the two images are identical in the paraxial sense. Notice that in Figure 4.11 we have placed the thin-lens equivalent at P rather than P', in order to account for rays that are not parallel to the axis: If the thin lens was placed at P', the ray height for ray 2 would not be equal to h; so the resulting ray deviation would be different in the two cases.

When writing the thin-lens formula for a general system, we need to know what plays the role of the power K, or EFL F. We can answer this question

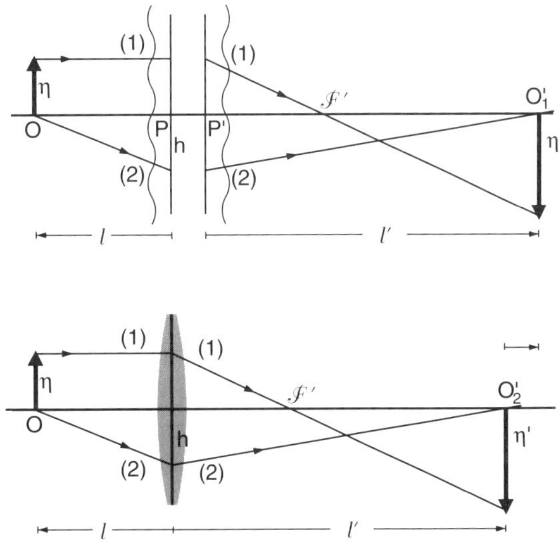

Figure 4.11. A general system and its thin-lens equivalent. The conjugate distances l, l' are the same in both cases. The image at O'_2 is identical with the image at O'_1, but displaced by a distance $(O'_2O'_1) = (PP')$.

by using an analogy with the thin lens: Using Eq. (4.2), which applies to a single surface as well as a thin lens, trace a ray with $u = 0$, to obtain $K = 1/F = -n'u'_\infty/h$, where the subscript ∞ indicates that the ray comes from the axial object point at infinity. For a general system, we can use a similar definition: The ray height refers now to the first surface (object space), while n' and u'_∞ refer to the last surface (image space). So we have

$$K = \frac{1}{F} = -\frac{n'u'_\infty}{h_1} \tag{4.13}$$

4.4.2. Power and Focal Lengths of a General System

The proof for all of the above statements hinges on the following assumptions: (1) that every point in the object space is imaged on a single point in the image space, or equivalently, that all rays emanating from a single object point will intersect at a single image point, and (2) that the image of an axial point is another axial point and the image of an extra-axial point is another extra-axial point, or alternatively, that the image of a plane perpendicular to the axis is another plane perpendicular to the axis. It should be clearly understood that these assumptions cannot be satisfied except in the paraxial approximation, where all wavefronts are spherical.

In Figure 4.12, we show the first and last surface of a general system. Take a ray parallel to the axis at a height h, which intersects the axis at point $\mathscr{F}'$ in the image space. (The case where the ray emerges parallel to the axis is treated separately in the following.) We then proceed to construct the plane $P'Q'$ through the intersection of the initial and final ray segments. Take a second

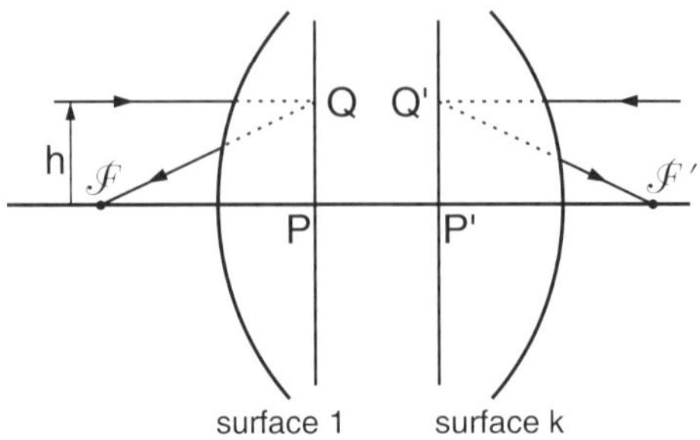

Figure 4.12. Existence of principal planes.

ray incident from the right but at the same height h; this ray will intersect the axis at point $\mathscr{F}$.* We proceed to construct the plane PQ in the same fashion as $P'Q'$; so $(PQ) = (P'Q')$. We then consider that the direction of propagation for ray 2 is reversed. Thus we now have two rays emanating from a point Q in the object space, and a point Q' in the image space. It follows that Q and Q' are conjugates, and since $(PQ) = (P'Q')$, the planes PQ and $P'Q'$ are conjugates of unit magnification. So we have shown that conjugate planes of unit magnification, called principal planes, exist for any system for which a focal point $\mathscr{F}'$ can be established at a finite distance.

Consider now any other pair of conjugate planes, as in Figure 4.13. We choose to measure the conjugate distances l and l' from the principal planes as shown. Assume that the segment $(AB) \equiv \eta$ is imaged on the segment $(A'B') \equiv \eta'$. From the diagram,

$$u'_\infty = -\frac{\eta - \eta'}{l'}$$

Dividing both sides by η and noting that $M \equiv \eta'/\eta$, we obtain

$$1 - M = -\frac{u'_\infty l'}{\eta} \tag{4.14}$$

The magnification formula (4.6) (or the optical invariant) gives

$$M = \frac{nu}{n'u'}$$

But from the figure, $u = -\eta/l$ and $u' = -\eta/l'$; therefore,

$$M = \frac{nl'}{n'l} \tag{4.15}$$

* Once we have accepted that the first ray will cross the axis in the image space, it follows that the second ray will also cross the axis somewhere in the object space. If it did not, that is, if it emerged parallel, then we would violate our assumption that a point on axis in the object space is imaged to a point on axis in the image space. This can be seen by reversing the direction of propagation of the second ray, in which case we would have two parallel rays, coming from the axial object point at infinity, intersecting at an extra-axial image point.

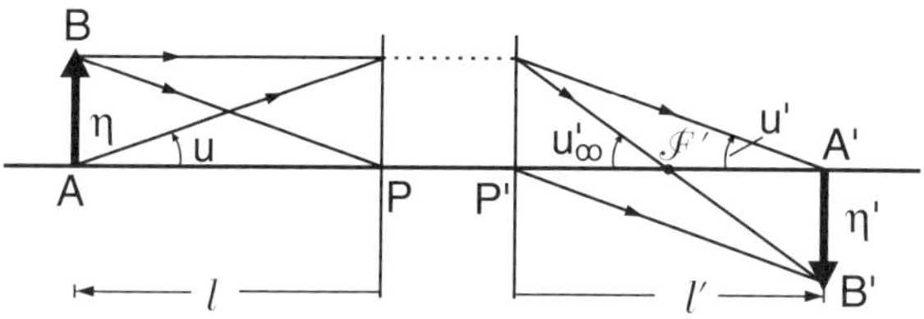

Figure 4.13. Conjugate equation with distances measured from the principal planes.

which is a formula identical with the thin-lens one, but with the conjugate distances measured from the principal points. Substitution of Eq. (4.15) into Eq. (4.14) yields, after some rearrangement of terms,

$$\frac{n'}{l'} - \frac{n}{l} = -\frac{n'u'_\infty}{\eta} \tag{4.16}$$

We note that the distance $(P'\mathscr{F}')$ is given by the ratio $-\eta'/u'_\infty$, which is the inverse of the ratio appearing on the right-hand side of Eq. (4.16). This distance is a constant, independent of the initial ray height η, if all rays from the axial object point at infinity are to meet at a single axial point $\mathscr{F}'$. We can identify the distance $(P'\mathscr{F}')$ with the secondary focal length f', exactly as we did with the thin lens. Therefore, Eq. (4.16) implies that the quantity on the right-hand side is a constant, and we call this constant the power of the system K, represented by Eq. (4.13). The equivalent (or effective) focal length is then given as $1/K \equiv F = f'/n'$, just as with the thin lens.

For the definition of power to be valid, we must also show that Eq. (4.16) holds also when tracing rays from right to left. By taking a ray parallel to the axis in the image space that crosses the axis at $\mathscr{F}$, we can employ reasoning identical to the preceding one and conclude that

$$\frac{n'}{l'} - \frac{n}{l} = \frac{nu_\infty}{\eta'} \tag{4.17}$$

and that the ratio u_∞/η' is constant, and can be set equal to $-f$. Therefore,

$$K = -\frac{n'u'_\infty}{\eta} = \frac{nu_\infty}{\eta'} \tag{4.18}$$

$$F = \frac{1}{K} = \frac{f'}{n'} = -\frac{f}{n} \tag{4.19}$$

The principal planes are often a source of great confusion for the beginner. Part of the confusion can usually be traced to an inadequate understanding of the concepts of object and image space. When we speak of the object-space principal plane, or any object-space quantity, we refer to a quantity that is encountered before any refraction. Similarly, an image-space quantity is encountered after all refraction. Thus a ray proceeding from any object-space location to a point on the first (or object-space) principal plane follows a single straight line. This holds irrespective of the physical location of the principal plane, that is, even if it is located somewhere inside or behind the system. It may be of some help at the beginning to imagine that once the principal planes have been established, the system has by some magic been removed, and replaced by those two planes. We can do this because we have already taken into account the refractive properties of the entire system in determining the principal planes.

4.4.3. Reference to an Arbitrary Set of Conjugates

The basic idea behind principal planes is that they provide a useful abstraction: However complex the system, it is replaced by only three parameters, the location of the principal planes and the power. The same abstraction can be achieved if we measure the conjugate distances not from the conjugates of unit magnification, but from the conjugates of some given reference magnification M_0. Thus, in Figure 4.14, suppose that O_0 and O'_0 are the reference conjugates of magnification M_0, and that O and O' are another arbitrary set of conjugates. We measure the distance L from O_0 and the distance L' from O'_0. Let the object height $O_0Q_0 = \eta_0$ and consider the ray Q_0Q parallel to the axis, which emerges at the image space forming an angle u'_∞ and intersects the image plane at Q'_0. Also consider the ray Q_0O, which will also cross at Q'_0 and further will intersect the axis at O', to define the image of O.

The optical invariant gives $H = nu\eta = n'u'\eta'$. Also, from Figure 4.14 $u = -\eta_0/L$ and $u' = -\eta'_0/L$, so that

$$\frac{n'\eta'_0\eta'}{L'} = \frac{n\eta_0\eta}{L}$$

or

$$M_0 M = \frac{nL'}{n'L} \tag{4.20}$$

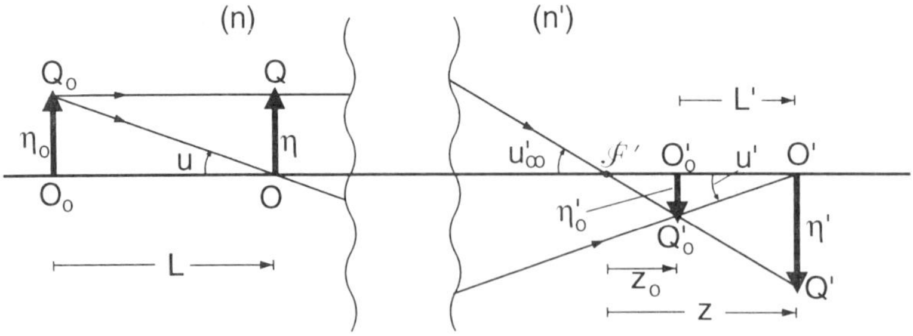

Figure 4.14. Use of an arbitrary set of conjugates as reference surfaces: the general conjugate formula.

Notice that this equation expresses not only the transverse magnification M, but also the longitudinal magnification L'/L [compare with Eq. (3.37)]. We also have:

$$u'_\infty = \frac{\eta' - \eta'_0}{L'} \rightarrow \frac{\eta'}{\eta} - \frac{\eta'_0}{\eta_0} = \frac{L'u'_\infty}{\eta_0}$$

since $\eta = \eta_0$. This is rewritten as

$$M - M_0 = \frac{L'u'_\infty}{\eta_0} \tag{4.21}$$

By eliminating M between Eqs. (4.20) and (4.21), and after a little algebra, we have

$$\frac{n'M_0}{L'} - \frac{n}{LM_0} = -\frac{n'u'_\infty}{\eta_0} = K \tag{4.22}$$

Both Eqs. (4.20) and (4.22) revert to the principal plane or thin-lens form if the magnification of the reference conjugates $M_0 = 1$. Equation (4.22) which may be called the general conjugate formula, represents the central theorem of Gaussian optics. It implies the existence of principal planes, defines the power of a system, and relates the conjugate locations from any convenient set of reference conjugates.

4.4.4. Afocal Systems

In proving the existence of principal planes, we assumed that a ray parallel to the axis in the object space will cross the axis somewhere in the image space. If the ray emerges parallel, then the point $\mathscr{F}'$ is at infinity, and so are the principal planes. In this case the principal planes are not useful since they cannot be used as reference points.

If the ray emerges parallel to the axis, $u'_\infty = 0$, and therefore $K = 0$. A zero-power system is called *afocal* or *telescopic*. For such a system, an object of height η is imaged to a segment of a constant height η', independent of conjugate location (see also Example 4.1). Thus an afocal system is characterized by its (constant) magnification, rather than power.

From Eq. (4.21), we have $M = M_0$ for an afocal system. Thus, if we establish any convenient set of conjugate reference planes, then the conjugate distances L and L', measured from the reference planes, are related by

$$L' = \left(\frac{n'}{n}\right) M_0^2 L \tag{4.23}$$

For example, consider the system of Figure 4.3. Since the focal point $\mathscr{F}_1$ is imaged onto the focal point $\mathscr{F}'_2$, we can choose these two points as the reference conjugates. The magnification M_0 is given by $-F_2/F_1$; so Eq. (4.23) can be applied to determine L' for any L.

4.4.5. Location of Principal Planes

It remains to specify the location of the principal planes from a given reference point within the system. We choose the first and last surfaces of the system, as they are normally the only physically accessible reference points. Through successive application of the refraction and transfer formulae, we can determine the final ray intersection height h_k and the image-space paraxial angle variable, u'_k. Call δ' the distance between the last surface and second principal plane, respectively. Figure 4.15 (ray 1) shows that the principal plane location is found from:

$$\delta' = \frac{h_1 - h_k}{u'_k}, \qquad u_1 = 0 \tag{4.24}$$

The corresponding distance of the first principal plane from the first surface is found by tracing a ray that is parallel to the axis in the image space (ray 2, Figure 4.15). Thus

$$\delta = -\frac{h_1 - h_k}{u_1}, \qquad u'_k = 0 \tag{4.25}$$

To avoid confusion, pay attention to the fact that all quantities in Eq. (4.24) are associated with ray 1, and all quantities in Eq. (4.25) are associated with ray 2 (Figure 4.15). The same symbols are used in both equations in order to avoid a cumbersome notation.

For a thin lens or a single surface $h_1 = h_k$; so $\delta = \delta' = 0$; that is, the principal planes coincide at the surface or lens. Notice that we now have a quantitative answer to the dilemma posed by Figure 2.10 (Chapter 2). The definition of power [Eq. (4.13)] predicts correctly that the black box system of Figure 2.10a is of negative power, while that of Figure 2.10b is of positive power (the sign of u'_k is opposite in the two cases).

A very useful and frequently encountered term is the *back focal length* F_b. This is the distance between the last surface of the system and the second focal point $\mathscr{F}'$. It is of practical significance, since it determines the clearance

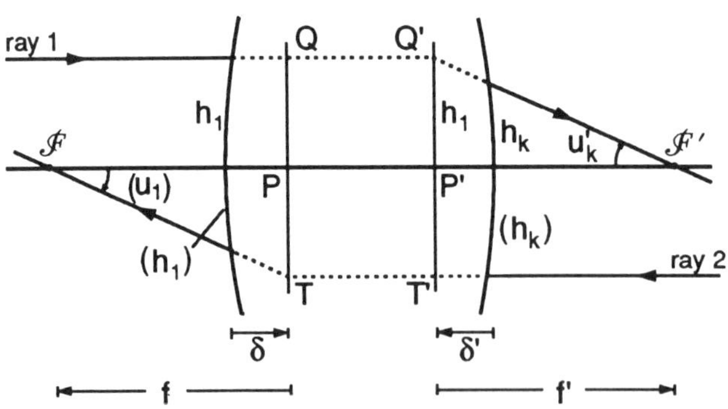

Figure 4.15. Principal plane location for the general system.

between the last surface and the image, if the image is at the focal plane. Figure 4.15 shows that

$$F_b = f' + \delta' \tag{4.26}$$

Similarly, we may define the front focal length F_f as the distance between the first surface and the first focal point. Thus

$$F_f = f + \delta \tag{4.27}$$

Example 4.3: The Cassegrain mirror. The concept of principal planes allows us to determine the focal length of the Cassegrain mirror of Example 4.2 (Figure 4.4). To determine the focal length, we need to know the focal point and the corresponding principal plane. The focal point was found to be at a distance $l'_2 = 200$ mm from the second surface. The image-space principal plane can be determined by the process described previously (Figure 4.16). From the figure [or from Eq. (4.13)], we have $F = -h_1/u'_2$. From Example 4.2, we have $u'_2 = 2h_1(c_1 - c_2 + 2dc_1c_2)$; so

$$F = -\frac{h_1}{u'_2} = -\frac{1}{2(c_1 - c_2 + 2dc_1c_2)} = 1 \text{ m}$$

Example 4.4: The telephoto lens. In photography, we often want to take a picture of a distant object that may subtend only a small angle at the camera (Figure 4.17a). Let us consider the camera lens to be a single thin lens for the moment; then from Figure 4.17a, we see that the image size on the film is $\eta' = \beta F$. So the larger the focal length F, the larger the image size and the more detail can be resolved. But since the film must be placed at the focal plane (object close to infinity), it follows that the camera or lens mount may have to be inconveniently long. Consider now the lens system of Figure 4.17b, which has a negative element (2) after the positive (1), at a distance d such that $F_1 > d > F_1 + F_2$ ($F_2 < 0$). By tracing the ray shown, we determine the

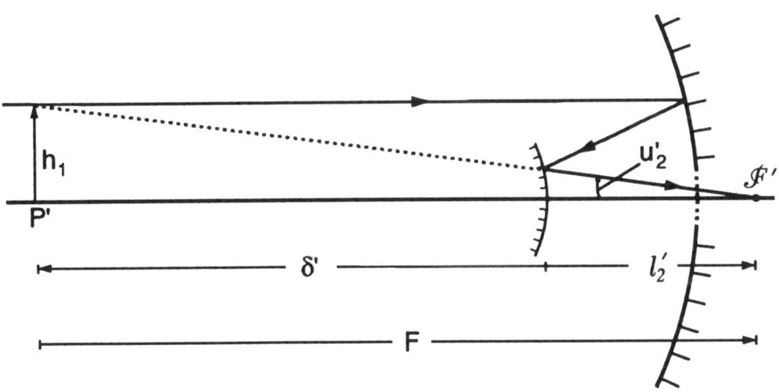

Figure 4.16. Focal length of the Cassegrain objective.

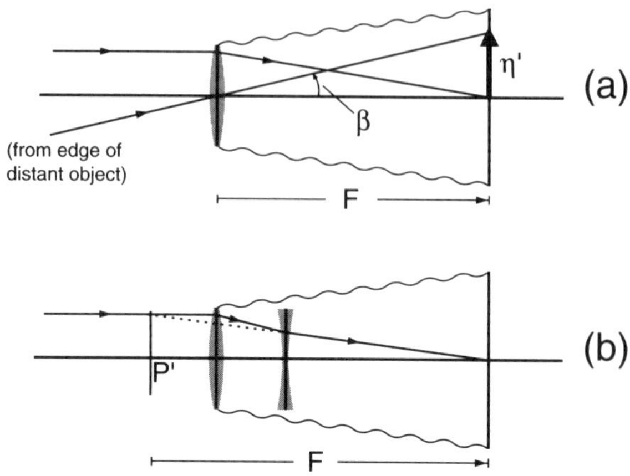

Figure 4.17. The telephoto lens in (b) has longer focal length than the lens in (a).

image-space principal plane to be in front of the positive element. Thus the focal length of the entire system ($P'\mathscr{F}'$) is larger than the distance between the front element and the focal plane. So a telephoto lens is more compact than a single thin lens of the same focal length.

Notice that the Cassegrain objective of the previous example achieves essentially the same effect, using mirrors instead of lenses. These last two examples illustrate clearly the importance of the principal planes, as the action of the telephoto lens or the Cassegrain objective cannot be understood without this concept.

4.5. POWER AND PRINCIPAL PLANES OF A SYSTEM OF TWO SEPARATED COMPONENTS

Consider now two systems, characterized by their principal planes, as in Figure 4.18, and by their powers K_1, K_2 or EFLs, F_1 and F_2. Let their separation be d, and let the refractive index of the medium between the two systems be μ ($n'_1 = n_2 = \mu$). To determine the power and the principal planes of this system, we need only trace a ray parallel to the axis ($u_1 = 0$). Thus we have:

$$n'_1 u'_1 - n_1 u_1 = -h_1 K_1 \rightarrow \mu u'_1 = -h_1 K_1 \tag{4.28}$$

$$n'_2 u'_2 - n_2 u_2 = -h_2 K_2 \rightarrow n'_2 u'_2 - \mu u'_1 = -h_2 K_2 \tag{4.29}$$

By adding these two,

$$n'_2 u'_2 = -h_1 K_1 - h_2 K_2 \tag{4.30}$$

The power of the system will be [from Eq. (4.13)]: $K = -n'_2 u'_2 / h_1 \rightarrow n'_2 u'_2 = -h_1 K$. Substituting into Eq. (4.30) and rearranging terms,

$$K = K_1 + \frac{h_2}{h_1} K_2 \tag{4.31}$$

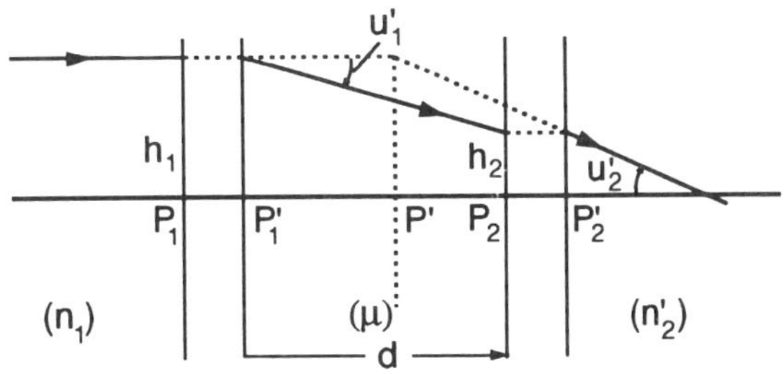

Figure 4.18. A system of two separated components.

The transfer formula (or Figure 4.18) gives: $h_2 = h_1 + du'_1 = h_1 - dh_1K_1/\mu$, from Eq. (4.28). So the ratio

$$\frac{h_2}{h_1} = 1 - \frac{dK_1}{\mu} \tag{4.32}$$

Substituting this value into Eq. (4.31), we have

$$\boxed{K = K_1 + K_2 - \frac{d}{\mu}K_1K_2} \tag{4.33}$$

We can now determine the location of the principal planes for the system of two components. From Eq. (4.24) we have:

$$\delta' = \frac{h_1 - h_2}{u'_2} = \frac{1 - h_2/h_1}{u'_2/h_1} = \frac{dK_1/\mu}{-K/n'_2}$$

where the last step follows from Eqs. (4.13) and (4.32). Setting $n'_2 \equiv n'$ to simplify the notation, we have

$$\delta' = -\frac{n'dK_1}{\mu K} \tag{4.34}$$

The corresponding distance for the first principal plane can be found by similar reasoning to be (with $n_1 \equiv n$):

$$\delta = \frac{ndK_2}{\mu K} \tag{4.35}$$

For a system of three or more components, it is cumbersome or impossible to derive analytical formulae for the location of the principal planes. We then follow the general procedure, which is to trace a ray parallel to the axis in the object space through the system and apply Eq. (4.24).

It is instructive to consider here the dilemma posed by Figure 2.11, where we had two black box systems with identical construction, giving seemingly different outputs. Clearly, Eq. (4.13) is not concerned with the bounds of the black box, and it will predict the same power K for both systems. It is a simple matter to predict the sign of K graphically: Taking ray 1 of Figure 2.11, produce the object-space ray segment until it intersects the image-space segment. The intersection is located somewhere just outside the black box, above the axis, defining the principal plane location. At that point, the ray bends upward. Such an effect would be produced by a negative thin lens placed at the principal plane; so the system power is negative. This holds for both black boxes, since the geometrical construction is identical. But the rays come to a real focus for system b, which can not happen with a negative thin lens. So we see that the reduction of the system to a thin lens is not valid in every aspect. The Gaussian equivalent form predicts correctly the location, size, and orientation of the image, but it does not predict the type of the image (real or virtual) without additional information. This is because the conjugate and magnification equations are not concerned with the type of the image. The additional information is provided by δ'; upon comparing δ' with l' we can determine whether the image is to the left or to the right of the last surface (virtual or real, respectively).

4.6. THICK LENSES: POWER AND LOCATION OF PRINCIPAL PLANES

We can now attack the problem of determining the power and principal planes of a thick lens, as in Figure 4.8, by considering it to be a system of two separated components (the two surfaces). The powers of the individual surfaces are $K_1 = c_1(\mu - n)$, and $K_2 = c_2(n' - \mu)$. Substitution into Eq. (4.33) will yield the desired result. We may simplify the expression by taking the thick lens to be in air, $n = n' = 1$. We then have from Eq. (4.33), after a bit of algebra:

$$\boxed{\frac{1}{F} = K = (\mu - 1)(c_1 - c_2) + \frac{dc_1c_2(\mu - 1)^2}{\mu}} \tag{4.36}$$

which gives the focal length or power of the thick lens as a function of the constructional parameters of the lens. Notice that the first term on the right-hand side represents the power of a thin lens with the same curvatures and refractive index as the thick lens. The second term is the one due to the lens thickness, and it becomes zero if $d = 0$, or if either c_1 or c_2 is zero (plano- lens). Another interesting case is a lens with $c_1 = c_2$: From Eq. (4.36), we see that its power is proportional to the thickness d, rather than zero.

The principal plane location for the thick lens can be determined from Eqs (4.34) and (4.35). Figure 4.19 shows the approximate location of the principal planes for typical thick lenses. Notice that for a plano- thick lens, one of the two principal planes is located on the curved surface. This is because either K_1 or K_2 is zero, and therefore either δ' or δ becomes zero. For example, take $K_2 = 0$, then $\delta = 0$, and $\delta' = -d/\mu$, since $K_1 = K$. Thus the thickness of a plano- lens has no effect on the power, but it affects the location of one principal plane. For a meniscus lens, both principal planes may be located outside the lens. When high accuracy is not paramount, the temptation is to treat as thin any lens that does not seem to be too thick. But this sometimes can lead to unexpected trouble, especially with thick menisci (see Problems 4.7 and 4.8).

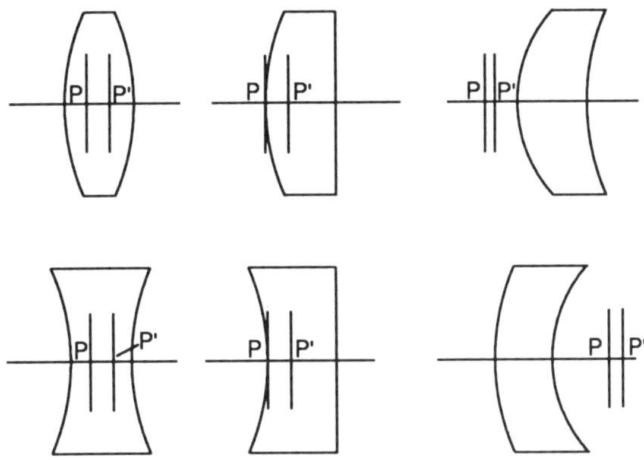

Figure 4.19. Approximate location of the principal planes for different types of thick lenses.

Example 4.5. We usually say that a lens that is thicker in the middle is positive, and a lens that is thicker at the edge is negative. This is true for lenses of moderate thickness, but it does not hold for very thick lenses. Take a lens (in air) with $c_1 > 0$ and $c_2 < 0$. Such a lens is thicker in the middle. From Eq. (4.36), we see that the two terms on the right-hand side are of opposite sign. Specifically, $(\mu - 1)(c_1 - c_2) > 0$, and $dc_1c_2(\mu - 1)^2/\mu < 0$. Thus if the second term becomes larger (in absolute value) than the first, the lens will have negative power. The corresponding condition for d is $d > \mu(c_1 - c_2)/[|c_1c_2|(\mu - 1)]$. When d is equal to the above value, the lens power is zero. The reader is encouraged to pursue this example further, by inserting numerical values, and to attempt a graphical explanation in terms of the principal plane location.

4.7. NODAL POINTS, MEASUREMENT OF FOCAL LENGTH

The nodal points of a system are axial conjugate points with the following property: *A ray crossing the axis at the first nodal point will emerge from the second nodal point parallel to itself*; that is, $u = u'$. The existence of nodal points can be easily established. Referring to Figure 4.20, the optical invariant gives $H = nu\eta = n'u'\eta'$, and for $u = u'$, we see that the nodal points are the conjugates with linear magnification equal to

$$M_N = \frac{\eta'}{\eta} = \frac{n}{n'} \tag{4.37}$$

Therefore, the nodal points coincide with the principal points when the lens has the same medium on both sides ($n' = n$, $M_N = 1$). The human eye is an example of a system where the nodal points and the principal points are not coincident: The object-space medium is air, but the image-space medium is a liquid called vitreous humour, which fills the eyeball and has a refractive index close to that of water (1.33).

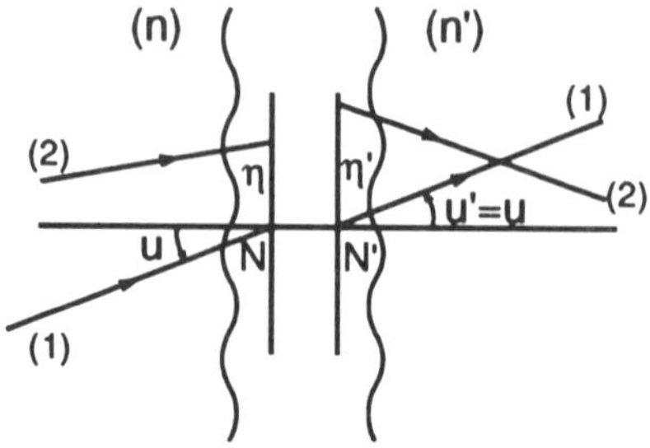

Figure 4.20. Nodal points.

The conjugate planes located at N and N' are often called nodal planes, but this terminology is somewhat unfortunate. For the principal planes, any point on the first plane will be imaged to a point on the second plane with $\eta' = \eta$. But for the nodal planes, only the axial points N and N' possess the property $u = u'$.

The principal, nodal, and focal points of a lens system are known collectively as the *cardinal points.*

For a thin lens (with $n = n'$) the principal and nodal points coincide at the lens. For a single surface, the principal points coincide at the apex, but the nodal points coincide at the center of curvature C, since any ray directed toward C will be undeviated. Similarly, the nodal points of a sphere coincide at the center of the sphere. If the sphere is thought of as a thick lens in air, the principal points must then coincide with the nodal points.

The properties of nodal points are used in the measurement of the focal length of thick or compound lenses through the *nodal slide.* The principle of the nodal slide is shown in Figure 4.21. Consider a collimated beam of light incident on the lens, and coming to focus at the focal point. When the lens is rotated about an arbitrary axis through a small angle, the image (which in the case shown would be a point) will, in general, move in a direction normal to the axis. But when the lens is rotated about N' (or P' for a lens in air), the image does not move. This is seen by taking the ray that crosses the new system axis at N: This ray will emerge from N' parallel to itself; that is, it will coincide with the previous system axis. Since all rays must go through the new image point, the new image point will still be located on the previous system axis, which means that it has not moved. The distance between the center of rotation (P') and the focal point is the focal length of the lens. The nodal slide is an

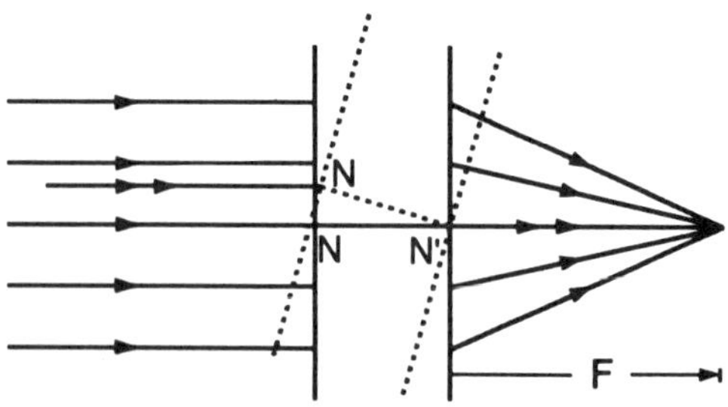

Figure 4.21. Principle of the nodal slide.

accurate and quick way of determining the principal point location, and hence the focal length of a thick or compound lens. It is, however, possible to measure the focal length without determining the principal points. One such way is to use the equation shown in Problem 3.17. Another way is to use Newton's form of the conjugate equation, to be shown in the following.

4.8. ADDITIONAL TOPICS IN GAUSSIAN OPTICS

4.8.1. Newton's Form of the Conjugate Equation

When the conjugate distances are measured from the focal points, the conjugate equation takes a simple form. Referring to Figure 4.22, we use $\mathscr{F}$ and $\mathscr{F}'$ as reference points (notice that the focal points are not conjugates), and measure the distances of a pair of conjugate planes at O and O' from $\mathscr{F}$ and $\mathscr{F}'$, respectively. Call $(\mathscr{F}O) = z$ and $(\mathscr{F}'O') = z'$. We then have

$$u'_{\infty} = \frac{\eta'}{z'} = -\frac{\eta}{n'F} \rightarrow z' = -\frac{n'\eta' F}{\eta} = -n'MF$$

and similarly, $z = nF/M$. By eliminating M,

$$zz' = -nn'F^2 \tag{4.38}$$

which is Newton's formula. It is often found convenient to use because of its simplicity. Notice that this formula does not apply to systems of zero power.

4.8.2. Imagery of a Tilted Plane

Consider a plane that is not perpendicular to the optical axis, but inclined at some angle θ, and a square $KLMN$ on that plane, with its center A on the axis (Figure 4.23). Let the optical system be characterized by its principal planes at P and P'. Point A is imaged at A' with a magnification M_1. Point B is the middle of the segment (KL), and (BC) is a perpendicular to the axis. Let the image of (CB) be $(C'B')$, with a magnification M_2, that is,

$$(C'B') = M_2(CB) \tag{4.39}$$

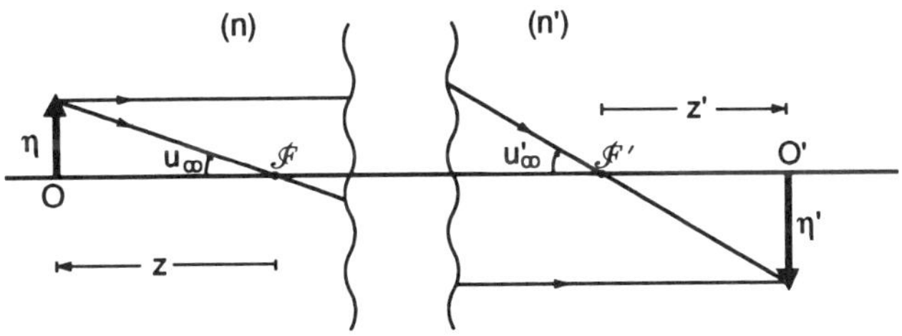

Figure 4.22. Newton's equation.

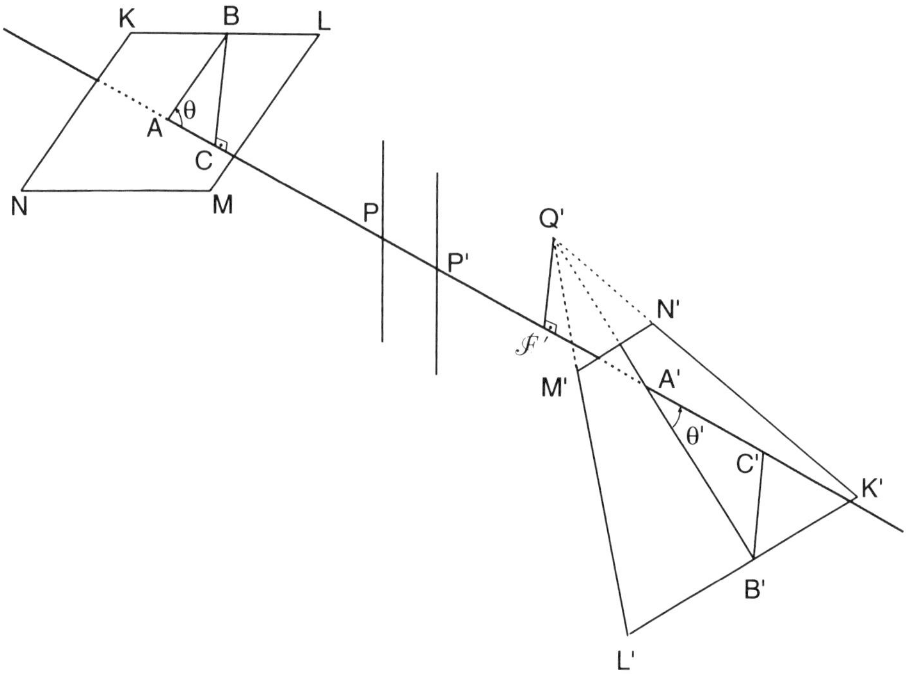

Figure 4.23. Imagery of a tilted plane.

The inclination of the object plane may be expressed as

$$\tan\theta = \frac{CB}{AC} \tag{4.40}$$

and that of the image plane as

$$\tan\theta' = \frac{C'B'}{A'C'} \tag{4.41}$$

The signs in these equations are consistent with taking the points A and A' as the local origin of coordinates. We then apply Eq. (4.20) to the two pairs of conjugates, (A, A') and (C, C'), to obtain

$$M_1M_2 = \frac{n(A'C')}{n'(AC)} \qquad \text{or} \qquad (A'C') = \frac{n'}{n}(AC)M_2M_1 \tag{4.42}$$

Using Eqs. (4.39) and (4.42) to substitute for $(C'B')$ and $(A'C')$ into Eq. (4.41), we obtain

$$\tan\theta' = \frac{n}{n'}\frac{CB}{AC}\frac{1}{M_1}$$

or, through Eq. (4.40), and setting $M_1 = \eta'/\eta$,

$$n\tan\theta\,\eta = n'\tan\theta'\,\eta' \tag{4.43}$$

so the product $n \tan\theta\, \eta$ is an invariant. Having determined the inclination of the image plane, we now need to find the shape of the image of the square. Let $\mathscr{F}'Q'$ be a perpendicular to the axis that is parallel to $(C'B')$, with Q' being on the image plane. Since Q' is also on the focal plane, the corresponding object point will be at infinity. If we consider the line (AB) as a ray, then the corresponding image-space ray is $(A'B')$, which passes through Q'. It follows that all lines (rays) parallel to AB in the object space, such as (KN) and (LM), will intersect at point Q' in the image space. The length of $(K'L')$ is simply $M_2(KL)$. Similarly, the segment (MN) is imaged to $(M'N')$ with a magnification M_3. Thus the image of the square $KLMN$ becomes a trapezoid.

4.9. SUMMARY OF GAUSSIAN OPTICS

In this chapter we have shown that the paraxial imaging properties of a general optical system comprising any number of refracting or reflecting surfaces are determined by only three parameters, the power K (or EFL, F) and the locations of the principal planes, δ and δ'. Given the system specification in terms of curvatures, separations, etc., we can determine δ and δ' by tracing two rays that are parallel to the axis in the object and image spaces, respectively. The same procedure is used to establish the power K. When the conjugate distances are measured from the principal planes, all thin-lens equations have been shown to apply to a general optical system.

The Gaussian properties of a system are summarized in Figure 4.24. The principal planes are at P and P', located at distances δ and δ' from the first and last surface, respectively. The equivalent focal length F equals the secondary focal length f' only if $n' = 1$; so it cannot be shown as a distance on this diagram.

The image of an arbitrary object can be determined by using the principal planes and applying the graphical raytracing rules of Chapter 2. This is shown in Figure 4.25. A ray parallel to the axis at a height $h_1 = \eta$ will emerge from the second principal plane at the same height and cross the axis at $\mathscr{F}'$. Similarly, a ray that crosses the axis at $\mathscr{F}$ and the first principal plane at $h_2 = \eta'$ will emerge from the other principal plane parallel to the axis and at the

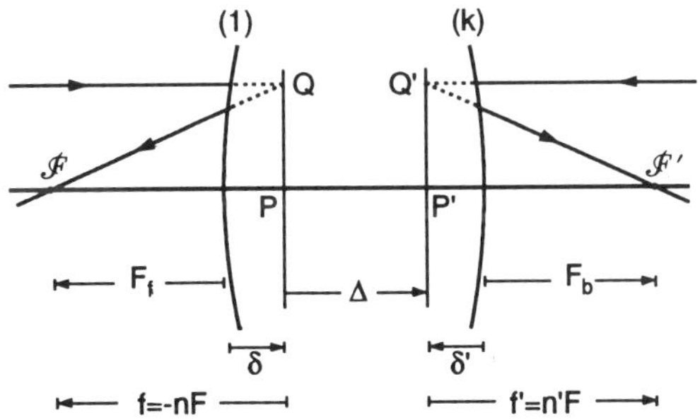

Figure 4.24. Gaussian properties of a system.

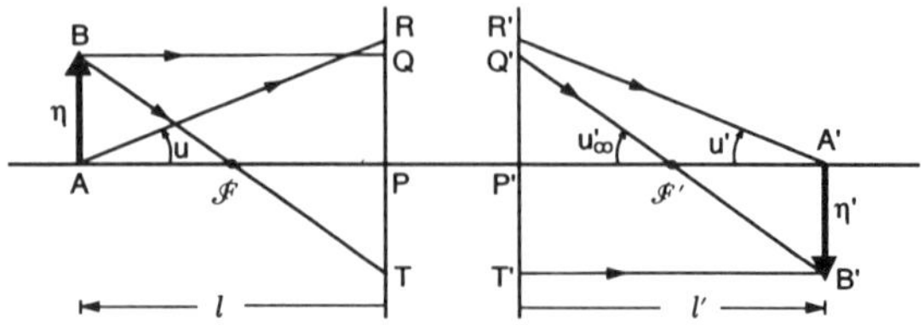

Figure 4.25. Graphical determination of the image.

same height. Also, if needed, the nodal points N,N' can be established as the conjugates of magnification n/n'. Any ray crossing the axis at N will emerge from N' undeviated.

It is critical that the student develop and cultivate the ability to draw generalized optical system diagrams in order to determine certain image- or object-space quantities, without having to resort to a detailed specification of the entire system. This activity represents the first step towards preliminary lens specification and design.

4.10. THE TWO-COMPONENT DESIGN PROBLEM

In Section 3.7 we examined the one-component design problem, and noted some of its limitations. The two-component problem can be stated in many equivalent ways, depending on the desired specifications. Our aim here is to determine the power and location of the individual components relative to the desired conjugates. We simplify the problem by considering the two components to be thin lenses in air, but they may equally well be more complex lenses in their Gaussian equivalent form. The following equations apply:

$$u'_1 - u_1 = -h_1 K_1, \qquad u'_2 - u'_1 = -h_2 K_2, \qquad h_2 = h_1 + d u'_1$$

$$M = \frac{u_1}{u'_2}, \qquad u_1 = -\frac{h_1}{l_1}, \qquad u'_2 = -\frac{h_2}{l'_2}$$

These six equations contain eleven unknowns, but if we proceed with the solution, we realize that only the ratio h_2/h_1 matters, rather than the two individual heights (see also Section 4.5); so the effective number of unknowns is ten. Therefore, we need to specify four of these ten in order to solve the two-component problem. (We may add other equations, for example, to include the throw T, but for every equation added we will also be adding another unknown.) Suppose, for example, that we know l_1, l'_2, d, and M. After a page or so of algebra, we obtain

$$K_1 = \frac{Ml_1 - l'_2}{Ml_1 d} - \frac{1}{l_1} \tag{4.44}$$

$$K_2 = \frac{1}{l'_2} - \frac{Ml_1 - l'_2}{dl'_2} \tag{4.45}$$

Knowledge of the number of parameters needed for an exact solution can save considerable time. When faced with a design problem that requires a two-lens solution, the reader can quickly determine whether the problem has or can be made to have a unique solution. But how does one decide that the problem needs a two-lens solution? The answer to this will become more obvious after the discussion of optical instruments in the next chapter. But sometimes, as in the following example, the need for a two-lens solution is clear.

Example 4.6: Surgeon's operating spectacles. You are an optics consultant, and your client is a brain surgeon who would like a pair of magnifying eyeglasses to use during surgery, to better see the delicate tissues of the brain. The initial problem specifications are: a magnification of $2\times$, an object distance of about 25 cm, and no allowance for additional spectacle correction (i.e., assume an emmetropic eye). To these we may add that the spectacles should be compact and lighweight, since they will be sitting on the surgeon's nose. We may also add that the location of the image should not be arbitrary; while the doctor may see a virtual image presented anywhere between -25 cm and infinity, it would be nice if the object and image were coincident; that is, $T = 0$ (any ambiguity in the location of the image could have dire consequences during brain surgery!).

The last requirement makes it clear that a single-lens solution is not possible. Let us then turn to the two-lens solution. We have ten unknowns and six equations. When the image is at the same place as the object, we can use the linear magnification as a visually relevant parameter; so $M = 2$. Also, set $l_1 = -25$ cm. Since the spectacles must be compact, we may set $d = 3$ cm as an upper limit, in which case, $l'_2 = -28$ cm. Thus we have specified four parameters, and can solve for K_1 and K_2 from Eqs. (4.44) and (4.45), to obtain $K_1 \cong 0.187\ \text{cm}^{-1}$ and $K_2 \cong -0.298\ \text{cm}^{-1}$. The corresponding focal lengths are $F_1 \cong 5.36$ cm, and $F_2 \cong -3.36$ cm (make a drawing). So we have obtained a preliminary design for the spectacles.

There are, of course, several other design questions to consider, such as the *field of view* (object size), the brightness of the image and the size of the lenses, and the quality of the image (aberration correction). The image quality specification will determine the exact curvatures of the lens surfaces, their thickness, and whether single elements will suffice. These questions are addressed in the following chapters, where we complete the discussion of this example.

The reader should notice the similarity between the operating spectacles design and the telescope described by Galileo in the opening quotation of this chapter.

In this example, we have encountered another set of interesting points: These are the self-conjugate points, or *Bravais points* (Longhurst, 1973), that is, points that are their own image. Such points do not exist for any arbitrary system, but they do exist for any two-lens system with one negative component.

PROBLEMS

1. Write a computer program to perform paraxial ray tracing, based on Eqs. (4.9) and (4.10). The input should be the surface curvatures, separations, and refractive indices. Also, you must specify h_1 and u_1 at the first surface. The output should be h_k and u_k' after the last refraction. (Option: Determine the image plane and the image height.)

2. Write a computer program that calculates the principal plane location δ, δ', the power of the system K, and the back focal length F_b. The input should be the system curvatures, refractive indices, and separations.

3. Consider a system of two thin lenses, separated by a distance equal to the sum of their focal lengths, $d = F_1 + F_2$ (Figure 4.3). By applying Eqs. (4.5), (4.8), and (4.9), show that the (transverse linear) magnification of this afocal system is constant, equal to $-F_2/F_1$.

4. A so-called Schwarzschild reflecting objective is shown in Figure 4.26. It comprises two mirrors with a common centre of curvature. (a) Show that the power of this system is given by $K = 2(c_1 - c_2)$. (b) For $F = 10$ mm and $d = -30$ mm, determine the radii of the mirrors, the location of the principal plane P', and the distance between mirror 1 and the image (O_1A'). (Note: This system is normally used with the conjugates reversed, as a reflecting microscope objective. Its advantage is its large working distance (O_1A') compared to a short focal length. A corresponding refractive microscope objective with similar focal length would normally have a working distance of only a few millimeters.)

Figure 4.26.

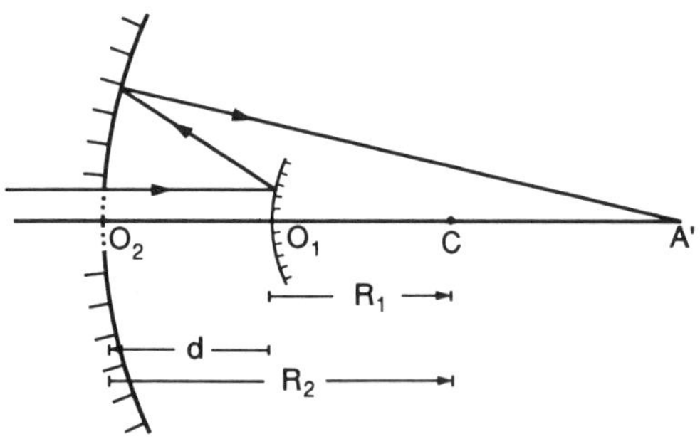

5. Consider a "reversed telephoto" lens (negative element before the positive). Sketch the system and determine approximately the location of the secondary principal plane P'. Why might this design be useful in making wide-angle (large-field-of-view) photographic lenses?

6. Derive Eq. (4.33) with $d < 0$ and draw Figure 4.18 for this case. (Note: The final equation is identical.)

7. A biconvex lens has radii $R_1 = 5$ cm and $R_2 = -3$ cm ($n = 1.5$). (a) Considering the lens as thin, determine its power. (b) If the lens is to have a size (diameter) of at least 2 cm, draw a scale drawing of the lens (use a compass!), and determine an appropriate thickness. Calculate the power of the resulting thick lens and compare the result with (a). Express your result as a percentage change in power from thin to thick.

8. Lens 1 is a negative meniscus with $R_{11} = 5$ cm, $R_{12} = 3$ cm. Lens 2 is a positive meniscus with $R_{21} = 2.8$ cm, $R_{22} = 6$ cm. The thickness is $d =$ 0.6 cm, and the refractive index is 1.5 (both lenses). (a) Determine their power and the location of the principal points. (b) Determine the power of the combination of these two lenses, in the following two cases: (i) surface R_{12} is in contact with surface R_{21}, and (ii) surface R_{22} in contact with R_{11}. Compare the two results and explain any difference.

9. A thick lens has $c_1 = c_2 = 0.3\ \text{cm}^{-1}$, $d = 0.5$ cm, $n = 1.5$. Determine its power and the location of the principal planes.

10. A thick lens comprises two concentric surfaces. Show that its power is given by $K = (n - 1)(c_1 - c_2)/n$. Where are the principal (nodal) points located? Determine the power of a sphere.

11. An optical system consists of two thin lenses of focal lengths $F_1 = 100$ mm, $F_2 = 25$ mm, separated by 140 mm. Sketch the arrangement, showing the path through the system of (a) a ray from infinity, (b) a ray from an object 833.3 mm to the left of the first lens (light travels →). Calculate the equivalent power of the total system and the positions of the principal points. Show these on a sketch. What is the separation of the principal points? Show the ray paths referred to above on a new diagram in which the system is replaced by its principal planes. Also show in this style the ray path for an object point for which $l = +167$ mm (measured from P).

12. Figure 4.27 shows the so-called *autocollimation method*, a quick method for determining approximately the focal length of a positive thin lens. A pinhole is imaged through the lens: The reflected rays form a sharp image of the pinhole only if it is located at the focal plane. Given that real lenses are thick, would the measurement accuracy depend on the type of lens (i.e., biconvex, planoconvex, meniscus)? If yes, is the orientation of the lens relevant?

Figure 4.27.

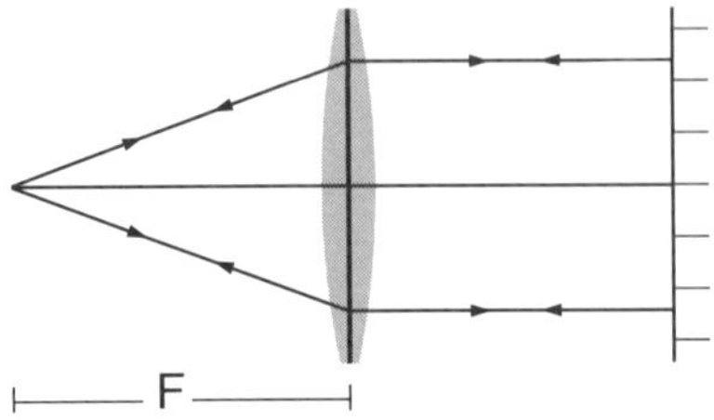

13. We wish to determine the focal length of a lens using the nodal slide technique. A collimating lens of focal length F_1 is used to generate a parallel beam of light. The test lens has focal length F_2. If the source is not exactly at the front focal point $\mathscr{F}_1$, then its image will not be exactly at $\mathscr{F}'_2$, thus leading to an incorrect result. If δz is the uncertainty in locating $\mathscr{F}_1$ and $\delta z'$ the corresponding error in F_2, show that $\delta z' = (F_2/F_1)^2\ \delta z$ (so that in order to minimize the errors, the focal length of the collimating lens must be longer than that of the test lens).

14. Two positive thin lenses are separated by a distance of 10 cm. The focal lengths are $F_1 = 10$ cm, $F_2 = 20$ cm. Determine the power of the combination and the location of both principal planes.

15. Two positive thin lenses are separated by a distance d. Determine graphically the sign of the power of the combination for the following two cases: (a) d slightly larger than $F_1 + F_2$, (b) d slightly smaller than $F_1 + F_2$.

16. A positive thin lens is followed by a negative thin lens at a distance d. Determine graphically the sign of the power of the combination for the following two cases: (a) d slightly larger than $F_1 + F_2$, (b) d slightly smaller than $F_1 + F_2$. Take $|F_1| > |F_2|$.

17. Show that the principal (nodal) points for an equiconvex lens ($n \cong 1.5$) that is not too thick are located approximately one-third of the lens thickness inside the lens. The lens is in air. (Hint: Use symmetry and the fact that the principal and nodal points are coincident; the derivation is not long.)

18. An equiconvex lens has curvatures $c_1 = -c_2 = 0.5\ \text{cm}^{-1}$. How thick must the lens be for its power to be zero?

19. An equiconcave lens has curvatures $c_1 = -c_2 = -0.5\ \text{cm}^{-1}$. If the lens is to have a diameter of 1 cm, what is the minimum edge thickness?

20. Show that it is not possible to produce an optical system comprising two separated thin lenses of finite power for which the principal points P and P' coincide.

21. A system has $n = 1$, $n' = 1.3$, $K = 0.1\ \text{cm}^{-1}$. Calculate the position of the nodal points relative to the principal points.

22. If a spectacle lens is to be placed in front of the eye in such a way that the equivalent power of the spectacle-plus-eye system is equal to that of the eye alone, calculate the position for the spectacle lens. Why should such a requirement be desirable? Treat the eye as a single thin lens, and discuss what happens to the principal points as the spectacle lens is inserted.

23. A thin lens of $F = 4$ cm is followed by another thin lens of $F = 2$ cm. The distance between the lenses is 4 cm. (a) Determine the location of both principal planes and the total system power. (b) Using the principal planes, determine the image location and magnification for an object located 8 cm in front of the first lens. (c) Determine the image location and magnification for an object that is located on the first lens.

24. A three-lens system has $K_1 = K_3$ and $d_1 = d_2 \equiv d$ in the usual notation. (a) Show that the location of the principal planes is independent of K_2. (b) Determine the location of the principal planes for $K_1 = 0.1\ \text{cm}^{-1}$ and $d = 3$ cm. (c) Show that the power of the system is given by

$$K = -(1 - dK_1)(2K_1 + K_2 - dK_1K_2)$$

(d) Sketch four different three-lens zero-power systems.

25. A system comprises a thin lens (power K_1) followed by a spherical mirror (power K_2). (a) Show that the principal plane location is independent of the power of the mirror, and determine the system power in terms of K_1, K_2, d. (b) Determine the nodal points of the system. (Note: Light passes through the lens twice.)

Putting It All Together

5

"...the only one [ocular] which I have gives a twentyfold enlargement, but the light is very weak. The reason for this is not unknown to me and I see how the intensity could be improved, but one hesitates to spend the money..."

Johannes Kepler, in a letter to Galileo, August 9, 1610. (From *Johannes Kepler: Life and Letters*, by Carola Baumgardt, Victor Gollancz Ltd., London, 1952.)

In this chapter, we complete the Gaussian theory by extending it to encompass the calculation of the EM energy transmitted through a lens system, and give examples of how to use the knowledge gained in order to set up a lens system that satisfies the first-order requirements of a given imaging problem.

5.1. STOPS AND PUPILS

Thus far we have ignored the fact that lenses are of finite size. Consider a single thin lens, as in Figure 5.1, and a point source at a distance l away. The size of the lens determines the maximum angle α, and therefore, the amount of light captured by the lens from the source. For a given lens size, we can control the amount of light by using a diaphragm of an appropriate diameter, called an *aperture stop*, in contact with the lens. If a diaphragm is not used, then the lens rim plays the role of the aperture stop. In both cases, we say that "the stop is at the lens."

Now move the diaphragm away from the lens (Figure 5.1b). Notice that in order to keep the angle α constant, we must also change the diameter of the stop; for example, as the stop moves toward the source, its diameter

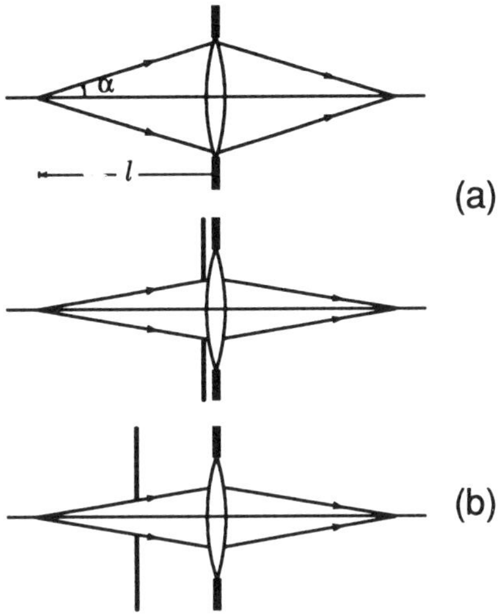

Figure 5.1. The action of an aperture stop for the axial ray pencil.

must decrease proportionately. When the stop is not at the lens, we call it "remote."*

5.1.1. Entrance and Exit Pupils

The most commonly encountered aperture stop is the one in camera objectives, which is the adjustable iris often seen inside the lens. Looking through the front of the lens, we see not the aperture stop itself, but its image through the lenses that precede it, called the *entrance pupil*. Similarly, the *exit pupil* is the image of the stop through the lenses that follow. From these definitions, it follows that the entrance and exit pupils are conjugates.

Some simple examples are shown in Figure 5.2. In Figure 5.2a, the entrance pupil coincides with the aperture stop, since there are no lenses in front of it. The exit pupil is then the image of the stop through the lens, shown in dashed lines. The entrance and/or exit pupil can be virtual or real, and located anywhere in principle. In both cases, we have drawn a single ray from the axial object point, passing through the edge of the entrance pupil or stop; since entrance and exit pupil are conjugates, this ray will also pass through the edge of the exit pupil. We call this ray the *paraxial marginal ray* (PMR). The name *marginal* derives from the fact that rays forming a larger angle with the axis than the PMR will be cut off, or *vignetted*, by the aperture stop.

For a general system, the role of the pupils is more clearly shown in Figure 5.3. For convenience, we take the entrance pupil to lie closer to the object

* A note on terminology: The term *lens aperture* is commonly used in place of lens size or diameter, or sometimes in place of the *numerical aperture*, defined in the next section. The term *aperture of a lens* normally indicates the size of the lens rather than the aperture stop. The term *lens semidiameter* also indicates the size of a lens and is perhaps a little less ambiguous; in what follows, we use the term semidiameter rather than radius, in order to avoid confusion between the lens size and the radii of curvature of the surfaces.

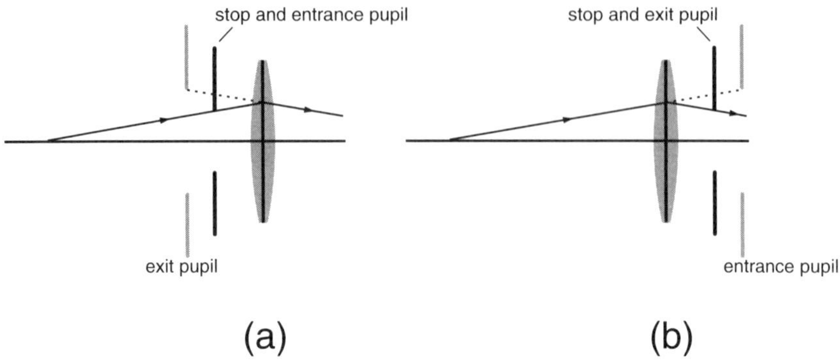

Figure 5.2. Examples of entrance and exit pupils for a thin lens with a remote aperture stop.

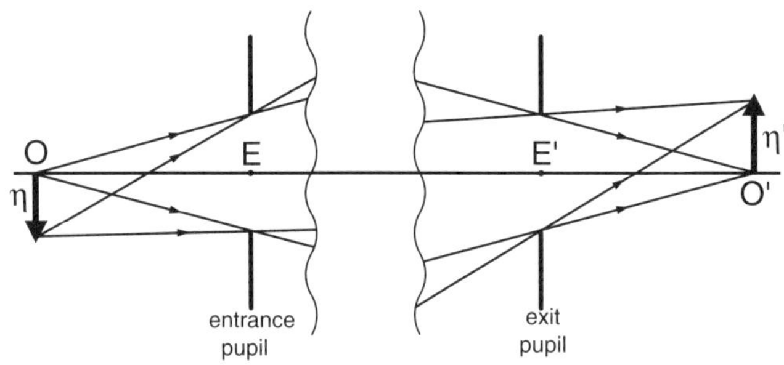

Figure 5.3. Entrance and exit pupils for a general system.

than the exit pupil. It is possible to have the exit pupil to the left of the entrance pupil, but this does not change any of the concepts. It is seen that the cones of light accepted by the system from points in the object have the entrance pupil as their common base; similarly, cones of light proceeding to the image have the exit pupil as a common base. If the aperture stop is varied in size, the bases of these light cones change proportionately, but there is no change in the object or image size.

Thus the entrance pupil represents the limiting window, or aperture, through which all object-space rays must pass; similarly, the exit pupil represents the limiting aperture through which all image-space rays must pass. This is the fundamental significance of the pupils.

The mere existence of a diaphragm does not automatically make it the aperture stop; it has to represent the limiting aperture. For example, consider the system of Figure 5.4a: a lens with an aperture in front of it, located at a distance $-2F/3$ in front of the lens. Let the lens diameter be D, and the diaphragm diameter be $D/2$. Assume that the object is at infinity; so rays from the axial object point are parallel to the axis. The maximum ray height is determined by the diaphragm diameter; so in this case, the diaphragm is the aperture stop (and entrance pupil). However, if the object is at $\mathscr{F}$ (Figure 5.4b),

Figure 5.4. Dependence of the pupil on the object location.

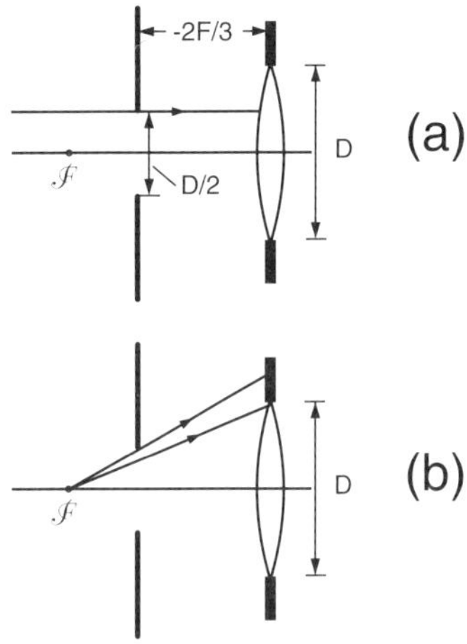

then the maximum ray angle is not determined by the diaphragm, but by the lens rim; in this case, the stop is at the lens.

A simple camera. A simplified schematic camera is shown in Figure 5.5a. The aperture stop coincides with the entrance pupil in this case, and the exit pupil can be found by producing backward the ray from the edge of the image. The two rays drawn are enough to specify the size and position of both the image and the pupils. The *field stop* indicated is an opening (in this case rectangular or square) that limits the image (field) size, hence the name. In a system with many lenses, a field stop can be placed at an intermediate image. The field stop is in a sense the opposite of the aperture stop; when its size is varied, the illumination is not affected, but the image size changes. (This change in image size is not a change in magnification, but simply a change in the physical limits of the image.) Figure 5.5b shows the same construction but with the stop behind the lens.

5.1.2. Numerical Aperture, *F* Number

Consider a lens system, characterized by its principal planes, as in Figure 5.6. Let the object be at infinity; so the incoming ray from the axial object point is parallel to the axis. The entrance and exit pupils are also indicated. If n' is the image-space refractive index, we define the numerical aperture (NA) of the lens to be

$$\boxed{\mathrm{NA} = n' \sin \alpha'} \tag{5.1}$$

In the paraxial approximation we can write $\sin \alpha' \cong \tan \alpha' \cong h_1/f' = h_1/n'F$, so that $\mathrm{NA} = n' \sin \alpha' = h_1/F$, or

$$\mathrm{NA} = \frac{h_e}{F} \tag{5.2}$$

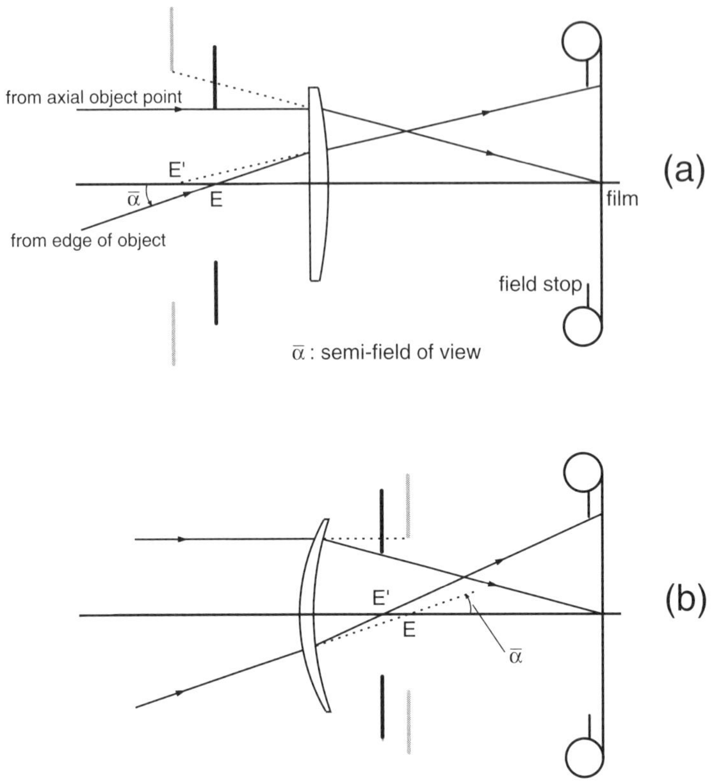

Figure 5.5. Schematic of two camera lenses. Pay close attention to the two rays drawn and the conjugates they define.

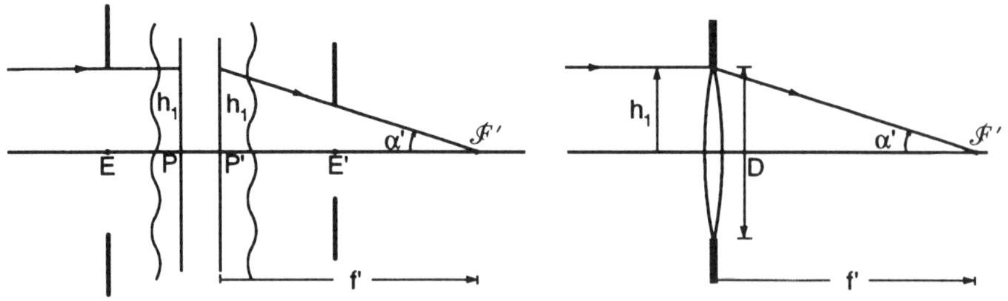

Figure 5.6. Numerical aperture and F number for a general system and a thin lens with the stop at the lens.

where h_e is the entrance pupil height and $h_1 = h_e$ for an infinite object. Normally, we should have written $\alpha' \cong -h_1/f'$, since the angle α' is negative as shown. However, no consistent sign convention is applied to the NA; so there is no reason to include the negative sign here.

For a single lens with the stop at the lens, $h_1 = D/2$, where D is the lens diameter, so

$$\mathrm{NA} = \frac{D}{2F} \tag{5.3}$$

The *relative aperture* or simply *F number* ($\mathcal{N}$ or F#) is defined as

$$\mathcal{N} = \frac{1}{2n' \sin \alpha'} = \frac{1}{2(\mathrm{NA})} \tag{5.4}$$

In the paraxial approximation, we have

$$\mathcal{N} \cong \frac{F}{D} \tag{5.5}$$

for a single thin lens, or

$$\mathcal{N} = \frac{F}{2h_e} \tag{5.6}$$

for a general system. In words, the F number is the EFL divided by the entrance pupil diameter. The notation F/x is often used to indicate $\mathcal{N} = x$; for example, $F/2$ means $\mathcal{N} \equiv \mathrm{F\#} = 2$. Both the NA and $\mathcal{N}$ are normally taken as positive quantities, independent of the sign of u' or α'. If the image, rather than the object, is at infinity, the definitions are identical, but involve object-space quantities. For example, the numerical aperture would be $\mathrm{NA} = n \sin \alpha$ or h'_e/F, etc.

It should be clearly understood here that the NA and the F number are properties of the light beam, rather than the optical system. The definitions make it clear that only the convergence angle of the beam matters. So a cone of rays has its own NA, without regard to the system that produced it. When a lens is said to have a certain NA or F#, the implication is that it can transmit a beam of maximum convergence (or divergence) angle equal to α' in the above equations. But a look at Figure 5.1 shows that, for a given lens size, the angle α' depends on the distance of the object from the lens. Thus for a lens system, both the NA and $\mathcal{N}$ are normally defined for an infinite object or image, though extensions to the definitions can be made for finite conjugates, as shown in the next section.

The F number and the NA are measures of the light-gathering ability of a lens for a given field size. This can be seen by considering two different camera objectives with the same $\mathcal{N}$ (that is, producing beams with the same convergence angle). From Eq. (5.6), the lens with the larger focal length must have a proportionately larger entrance pupil to maintain the same $\mathcal{N}$; so it collects more light. However, it also produces a proportionately larger magnification;

so the available light is spread over a larger area. In the photometry Section 5.3, we show in detail that those two effects cancel each other exactly. Thus the film (image) illuminance is determined by the lens F number, not the entrance pupil diameter.

In the paraxial approximation, the angle $\alpha' \cong \sin \alpha' \cong u'$, the PMR angle variable. Therefore, we can write the numerical aperture as

$$\mathrm{NA} = n'u' \tag{5.7}$$

showing more clearly that the PMR determines the image illuminance as well as the image location. The validity of Eq. (5.7) can be extended well beyond the paraxial region, as discussed in more detail in Section 5.2.

5.1.3. Depth of Focus, Depth of Field

Often, the final image is formed on some image recording device, such as photographic film. How accurately must the film be positioned with respect to the image plane? We shall show that the accuracy depends on two parameters: the size of the minimum resolvable detail (resolution limit) of the film and the F number.

In Figure 5.7, let us consider for simplicity a single thin lens, with the stop at the lens (the general case is treated in the fine print section). If the object is not at infinity, the image will be at a distance l' away from the lens. Let ρ be the resolution limit of the film; we can see that the film may be positioned anywhere within the segment $\delta z'$ with no loss of image clarity. Similar triangles give

$$\frac{D/2}{l'} = \frac{\rho/2}{\delta z'/2} \qquad \text{or} \qquad \delta z' = \frac{2\rho l'}{D}$$

If the object is at infinity, then $l'/D = F/D = \mathcal{N}$; so

$$\delta z' = 2\rho\mathcal{N} \tag{5.8}$$

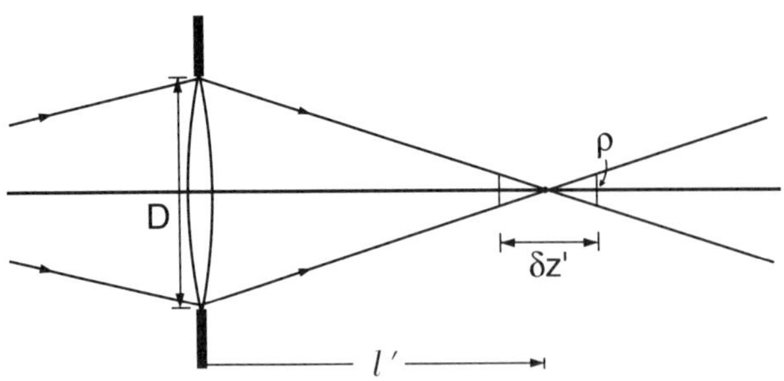

Figure 5.7. Depth of focus for a thin lens with the stop at the lens (arbitrary object location).

If the object is not at infinity, then from Eq. (3.32) we have $l' = F(1 - M)$; so

$$\delta z' = \frac{2\rho F}{D}(1 - M) = 2\rho \mathcal{N}^* \tag{5.9}$$

where

$$\mathcal{N}^* = \mathcal{N}(1 - M) \tag{5.10}$$

the effective F number for finite conjugates. If both conjugates are real, then $M < 0$; so $\mathcal{N}^* > \mathcal{N}$. For camera lenses, we usually have $|M| \ll 1$, so the resulting correction is not large. However, an enlarging or closeup lens can operate at around $M \cong -1$, in which case $\mathcal{N}^* = 2\mathcal{N}$. The smaller number ($\mathcal{N}$) is normally inscribed on the lens; so the casual user might think that the lens will transmit a beam of divergence angle $\sin^{-1}(1/2\mathcal{N})$. Indeed the lens can transmit such a beam but only with the object at infinity; however, at $M \cong -1$ the *beam* F number is twice the value inscribed on the lens (and the corresponding beam divergence angle is only half of the above value).

The image of the segment $\delta z'$ in the object space establishes the depth of field, δz. Objects located anywhere inside the depth of field will appear to be equally in focus (that is, the film cannot tell the difference). If the lens is focused so as to place the film a short distance past the focal plane at $\mathcal{F}'$, the useful depth of field can be maximized. The corresponding object distance (conjugate to the film) is called the *hyperfocal distance*. When the lens is focused at the hyperfocal distance, the depth of field extends to infinity.

The two pictures shown in Figure 5.8 demonstrate the effect of changing the F number on the depth of field. Notice that for the picture taken with the high F number, most objects appear in focus. The logical conclusion of this reasoning is provided by the pinhole camera, examined in Chapter 1; its F number is so high that virtually all objects appear in focus, irrespective of their distance (so long as it is many times the pinhole size). If we consider the pinhole to be a lens with a focal length equal to the length of the box, then we can express the optimum pinhole size in terms of the F number. Analysis (or experiment) shows that the optimum aperture is around $F/500$.

Equation (5.9) applies only for a thin lens with the stop at the lens and cannot be extended to a general system; for example, it predicts that $\delta z' = 0$ if $M = 1$, that is, at the principal planes. This is clearly not the case. We derive in the following a general formula. In Figure 5.9 we have drawn a general system characterized by its principal planes, and also the entrance and exit pupils. From the figure,

$$\delta z' = \frac{2\rho/2}{-\tan\alpha'} = -\frac{\rho}{\tan\alpha'} \tag{5.11}$$

We may take $\delta z'$ to be positive, since it is not measured from any surface or reference plane. We also have

$$\tan\alpha' = -\frac{h'_e}{l' - \bar{l}'} \tag{5.12}$$

(a)

(b)

Figure 5.8. Demonstration of the effect of the F number on the depth of field. Picture (a) was taken at $F/4$ and 1/1000 s shutter speed. Picture (b) was taken at $F/22$ and 1/30 s shutter speed. In the second case, objects appear sharp almost independent of distance, while the moving water of the waterfall appears blurred due to the slow shutter speed. In the first case, only the rock (left foreground) and the waterfall are in sharp focus, while the fast shutter speed has also arrested the movement of the water.

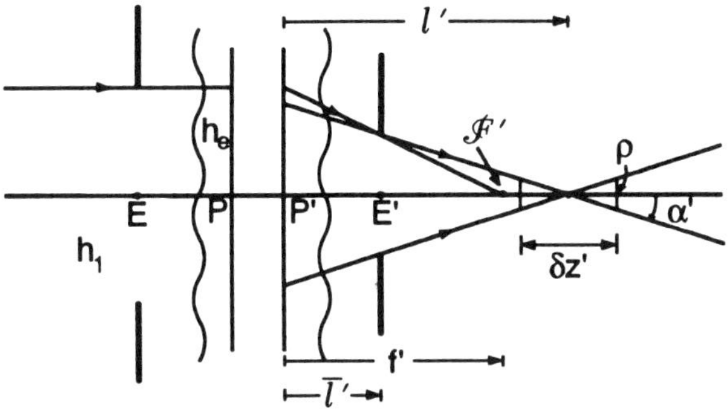

Figure 5.9. Depth of focus for a general system.

Consider now the ray that is parallel to the axis in the object space and passes through the edge of the entrance pupil. This ray will also pass through the edge of the exit pupil and intersect the axis at $\mathscr{F}'$. Similar triangles give

$$\frac{h_e}{F} = \frac{h'_e}{F - \bar{l}'} \qquad \text{or} \qquad \bar{l}' = F\left(1 - \frac{h'_e}{h_e}\right)$$

taking $f' = F$ for simplicity. [This last equation can also be derived immediately by applying Eq. (3.32) to the pupil conjugates.] Substitution of this value into Eq. (5.12) and using again $l' = F(1 - M)$ gives

$$\tan \alpha' = \frac{h_e}{F} \frac{1}{M/\bar{M} - 1}$$

where $\bar{M} = h'_e/h_e$ the *pupil magnification*. Substituting this value into Eq. (5.11) and using Eq. (5.6), we obtain

$$\delta z' = 2\rho\mathscr{N}\left(1 - \frac{M}{\bar{M}}\right) \tag{5.13}$$

This equation is correct in all cases except when the entrance pupil is at infinity ($\bar{M} = 0$). But in that case, $\bar{l}' = F$; so we can substitute $\tan \alpha'$ directly from Eq. (5.12) into Eq. (5.11), to obtain

$$\delta z' = \frac{\rho M F}{h'_e} = \frac{\rho M}{\tan \bar{\alpha}} \tag{5.14}$$

where $\tan \bar{\alpha}$ is the angular size of the entrance pupil. The reader should note that the depth of focus depends on the resolution limit, which is a loosely defined concept. Therefore, the calculation of the depth of focus is inherently approximate.

The concept of depth of focus is important also when considered independent of any image recording device. It then becomes a means of characterizing how rapidly the image blurs with defocus. The tolerable defocus is set not by the recording device resolution, but by the *Rayleigh quarter-wave criterion*. This states that if the path difference between the rays forming the image is less than $\lambda/4$, a near-perfect image results. Application of this criterion leads to the formula

$$\delta z' = \frac{\lambda}{n'u'^2} \tag{5.14a}$$

which is also often stated as $\lambda/(\mathrm{NA})^2$ for $n' = 1$. Justification of this formula may be found in Appendix 4, Eq. (A4.8) by setting $\delta W_{20} = \lambda/4$, but this requires prior understanding of the aberration Chapters 7–9.

5.1.4. Pupils: Off-Axis Imagery

As Figure 5.10 shows, a remote stop has a different effect on the rays from an extra-axial object point than from an axial one. When the stop is at the lens, ray pencils from both the axial and the extra-axial object points pass through the same (middle) part of the lens. With a remote stop, the ray pencil from the axial object point illuminates a different part of the lens than the extra-axial point; the extra-axial ray pencils pass eccentrically through the lens. This eccentricity can be used to control the aberrations of a lens, as we show in some detail in Chapter 8. Therefore, apart from illumination control, the aperture stop is also used for aberration correction.

It is also important to notice that the pupil shape for extra-axial object points is not circular. In the simplest case, it will be elliptical (imagine a circle seen from an angle). But it can also have a much more complicated shape, because *vignetting* can occur at surfaces other than the stop. This effect is easily seen if you have a camera objective handy with a variable stop: All you need to do is to look through the lens off-axis at an increasing angle, for various stop diameters. The effect is also illustrated in Figure 5.11, which shows a common form of a photographic objective, called a *Cooke triplet*. Vignetting at a surface other than the stop has a detrimental effect on the uniformity of the image illuminance, but it is often preferred to the large aberration that might result if the extreme rays were allowed to pass.

5.1.5. Pupil Matching

If you look through a visual instrument such as a telescope or microscope and move your head back and forth along the axis, it will be obvious that there is a best position for your eye, where the whole field appears uniformly illuminated. When that happens, you have positioned your eye pupil at the exit pupil of the instrument. Since the pupil represents a common window for all

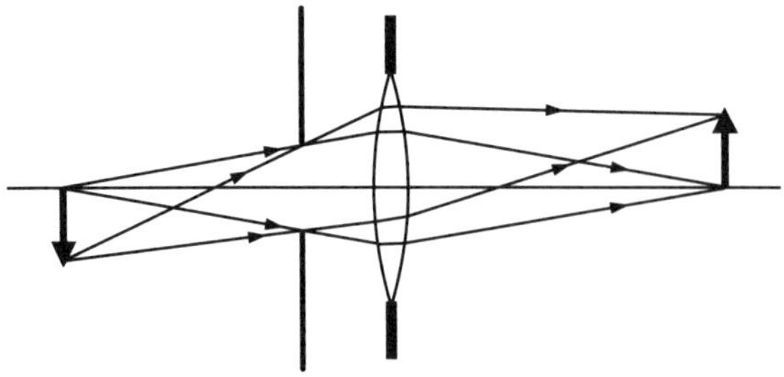

Figure 5.10. The action of the aperture stop for an extra-axial ray pencil.

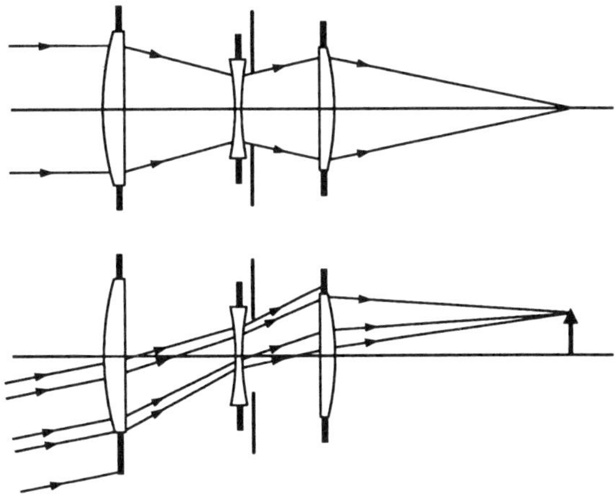

Figure 5.11. Illustration of vignetting at a surface other than the stop. For the axial pencil, there is no vignetting except at the stop. Look carefully to see where each ray is vignetted for the extra-axial pencil, and locate the pupil ray. Notice that the $F\#$ of the extra-axial image forming pencil is higher than that of the axial pencil. Also notice that if the aperture stop is "stopped down" sufficiently, then it can become again the limiting pupil even for the extra-axial pencil.

rays, you are then also collecting the maximum amount of light that your eye pupil will allow. If you move away from that position, the rays will spread more, and your eye will therefore collect less light for the same pupil diameter. This is illustrated in Figure 5.12, where the exit pupil of an inverting telescope is shown. It is easy to see the exit pupil by using a pair of binoculars and a piece of white paper: Point the binoculars at a light source, and observe a bright circle formed just after the eyelens on the paper—that is the exit pupil.

"Pupil matching" means that when two instruments or lenses are cascaded (e.g., telescope and eye), the exit pupil of the first must match the entrance pupil of the second, in location and size. This will guarantee that all the light from the first stage will also go through the second stage, and that the final image will appear uniformly illuminated.

5.1.6. Paraxial Marginal and Pupil Rays

As we saw in the camera example, we can learn all about the conjugate location and size as well as the pupil location and size by tracing two rays through an optical system. The first ray starts from the axial object point and

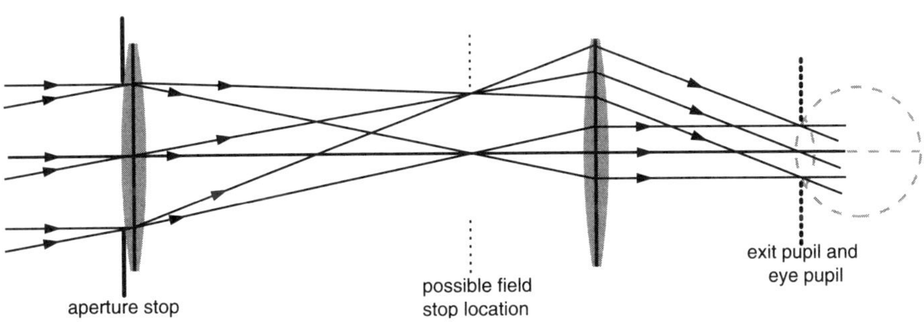

Figure 5.12. Pupil matching between an inverting visual telescope and the eye.

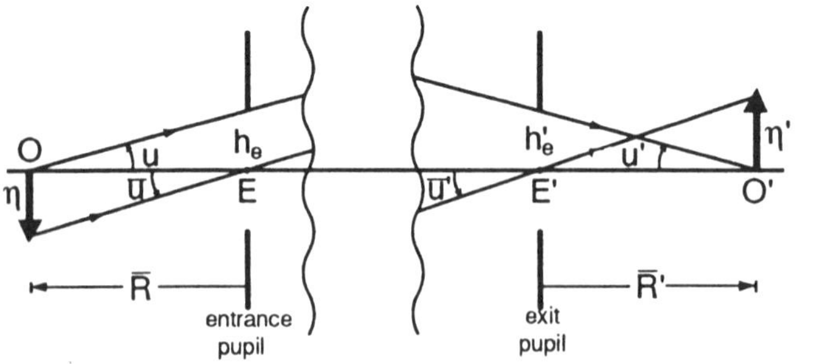

Figure 5.13. Paraxial marginal ray and paraxial pupil ray for a general system. Notice that even in such an abstract diagram, the quantities indicated must still satisfy the optical invariant $nu\eta = -nh_e\bar{u} = -n'h'_e\bar{u}' = n'u'\eta'$ [see Eq. (5.18)].

goes through the edge of the entrance pupil: It is called the *paraxial marginal ray* (PMR). Clearly, it will also go through the edge of the exit pupil (and the aperture stop), and will cross the axis at the image plane. The second ray starts from the edge of the object and goes through the middle of the entrance pupil: It is called the *paraxial pupil ray* (PPR, often also called the *chief ray* or *principal ray**). This ray will cross the axis at the aperture stop; in the image space, it will go through the middle of the exit pupil and will cross the image plane at the edge of the image. Thus the PMR determines the location of the image and the size of the pupils, while the PPR determines the location of the pupils and the size of the image (Figure 5.13).

The PPR can be traced through any multielement system using exactly the same paraxial refraction and transfer equations that we developed in Chapter 4 [Eqs. (4.9) and (4.10)]. However, the ray heights and angles associated with the PPR provide interesting information; so we give them special names. We will use a bar to denote any quantities associated with the PPR. Therefore, we have, for PPR refraction and transfer, respectively:

$$n'_i\bar{u}'_i - n_i\bar{u}_i = -\bar{h}_i c_i(n'_i - n_i) \tag{5.15}$$

$$\bar{h}_{i+1} = \bar{h}_i + d_i\bar{u}'_i \tag{5.16}$$

Similarly, the entrance pupil distance from the first surface is $\bar{l}_1 = -\bar{h}_1/\bar{u}_1$, or from the first principal plane, $\bar{l} = -\bar{h}/\bar{u}$. Image-space quantities are again primed, as before.

Figure 5.14 shows a two-lens system, with the entrance and exit pupils, and the aperture stop (object size and location are also given). The figure contains

* The terms *chief ray* and *principal ray* are more common in the literature than *paraxial pupil ray*. We feel that a departure from the more common names is justified, in order to draw attention to the special nature of this ray. In aberration theory, it is known that the exact location of the pupil is often ill-defined. The term *chief* or *principal* ray is then used to indicate the central ray of an off-axis pencil, and there are several chief rays, for different field points. However, there is only one PPR, because the paraxial approximation removes any ambiguity about the pupil location.

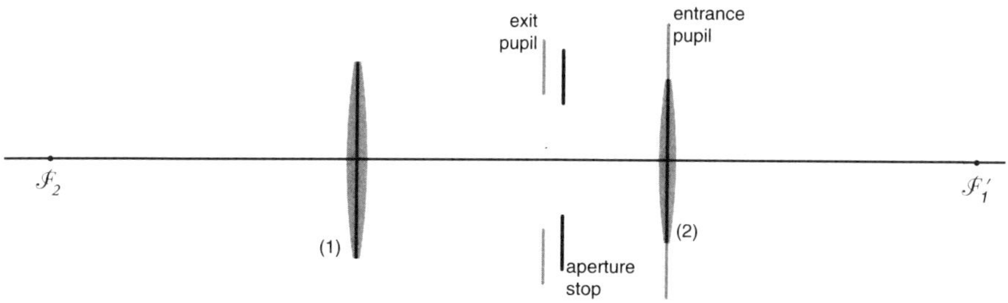

Figure 5.14. From the information shown, determine graphically the location and size of the entrance and exit pupils by tracing graphically the PMR and the PPR. The lenses are considered thin and large enough to avoid vignetting. Take $u = 0$, $\bar{u} = 0.2$ (object at ∞).

enough information for a precise graphical determination of the image location and size, through tracing the PPR and PMR. This should be attempted as an exercise (solution at the end of this chapter, Figure 5.32).

5.1.7. How to Find the Stops

In a multilens system, the role of the aperture stop is played either by an obvious physical stop or a lens, the rim of which limits the incoming pencil of rays. To determine the aperture stop in a system where there is no obvious physical stop, we trace a ray from the axial object point and find the intersection heights h_i with all surfaces. If r_i are the corresponding surface (or lens) semidiameters, then the ratios h_i/r_i are all less than unity if the ray is to go through all lenses. The surface (or lens) with the largest ratio $|h_i/r_i|$, that is, the one where the ray is proportionately closest to its rim, is the aperture stop. Notice that the result of these calculations is independent of the initial ray angle; any ray from the axial object point will do, provided it does not miss any surface. This is due to the linearity of the paraxial ray-tracing equations.

A field stop is placed at the final or intermediate image locations; so it should be easy to spot. However, often there is no specific field stop in the system, and the role of field stop is played again by the rim of some lens. The procedure for determining the field stop is the same as that for the aperture stop, only this time we trace a ray from the axial pupil point. The lens (surface) with the highest ratio $|\bar{h}_i/r_i|$ will play the role of the field stop.

The preceding procedures are very useful in understanding the action of the different elements within a system, especially if they do not happen to be located clearly at a field stop or an aperture stop position. When faced with a multielement system, it is important to be able to deduce which elements affect primarily the field size and which affect primarily the illumination.

Inappropriate positioning of a diaphragm within a system can mean that the diaphragm will play partly the role of the aperture stop, and partly the role of the field stop. An example is provided by Figure 5.4b. The aperture shown plays the role of a field stop, if the object size η is larger than $3D/8$. However, if the diaphragm diameter is reduced beyond $D/3$, then it begins to limit also the angle of the rays from the axial object point; so it becomes an aperture stop as well. Remember always that if the stops are properly positioned, then the field stop changes the image size but has no effect on illumination, while the aperture stop changes the illumination but has no effect on the image size.

5.1.8. Size of a Lens or Surface (Clear Aperture)

The desired diameter of an element is set by the requirement that the marginal ray from the edge of the object must pass. This is shown in Figure 5.15, for a simple case of a thin lens with a remote stop. The marginal ray of importance is the one that passes closest to the lens rim. An exact calculation of the height of that ray is complicated, but fortunately, the use of the paraxial approximation gives us a good starting value. In the paraxial approximation, the angle between the PMR and the axis is the same as the angle between the PPR and the extreme marginal ray. Therefore, this ray intercepts the lens at a height equal to $|h| + |\bar{h}|$. The absolute values are necessary to make the result independent of the sign of h and $\bar{h}$, which depend on the arbitrary choice of ray (e.g., above or below the axis). Thus the desired lens (surface) diameter is given to a first approximation by

$$D = 2(|h| + |\bar{h}|) \tag{5.17}$$

5.1.9. Two-Ray Forms of *H*

The forms $nu\eta$ and $-nh\bar{u}$ [Eqs. (4.11) and (4.12)] can be seen as special cases of the general two-ray form of the optical invariant, which involves both PMR and PPR quantities. The two-ray form(s) of H are very useful in practice. Consider the paraxial refraction equations for the PMR and the PPR

$$n'u' - nu = -hK$$

$$n'\bar{u}' - n\bar{u} = -\bar{h}K$$

Multiplying the first equation by $\bar{h}$, the second by h, subtracting, and rearranging terms, we obtain:

$$n'(u'\bar{h} - \bar{u}'h) = n(u\bar{h} - \bar{u}h)$$

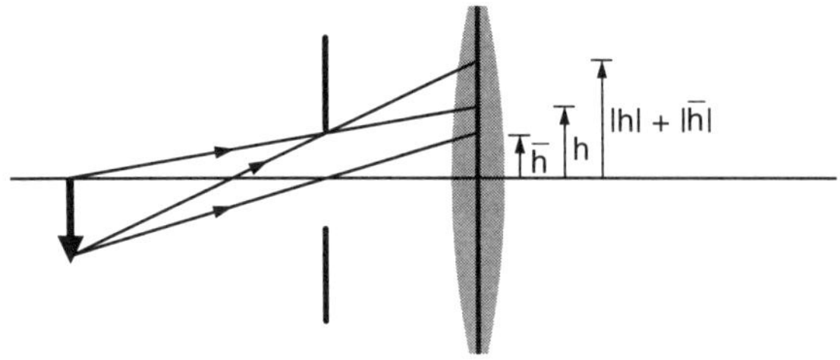

Figure 5.15. Minimum clear aperture of a lens to avoid vignetting.

therefore, the quantity $n(\bar{h}u - h\bar{u})$ is invariant upon paraxial refraction. (The reader can verify that it is also invariant upon paraxial transfer.) At a field stop (or at the object), we have $h = 0$, $\bar{h} = \eta$, while at an aperture stop, we have $\bar{h} = 0$, $h = h_e$; that is, we obtain Eqs. (4.11) and (4.12). Therefore, our invariant form is the same as the optical invariant H

$$H = n(\bar{h}u - h\bar{u}) \tag{5.18}$$

It is left as a problem to also show that

$$H = \frac{nn'(u'\bar{u} - u\bar{u}')}{K} \tag{5.19}$$

where K is the system power. This form can be used for computing K without having to trace an additional ray parallel to the axis.

Example 5.1. Calculate the optical invariant H, in the following two cases: (1) a microscope objective, with NA = 0.16 (object space), magnification $10\times$, and image diameter 20 mm, (2) a telescope with an entrance pupil diameter of 50 mm, $\pm 3°$ field of view, and entrance pupil located at the first surface.

Conventionally, we choose the PMR and the PPR to be such that h_1 is positive and H is negative. If the object is real ($l < 0$), this means that $u_1 > 0$, and, therefore, $\eta < 0$. Most of the diagrams in this book follow this convention, but it is only a convention, with no special significance. However, in calculations it is helpful to maintain a consistent convention. Thus we have:

1. $\mathrm{NA} = n \sin\alpha = nu = 0.16 \rightarrow u = 0.16$. Also, $M = nu/n'u' = -10 \rightarrow n'u' = -0.016$. So $H = n'u'\eta' = -0.016 \times 10 = -0.16$ mm (notice that $\eta' > 0$ because $M < 0$ and $\eta < 0$).
2. Since the object is at infinity, we must employ an alternative form of H. We can use Eq. (4.12), which is a special case of Eq. (5.18): $H = n(\bar{h}u - h\bar{u}) = -nh\bar{u}$ for $\bar{h} = 0$ (stop at the lens). Take $h = 25$ mm (>0) and $\bar{u} < 0$. When the field of view is given in degrees, we show in the next section that the correct conversion to paraxial angle variable is through the tangent of the angle. Thus $\bar{u} = \tan(3°) = 0.0524$, and $H = -25 \times 0.0524 = -1.31$ mm.

Example 5.2. A photographic objective has an infinite object. You are given the focal length, the entrance pupil diameter, and the semifield of view. Calculate the following quantities: h_1, H, η'_k, u'_k (in the usual notation).

The given quantities are F, h_e, and $\bar{u}_1$. Also, since the object is at infinity, $u_1 = 0$; so $h_1 = h_e$. We then have:

$$H = -h_1\bar{u}_1, \qquad n_1 = 1$$
$$u'_k = -h_e/F \qquad \text{(see Figure 5.6)}$$
$$\eta'_k = H/u'_k = F\bar{u}_1$$

So all required quantities have been expressed in terms of the given ones. Notice that when one conjugate is at infinity, the focal length represents a sort of "hybrid" magnification of the system (in this case, the linear size of the image η'_k divided by the angular size of the object $\bar{u}_1$).

Example 5.3. A two-lens system has $F_1 = 5$ cm, $F_2 = -2$ cm, $d = 2.5$ cm, $l_1 = -20$ cm, and lens diameters $D_1 = 2$ cm, $D_2 = 0.6$ cm. Determine the optical invariant for this system.

Notice that in this problem the object size is not given, but we can determine the maximum allowable object size from the given lens diameters. The first task is to determine the aperture stop. Trace a ray from the axial object point, with any value of u_1, for example, $u_1 = 0.01$. We have:

$$h_1 = -u_1 l_1 = 0.2 \text{ cm}, \qquad u'_1 = u_1 - h_1/F_1 = -0.03$$

$$h_2 = h_1 + u'_1 d_1 = 0.125 \text{ cm}$$

so

$$h_1/D_1 = 0.1 \text{ cm}, \qquad h_2/D_2 = 0.208 > h_1/D_1$$

therefore, lens 2 is the stop (and exit pupil). The PMR will have a height $h'_e = D_2/2 = 0.3$ cm at the stop. The PPR will cross the axis at the second lens; so lens 1 must play the role of the field stop. The PPR will pass undeviated through lens 2; so $\bar{u}'_2 = \bar{u}'_1$ and $\bar{l}'_1 = d$. In order to determine the field size, we might be tempted to draw a ray from the middle of the second lens to the edge of the first lens (make a drawing!). However, this is not correct, because lens 1 is not exactly at a field stop position (i.e., it is not at an intermediate image where h would be zero), even though it limits the field. Therefore, we must use Eq. (5.17):

$$|\bar{h}_1| = \frac{D_1}{2} - |h_1|$$

where h_1 is the PMR height at the first lens. This can be determined from the previous calculation: If h_2 is increased from 0.125 to 0.3, then h_1 will increase proportionately to $0.2 \times 0.3/0.125 \rightarrow h_1 = 0.48$ cm. Therefore,

$$|\bar{h}_1| = 0.52 \text{ cm} \qquad \text{and} \qquad \bar{h}_1 = -0.52 \text{ cm}$$

so that

$$\bar{u}'_1 = \bar{u}' = -\frac{\bar{h}_1}{\bar{l}'_1} = 0.52/2.5 = 0.208$$

$$H = -h'_e \bar{u}' = -0.3 \times 0.208 \text{ cm} = -0.0624 \text{ cm}$$

In this example, $\bar{h}_1$ was taken as negative in order to make $H < 0$, as discussed in Example 5.1. The field (object) size can be determined by completing the PPR ray trace in the object space. It is to be noted that the object height determined through this procedure represents the maximum object height for which the marginal rays from the extreme object point pass through the lens with no vignetting. If vignetting were allowed, then the maximum field angle would be given by setting $|\bar{h}_1| = D_1/2$. This discussion shows the usefulness of a properly located field stop in controlling the illumination at the edge of the field.

5.2. SIGNIFICANCE AND USE OF THE MARGINAL AND PUPIL RAYS

5.2.1. Connecting Paraxial and Finite Optics

In all the ray-tracing problems we have given so far, we have used finite values for the paraxial quantities h, u, etc. Yet in the paraxial approximation, these quantities are assumed to be infinitesimal. In other words, during the Gaussian design stage rays are traced according to the laws of paraxial optics even though the approximation $\sin\theta \cong \theta$ may not be closely satisfied. Why are we allowed to do that? In order to use the paraxial equations with finite quantities, we must make sure that the two conditions to be stated are satisfied; we can then show that our use of the paraxial equations will not lead to large errors but rather will provide quite accurate results for systems without a lot of aberrations.

The two conditions under question relate the values of the field of view (object size) and the aperture of a system to the corresponding PMR and PPR angle variables as follows. The aperture angle α is related to the PMR angle variable by

$$\boxed{u = \sin\alpha} \tag{5.20}$$

The semifield of view $\bar{\alpha}$ of a system is related to the PPR angle variable by

$$\boxed{\bar{u} = \tan\bar{\alpha}} \tag{5.21}$$

It should be clearly understood that the angles α and $\bar{\alpha}$ represent the desired aperture and field of the system, and are thus finite angles, for which the paraxial approximation may not hold. The preceding two equations represent the link between Gaussian optics and the real world. They must be used whenever one needs to convert finite angle values to the corresponding paraxial angle variables, to provide, for example, the opening data for ray tracing. Thus if we say that the field of view of a system is $\pm 20°$, we must set $\bar{u} = \tan 20°$.

The choice of sine or tangent is not arbitrary. Let us first examine the geometrical significance of these choices. In Figure 5.16 we show the object and entrance pupil of a system and the wavefront emanating from A and arriving at E. The PMR cuts this wavefront at point C. Let the radius of the wavefront be $\bar{R}$ (measured from E), and h_e the entrance pupil radius. We then have $u = -h_e/\bar{R}$ paraxially. Also, from the figure, $\sin\alpha = -(DC)/(CA) = -(DC)/\bar{R}$. Therefore, if we set $u = \sin\alpha$, we must also have $h_e = (DC)$. This means that the paraxial height variable h must be interpreted as the height at which the ray cuts the wavefront touching the surface at the apex. For the PPR, we have $\bar{u} = \eta/\bar{R}$ paraxially, which is consistent with $\bar{u} = \tan\bar{\alpha} = (AB)/(EA) = \eta/\bar{R}$.

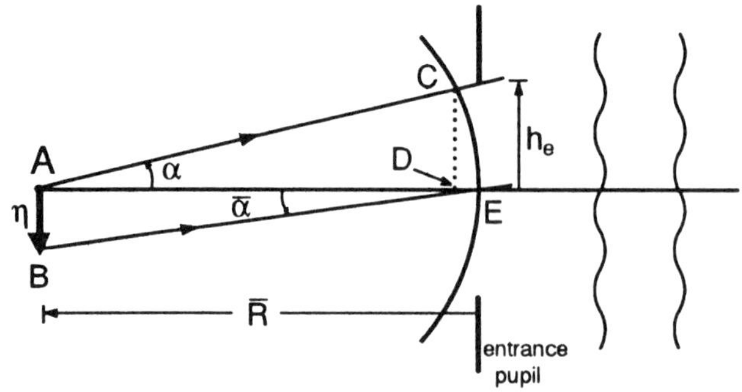

Figure 5.16. Geometrical significance of the choices $u = \sin \alpha$ and $\bar{u} = \tan \bar{\alpha}$.

The justification for the choices of Eqs. (5.20) and (5.21) is governed by the following reasoning. In order for the paraxial calculations to be meaningful, we must ensure that the convention used for converting the finite angles to paraxial quantities is as valid in the object space as it is in the image space. In other words, if we set $u = \sin \alpha$ in the object space, then the relation $u' = \sin \alpha'$ must also hold. This is by no means guaranteed in advance, because u' and α' are computed in different ways: u' is computed paraxially, but α' is computed through finite, or exact, ray tracing (i.e., using the exact form of Snell's law, as in Appendix 3). So we are attempting to impose a rather demanding requirement.

By choosing to link paraxial and real quantities through Eqs. (5.20) and (5.21), we ensure that there is a fixed correspondence between them. But *this correspondence holds only in the object and image spaces of systems in which the aberrations are reasonably well corrected.* However, even for such systems the aberrations are not corrected in intermediate spaces. In those spaces, the discrepancy between paraxial and real quantities may be very large, independent of the convention adopted for linking the two.*

Although the connection between paraxial and finite optics is very rarely discussed in the literature, it is more than mere pendanty. For example, the F number is sometimes defined as $1/2n'u'$, with $u' = \tan \alpha'$. This can lead to serious errors for "high-speed" (high-NA) systems such as microscope objectives.

The conditions $u' = \sin \alpha'$ and $\bar{u}' = \tan \bar{\alpha}'$ arise from two theorems, to be discussed in the following. For this discussion, some understanding of aberration theory is needed; so the next two sections may be omitted on first reading.

5.2.2. The Sine Condition

The sine condition is strictly stated as follows: For a system with no spherical aberration, if the relation $u'/u = \sin \alpha'/\sin \alpha$ is satisfied, then there is no coma

---

* For a complete discussion, see H. H. Hopkins: "The Nature of the Paraxial Approximation," *J. Modern Optics* **38**, 427–45 (1991).

or other aberrations linear in the field. Proof of this strict form will take us too far into aberration theory; so we will satisfy ourselves with a weaker version, which is stated as follows: For a system with perfect imagery near the axis, the relation $n \sin\alpha\ \eta = n' \sin \alpha'\eta'$ holds. This relation can be seen as the finite form of the optical invariant, $nu\eta = n'u'\eta'$, and suffices to justify the choice $u = \sin \alpha$.

In Figure 5.17, consider the entrance and exit pupils at E and E', respectively, and a small object $(AB) = \eta$. We start by assuming that it is possible for an optical system of finite aperture to image perfectly an infinitesimally small object, and then derive a necessary condition for this. Thus η is taken as infinitesimal, but the angle α as finite. (This represents a step up from the paraxial assumption, in which both the aperture and the field are infinitesimal.) The ray from A through P will emerge at some point P' at the exit pupil, and will cut the axis at A', the image of A. The optical path length from A to A' is the same along any ray; so

$$[AP\cdots P'A'] = [AE\cdots E'A'] \tag{5.22}$$

Consider now the ray from B to P. It will emerge at the exit pupil at a point Q' and intersect the image plane at B'. Let us drop two perpendiculars BB_0 and $A'A'_0$. Because η and η' are infinitesimal, we may consider BB_0 to be normal to either AP or BP or both, and similarly for $A'A'_0$. We then have

$$[BP\cdots Q'A'_0] = [B_0P\cdots P'A'] \tag{5.23}$$

The last equation states that the path difference between two nearby rays is zero (Section 1.5).

Finally, let us also consider the ray from B to E; it will emerge from a point D' and intersect the image plane at B', on the assumption of perfect imagery. Thus, for imagery between B and B',

$$[BP\cdots Q'B'] = [BE\cdots D'B'] = [AE\cdots E'A'] \qquad \text{for } \eta \to 0$$

Therefore, from Eq. (5.22), $[AP\cdots P'A'] = [BP\cdots Q'B']$, or

$$[AB_0] + [B_0P\cdots P'A'] = [BP\cdots Q'A'_0] + [A'_0B']$$

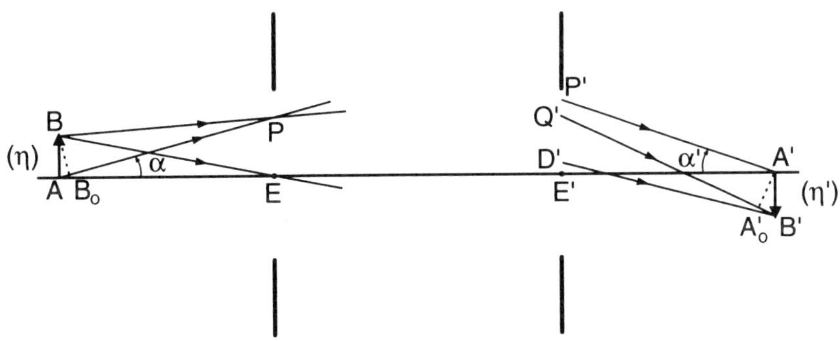

Figure 5.17. The sine condition and the finite form of the optical invariant.

so

$$[AB_0] = [A'_0B'] \tag{5.24}$$

by virtue of Eq. (5.23). The figure gives $(AB_0) = \eta \sin \alpha$, and $(A'_0B') = \eta' \sin \alpha'$, so from Eq. (5.24) and the definition of optical path,

$$n \sin \alpha\eta = n' \sin \alpha'\eta' \tag{5.25}$$

The paraxial form of the optical invariant gives $nu\eta = n'u'\eta'$, so from Eq. (5.25) we have $u/u' = \sin \alpha/\sin \alpha'$. Therefore, if we choose to represent the PMR angle variable through $u = \sin \alpha$, we will also have $u' = \sin \alpha'$ under the preceding conditions of perfect (or near perfect) imagery near the axis.

5.2.3. The Tangent Condition

In Figure 5.18, consider a general system with the entrance and exit pupils at E and E', respectively. We take two object heights, η_1 and η_2, and the two corresponding image heights η'_1 and η'_2. From the figure we have

$$\frac{\eta_1}{\eta_2} = \frac{\tan \bar{\alpha}_1}{\tan \bar{\alpha}_2} \quad \text{and} \quad \frac{\eta'_1}{\eta'_2} = \frac{\tan \bar{\alpha}'_1}{\tan \bar{\alpha}'_2} \tag{5.26}$$

The requirement $\eta'_1/\eta_1 = \eta'_2/\eta_2$ means that the magnification is constant across the field; that is, there is no distortion. We then have from Eq. (5.26),

$$\frac{\eta'_1}{\eta_1}\frac{\eta_2}{\eta'_2} = \frac{\tan \bar{\alpha}'_1}{\tan \bar{\alpha}_1}\frac{\tan \bar{\alpha}_2}{\tan \bar{\alpha}'_2} = 1$$

or

$$\frac{\tan \bar{\alpha}'}{\tan \bar{\alpha}} = \text{constant} \tag{5.27}$$

as a requirement for the absence of distortion. The value of the constant can be found from the figure as

$$\frac{\tan \bar{\alpha}'}{\tan \bar{\alpha}} = \frac{\eta'/\bar{R}'}{\eta/\bar{R}} = M\frac{\bar{R}'}{\bar{R}} = \frac{\bar{u}'}{\bar{u}} \tag{5.28}$$

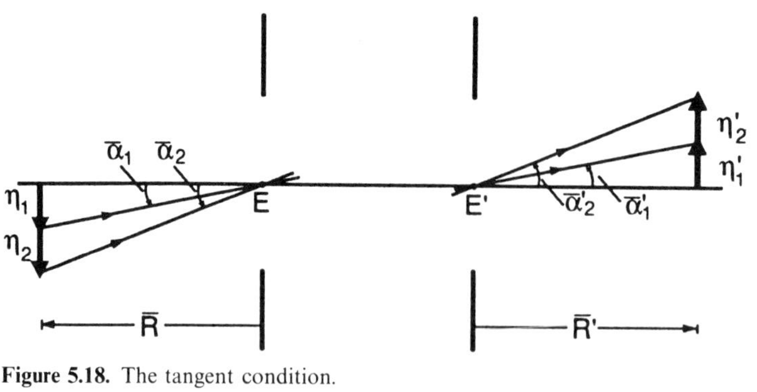

Figure 5.18. The tangent condition.

In drawing Figure 5.18, we have assumed that both rays will emerge from point E'. This is only true if there is no spherical aberration for imagery between the pupil planes. In practice, the spherical aberration of the pupil may be large; so the tangent condition [Eq. (5.27)] is not as useful as the sine condition. However, for our purposes, the choice $\bar{u} = \tan \bar{\alpha}$ may also be justified on purely geometrical grounds, as noted previously.

5.2.4. Gaussian Predesign

The PMR and PPR concepts facilitate the specification of a preliminary design that will have the required Gaussian properties, that is, location, size, and illumination of image. When faced with the problem of specifying a lens for a given task, it is best to think in terms of the PMR and PPR. As soon as these two rays can be traced through the system, the Gaussian predesign is complete. The following two examples illustrate how to think in terms of the PMR and PPR.

Example 5.4. Consider the operating spectacle problem (Example 4.6) of Chapter 4: Design a pair of magnifying spectacles, with the following initial specifications (1) $M = +2$; (2) the image plane should coincide with the object, at about 25 cm in front of the observer; (3) field of view "as large as possible" (typical of the way a surgeon might specify the spectacles!).

In Figure 5.19 we have drawn a ray from the axial object point through the system. This ray would be the PMR if we knew the aperture size. Therefore, we must determine the pupils. There is no obvious aperture stop at first sight, but since this is a visual system, the exit pupil must coincide with the eye pupil; so we may take the eye pupil to be both the aperture stop and the exit pupil of the spectacles. Take 4 mm eye pupil diameter to be typical of indoor lighting; so $h'_e = 0.2$ cm. We may also assume a comfortable distance of, say, $\bar{l}'_2 = 1$ cm between the negative element and the eye. Therefore, we now have the exit pupil location and size, and we can proceed to determine the entrance pupil

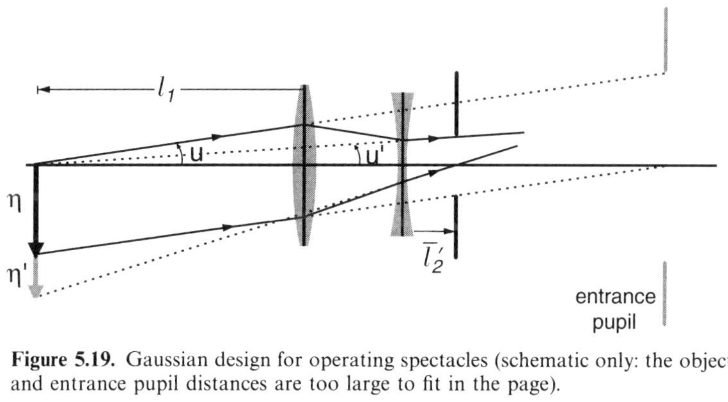

Figure 5.19. Gaussian design for operating spectacles (schematic only: the object and entrance pupil distances are too large to fit in the page).

by imaging the exit pupil through the system. Using the values of Chapter 4 ($K_1 = 0.187\ \text{cm}^{-1}$, $K_2 = -0.298\ \text{cm}^{-1}$, $d = 3$ cm, we obtain $\bar{l}_1 = 12.8$ cm, $h_e = 0.522$ cm, $h_1 = 0.345$ cm, $h_2 = 0.193$ cm. So the PMR tracing is complete, giving us all the first-order information about the location of the image and the illumination.

We then need to determine the size of the image (object) by tracing the PPR. The size of the field will be limited by a field stop. We can determine the existence of a field stop by following the procedure outlined in the Section 5.1, that is, by tracing any ray from the axial pupil point and finding the ratios $\bar{h}_i/r_i$, but we do not know yet the lens diameters. Therefore, we must first choose a field size arbitrarily, then trace the PPR and determine the lens sizes from Eq. (5.17). Take, for example, $\eta = -5$ cm, then Figure 5.19 gives $\bar{u}_1 = 0.132$, and $\bar{h}_1 = -1.69$ cm. Upon completing the ray tracing, and applying Eq. (5.17), we obtain $D_1 = 4.07$ cm, and $D_2 = 1.08$ cm for the lens diameters.

At this point, we should probably halt the design, because the size of the first lens is rather large and might prove uncomfortable. The major contributor to the size is the PPR height; this can be reduced by restricting the field of view. If we halve the field ($\eta = -2.5$ cm), the PPR heights will also be halved, but PMR quantities will not be affected. For this case, then, we have $D_1 = 2.38$ cm and $D_2 = 0.73$ cm. Since we have a $2\times$ magnification, the total field seen by the observer is 10 cm wide at 25 cm away. This is not too narrow, but if desired, one can give the first lens a rectangular (instead of circular) shape, trading the reduced field of view along the direction of the short side of the rectangle for reduced weight. This completes the Gaussian predesign for this lens. The question of image quality will be addressed in Chapters 7–9, and the design is completed in Chapter 10.

Example 5.5. It is required to specify a system that will be used for viewing a distant object. The initial specifications are as follows:

1. Object at infinity;
2. Image vergence (image distance) variable from +1D to −3D approximately.
3. Magnification: ± 1 (either choice acceptable);
4. Provide means for varying the exit pupil diameter easily;
5. Apparent image size should not change as the image vergence is varied.

Notice that requirement (3) does not refer to the linear magnification, but to the *angular magnification*, $\bar{u}'/\bar{u}$. Similarly, requirement (5) does not refer to the linear size of the image, but to the angle that it subtends at the eye. Requirements (2) and (5) make it clear that we cannot have a single lens, since the latter produces an inevitable magnification change when the image shifts ($M = l'/l$). Therefore, we need at least a two-lens system. A starting point can be provided by the telescope design of Figure 5.12. From Figure 5.12 we see that the exit

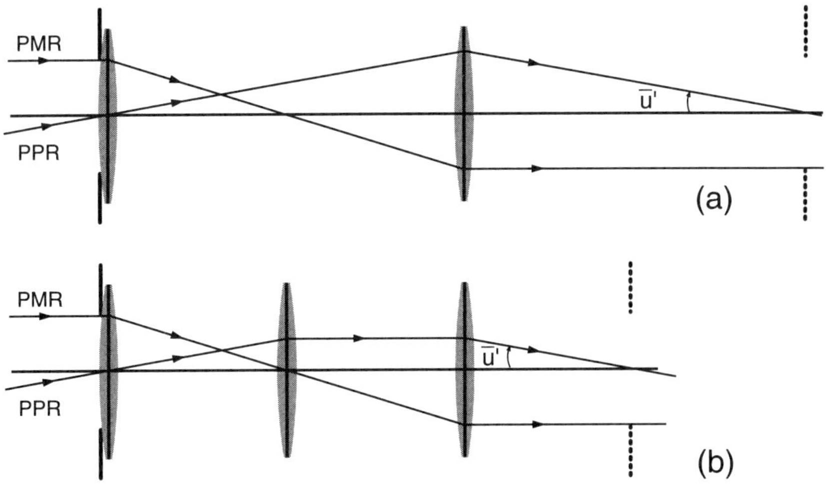

Figure 5.20. Gaussian design of a telescopic viewer that permits defocus but maintains constant apparent image height (b).

pupil is real (i.e., behind the last lens), which is necessary for pupil matching with the eye. We also see that the exit pupil size can be varied easily by placing an iris diaphragm in contact with the first lens (entrance pupil). So we may use the system of Figure 5.20a as a starting point. We have drawn the PMR and PPR with as much information as we have up to this point.

This system allows defocusing of the image by changing the distance between the two lenses. Unfortunately, the angular size of the image will vary as a result. However, the angular size of the image is determined by the PPR angle $\bar{\alpha}'$ (or $\bar{u}'$). This immediately suggests a solution: If we could arrange for the PPR to be parallel to the axis before the second lens (eyelens), the PPR angle, and thus the apparent image size in the eye space, would remain constant even as the eyelens distance was varied (to achieve the required defocusing). Thus we need a lens that would change the PPR angle in the intermediate space, but would have no effect on the PMR. After a bit of thought, we may come up with the solution shown in Figure 5.20b. If the lens is placed at the intermediate image location (hence the name *field lens*), the PMR will not be deviated, but the PPR will. The focal length of the field lens will then be approximately equal to that of the other two lenses. Clearly, the system as drawn can be made with lenses of any focal length. A focal length of 50 mm may be chosen as common enough to guarantee easy selection of lenses. The final choice of focal length will be driven by the tolerable field curvature, as discussed in Chapters 8 and 9.

5.2.5. The Two-Lens System with Fixed Pupil Positions

So far, we have examined the Gaussian predesign with a single-lens solution (Section 3.7), and a two-lens solution, but without taking the pupils into account

(Section 4.10). It is possible that information about the pupils is so critical that it must be taken into account at the very beginning of the design. This is true of any lens system that requires pupil matching, such as the telescope design of the previous example. Other examples are image relay lenses or the so-called Fourier transform lenses, used in optical spatial filtering experiments. There are also several counter examples, such as the system of Example 5.1 (match only exit pupil) or photographic objectives (no matching).

A typical specification for a system with fixed pupil positions (in air) is: l_1, l'_2, M, $\bar{l}_1$, $\bar{l}'_2$ and two of u, η, H. From the following equations, all of the Gaussian parameters are found:

$$h_1 = -u_1 l_1, \qquad \bar{u}_1 = -\frac{\eta}{\bar{l}_1 - l_1}, \qquad \bar{h}_1 = -\bar{u}_1 \bar{l}_1$$

$$h_2 = h_1 \left(\frac{l'_2}{l_1}\right)\left(\frac{1}{M}\right), \qquad \bar{h}_2 = \frac{H}{h_2(1/\bar{l}'_2 - 1/l'_2)}$$

The solution is:

$$d_{12} = \frac{h_1 \bar{h}_2 - \bar{h}_1 h_2}{H} \tag{5.29}$$

$$K_1 = \frac{1 - \bar{h}_2/\bar{h}_1}{d_{12}} - \frac{1}{\bar{l}_1} \tag{5.30}$$

$$K_2 = \frac{1}{\bar{l}'_2} - \frac{\bar{h}_1/\bar{h}_2 - 1}{d_{12}} \tag{5.31}$$

5.2.6. The y–$\bar{y}$ Diagram

This diagram takes its name from the paraxial height variables h and $\bar{h}$, which are frequently also called y and $\bar{y}$. As the name suggests, it is a plot of the PMR ray height versus the PPR ray height, with each point on the plot representing one surface.* Figure 5.21 shows the y–$y$bar (or h–$h$bar) diagram for the lens system of Figure 5.20b.

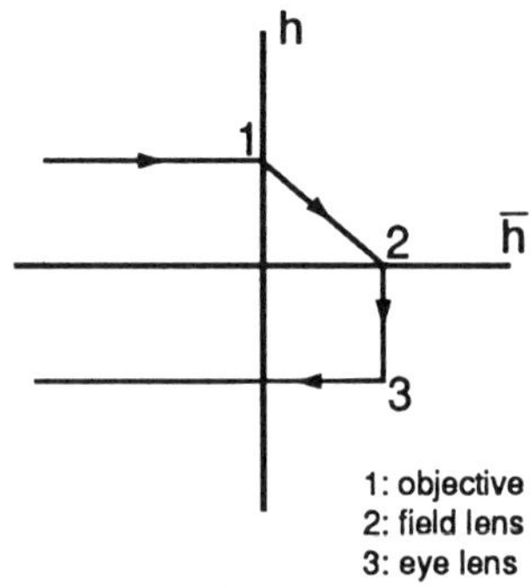

Figure 5.21. The y–$\bar{y}$ diagram for the system of Figure 5.20(b).

* E. Delano: "First order design and the Y–$\bar{Y}$ diagram," *Appl. Opt.* **2**, 1251 (1963).

An alternative way of thinking about the diagram is to consider it as the end view of a single skew ray that has the PMR as its projection on the x–y plane, and the PPR as its projection on the x–z plane. (A skew ray is a ray that does not cross the optical axis.) The diagram can be useful in visualizing vignetting or clear aperture constraints. Since it combines to some degree the properties of both the PMR and the PPR, an experienced practitioner may favor its use as a shortcut. However, it should be clear that the diagram does not contain any more information than a ray sketch of the PMR and the PPR through the system; rather, it contains less information; so its use is to a large extent a function of personal preference.

5.3. LIGHT FLUX TRANSMISSION THROUGH OPTICAL SYSTEMS

The concepts of the entrance and exit pupil allow us to calculate the amount of light passing through an optical system. In order to do that, we must first review some of the fundamental concepts of radiometry and photometry; we will then apply them to lens systems.

5.3.1. Radiometry versus Photometry

Photometry is concerned with measurement of the visible part of the EM spectrum, whereas radiometry considers the whole spectrum. In the case of photometry, the wavelength sensitivity of the human eye is taken into account as a weighting factor. Table 5.1 and Figure 5.22 show the accepted standard response of the human eye, under photopic (daylight) conditions. The peak sensitivity occurs at 555 nm, a convenient number to remember. The official name for the sensitivity function is "spectral luminous efficiency," V_λ. The inverse of this curve gives the number of watts of radiant power at any given wavelength that is required to produce a constant sensation of brightness.

The unit of radiant power is the watt, while the unit of luminous power (usually called luminous flux) is the lumen. The watt (lumen) is the only fundamental unit in radiometry (photometry), in the sense that all other units are defined in terms of watts (lumens), areas, and solid angles. Thus photometric and radiometric units can be converted into each other if we know the conversion factor between watts and lumens. The value of this factor is given in the third column of Table 5.1. These values represent a scaled version of the V_λ curve, with a scaling factor $K_m \cong 683$ lm/W.

One concept that is peculiar to photometry is that of the luminosity of a source. This is defined as the ratio of total luminous power to total radiant power, integrated across the whole spectrum,

$$K = \frac{\int V_\lambda W_\lambda \, d\lambda}{\int W_\lambda \, d\lambda} \tag{5.32}$$

The source luminosity is a measure of its efficiency in converting the input energy to visible radiation. Typical values for broadband sources are 2–10%.

TABLE 5.1.
Photopic spectral luminous efficiency* and luminous to radiometric conversion

λ (nm)	$V(\lambda)$	$K(\lambda)$ (lm/W)
410	0.0012	0.82
420	0.0040	2.73
430	0.0116	7.92
440	0.0230	15.7
450	0.0380	26.0
460	0.0600	41.0
470	0.0910	62.2
480	0.1390	94.9
490	0.2080	142
500	0.3230	221
510	0.5030	344
520	0.7100	485
530	0.8620	589
540	0.9540	652
550	0.9950	680
(555)	(1.0000)	(683)
560	0.9950	680
570	0.9520	650
580	0.8700	594
590	0.7570	517
600	0.6310	431
610	0.5030	344
620	0.3810	260
630	0.2650	181
640	0.1750	120
650	0.1070	73.1
660	0.0610	41.7
670	0.0320	21.9
680	0.0170	11.6
690	0.0082	5.60
700	0.0041	2.80
710	0.0021	1.43
720	0.0010	0.68

* Restricted in the range 410–720 nm and to fourth decimal place; the standard curve extends from 360 to 830 nm.

5.3.2. Radiometric (Photometric) Quantities and Units

Before examining the various quantities, we review the concept of solid angle.

The solid angle may be defined as a surface formed by all rays joining a point to a closed curve. For simplicity, we may take the curve to be a circle or a rectangle. The solid angle is then measured as the area intercepted at the surface of a sphere by a cone or pyramid with its apex at the center of the sphere,

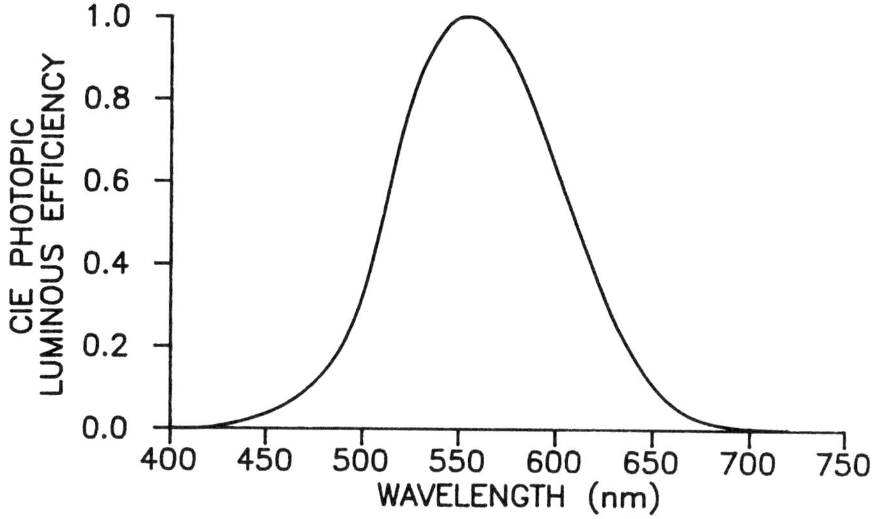

Figure 5.22. The photopic spectral luminous efficiency function, V_λ plotted against wavelength λ.

divided by the square of the radius of the sphere (Figure 5.23).

$$d\Omega = \frac{dA}{r^2} \tag{5.33}$$

For a right circular cone of half-angle θ, the solid angle can be shown to be

$$\Omega = 2\pi(1 - \cos\theta) = 4\pi \sin^2\left(\frac{\theta}{2}\right) \tag{5.34}$$

or

$$\Omega = \pi\theta^2 \tag{5.35}$$

if θ is small. The unit of solid angle is the steradian (sr). A complete sphere has 4π sr.

We may now define the following quantities (photometric quantities are in parentheses).

Figure 5.23. Definition of solid angle.

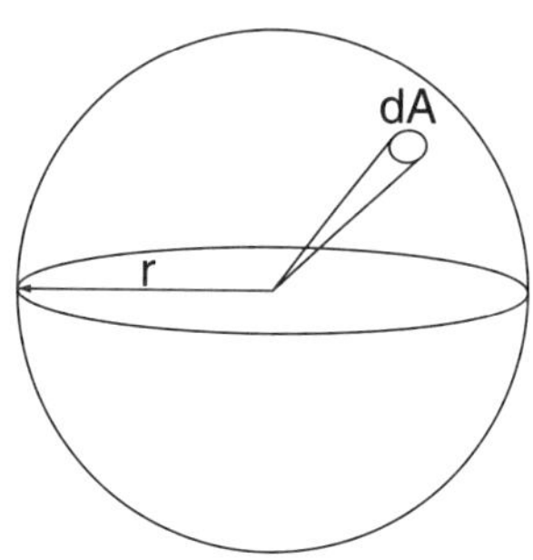

Radiant (Luminous) Intensity of a Source *I*. I (or simply intensity) is the emitted power per unit solid angle

$$I = \frac{d\Phi}{d\Omega} \tag{5.36}$$

Unit: watt/sr (photometric: 1 candela = 1 cd = 1 lm/sr).

Although the lumen is perhaps a more fundamental unit conceptually, it is in fact defined in terms of the candela, which is considered a fundamental SI unit (along with the meter, second, etc.). Several definitions of the candela have been used at different times. The one currently in use is as follows: "the candela is the luminous intensity, in a given direction, of a source that emits monochromatic radiation of frequency 540×10^{12} Hz and that has a radiant intensity in that direction of (1/683) watt per steradian." From this, the lumen is then defined as the flux emitted in one steradian by an isotropic point source having a luminous intensity of one candela.*

Radiant (Luminous) Emittance or Exitance of a Source *M*. M is the power emitted per unit surface area of the source.

$$M = \frac{d\Phi}{dA} \tag{5.37}$$

Units: W/m^2 (lm/m^2). A 50 W tungsten filament has the same emittance as a 100 W filament, since the additional output is achieved by increasing the filament area.

Radiance (Luminance) of a Source *L*. L is the emitted flux per unit solid angle, per unit projected area of the source. "Projected area" means normal to the direction of observation.

$$L = \frac{d^2\Phi}{d\Omega \cos\theta \, dA} \tag{5.38}$$

where θ is the angle between the normal to the plane of the source and the direction of observation (Figure 5.24). Unit: W/sr/m^2 (cd/m^2). L is in general a function of the direction of observation. A source for which L is independent of direction is called a *Lambertian source*. There is no such thing as a truly Lambertian source, though some diffusing surfaces may approximate Lambertian behavior over a restricted range of angles (say $\pm 40°$ about the normal). However, the Lambertian assumption provides usually a good approximation for diffusely illuminated objects.

Irradiance (Illuminance) of a Receiving Surface *E*. E is the power per unit receiving area.

$$E = \frac{d\Phi}{dA} \tag{5.39}$$

* CIE Publication No. 18.2 (TC-1.2). "The basis of physical photometry," 1983.

Figure 5.24. Definition of radiance (luminance).

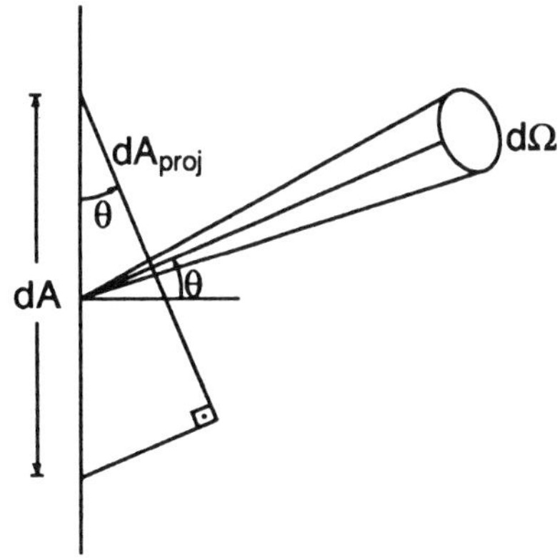

Units: W/m^2 (1 lm/m^2 = 1 lux). Notice that this is dimensionally the same quantity as the emittance of a source. No confusion should arise since the distinction between a source and a receiving screen should be obvious. The term *lux* is reserved for illuminance and is not used for the source emittance M.

Numerical examples of luminance and illuminance are given in Tables 5.2 and 5.3.

5.3.3. Flux Emitted into a Cone by a Lambertian Source

Suppose we have a Lambertian source, or an illuminated surface that approximates Lambertian behavior. In the latter case, the total reflected or transmitted flux will be related to the incident flux Φ_i through

$$\Phi_R = \Phi_i R \qquad \text{or} \qquad \Phi_T = \Phi_i T$$

where R is the reflectivity and T the transmissivity of the surface. We want to find the flux emitted within the solid angle enclosed by a cone of half-angle α.

TABLE 5.2.
Typical sources

Source	Input power (W)	Size (mm)	Intensity (cd)	Luminance (10^6 cd/m^2)
Tungsten	750	10 × 9	1550	17
Tungsten–halogen	250	5.5 × 4	680	30
Compact Hg iodide	400	9 × 5	36 000	80
Compact Xe	250	3 × 2	660	110
High-pressure Hg arc	250	4 × 2	1600	250
Carbon arc	2000	5 × 5	7500	300
High-pressure Hg arc	1000	6.5 × 2.5	6500	400
Compact Xe	2000	4.5 × 4	21 600	1200
Sun				1600

TABLE 5.3
Typical illuminance levels (adapted from the *RCA Electro-Optics Handbook*)

	(lux)	
Direct sunlight	10^5	
Daylight	10^4	
Overcast day	10^3	
Very dark day	10^2	(photopic)
Twilight	10^1	
Deep twilight	10^0	
Full moon	10^{-1}	(mesopic)
Quarter moon	10^{-2}	
Starlight	10^{-3}	(scotopic)
Moonless overcast night	10^{-4}	
Artificial illumination	100–2000	

(We could then, for example, predict the irradiance on a photodetector that subtends the above solid angle at the source.) An alternative statement of the problem is this: If we know the irradiance (illuminance) at a receiving surface, how do we find the radiance (luminance) of the surface, considered as a source?

Consider then a small source, area dA, and let $d^2\Phi$ be the flux radiated inside the incremental solid angle $d\Omega$ (Figure 5.25). We then have, from the definition of radiance:

$$d^2\Phi = L\,dA\cos\theta\,d\Omega$$

where L is the source radiance, and dA the source area. The solid angle $d\Omega$ is (Figure 5.26)

$$d\Omega = \frac{dS}{r^2} = \frac{r\,d\theta\,r\sin\theta\,d\phi}{r^2} = \sin\theta\,d\theta\,d\phi$$

Substituting

$$d^2\Phi = L\,dA\cos\theta\sin\theta\,d\theta\,d\phi = (L\,dA/2)\sin 2\theta\,d\theta\,d\phi.$$

Therefore, the flux radiated into a cone of half-angle α is*

$$d\Phi = \frac{L\,dA}{2}\int_0^{2\pi}\int_0^{\alpha}\sin 2\theta\,d\theta\,d\phi$$

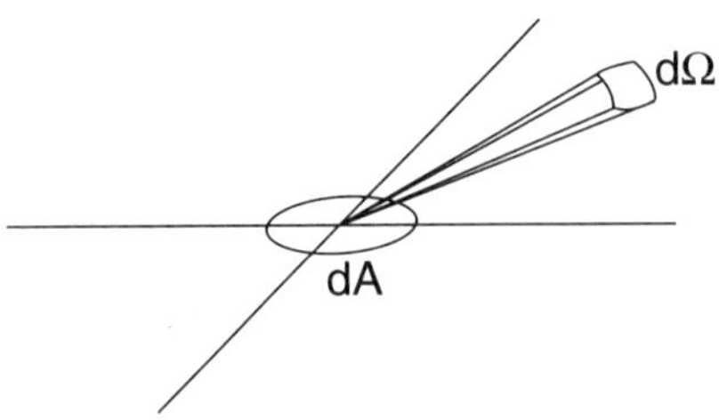

Figure 5.25. Small source area dA radiating flux into an incremental solid angle $d\Omega$.

* Notice that the Lambertian assumption (constancy of L) allows us to take L out of the integral. For a non-Lambertian source, the calculation is more complicated.

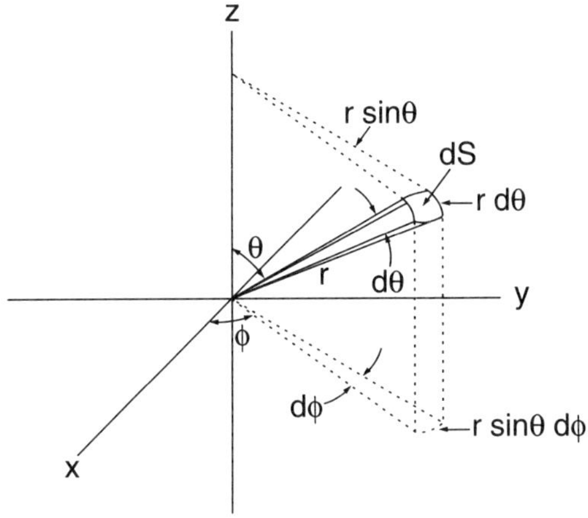

Figure 5.26. Calculation of the elemental solid angle.

or

$$\boxed{d\phi = \pi L\, dA \sin^2 \alpha} \tag{5.40}$$

The flux radiated into a complete hemisphere is found by setting $\alpha = \pi/2$,

$$d\Phi = \pi L\, dA \tag{5.41}$$

The radiant exitance of a source is $d\Phi/dA$, so from Eq. (5.41) we find the radiant exitance of a Lambertian source as

$$\boxed{M = \pi L} \tag{5.42}$$

Conversely, if we have E watts/m^2 (or lux) incident on a diffuse receiving surface, the radiance of the surface (considered as a source) can be found from

$$L = \frac{M}{\pi} = \frac{ER}{\pi}$$

$$L = \frac{M}{\pi} = \frac{ET}{\pi}$$

for a reflecting or a transmitting surface, respectively. Notice that this result is not the same as dividing the emitted flux by the solid angle within which the flux is distributed (which would be 2π for a hemisphere). In photometry, the average reflectivity of a typical scene is taken as 0.18.

While the above relations were derived for a source element dA, they hold true to a good approximation when the source has a finite size.

5.3.4. Flux Collected by a Lens

Consider a small Lambertian source of area ΔA and radiance L, centered on the optical axis of a lens. The entrance pupil of the lens subtends an angle

Figure 5.27. Flux collected by a lens.

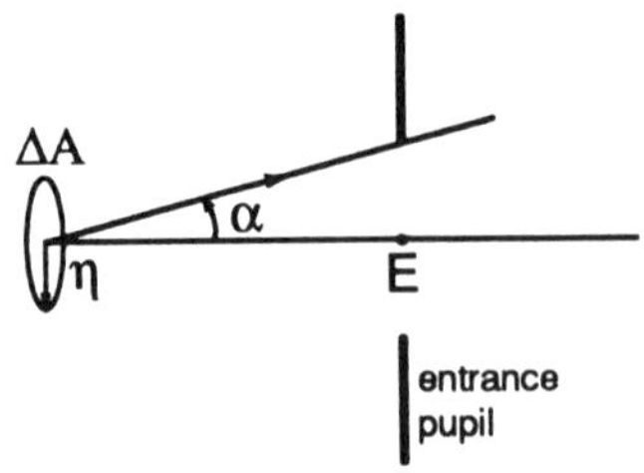

2α at the source (Figure 5.27). The flux radiated into the solid angle defined by the rim of the entrance pupil is, from Eq. (5.40),

$$\Delta\Phi = \pi L\, \Delta A \sin^2 \alpha$$

If the source is a disk of radius η, this equation is written as

$$\Delta\Phi = \pi^2 L\eta^2 \sin^2 \alpha \tag{5.43}$$

If the choice $u = \sin\alpha$ is made for the PMR angle variable, we can rewrite this equation as:

$$\boxed{\Delta\Phi = \left(\frac{\pi}{n}\right)^2 LH^2} \tag{5.44}$$

where $H = nu\eta$ the optical invariant. Thus *the flux collected by a lens is proportional to the square of the optical invariant.* Qualitatively, this is because H is a measure of how much light is accepted from each object point (characterized by u) times the number of such object points (characterized by η).

5.3.5. Irradiance of an Image

Suppose we are taking a photograph of an object, which for convenience we consider to be a small, circular Lambertian source, of area ΔA and radius η. The flux collected by the lens, $\Delta\Phi$, will be $\pi^2 L\eta^2 \sin^2\alpha$, from Eq. (5.43). Assuming that the lens has a transmission factor τ, then flux equal to $\tau\,\Delta\Phi$ will be transmitted through the lens and be spread over the image area $\pi\eta'^2$. Therefore, the image irradiance will be

$$E = \frac{\tau\, \Delta\Phi}{\pi\eta'^2} \tag{5.45}$$

For a reasonably well-corrected system, we have from Eq. (5.25), $H = n \sin\alpha\eta = n' \sin\alpha'\eta$. Therefore, using also Eq. (5.44),

$$E = \pi\tau L \left(\frac{n'}{n}\right)^2 \sin^2 \alpha' \tag{5.46}$$

In terms of the F number $\mathcal{N} = 1/2n' \sin\alpha'$, this equation becomes:

$$E = \frac{\pi\tau L}{n^2} \frac{1}{4\mathcal{N}^2} \tag{5.47}$$

Therefore, the image irradiance is given by the source radiance, the F number of the lens, and the lens transmission factor. In the infrared community the G number is sometimes defined (for $n = 1$)

$$\frac{L}{E} = \frac{4\mathcal{N}^2}{\tau\pi} \equiv G \text{ number} \tag{5.48}$$

as a direct ratio of the scene radiance and the image irradiance.

The $1/\mathcal{N}^2$ dependence of E from Eq. (5.47) helps explain the full stop numbers on a camera lens: To halve the irradiance, we must multiply the F number by $\sqrt{2}$. Thus $1.4 \cong \sqrt{2}$, $2 \cong 1.4 \times \sqrt{2}$, $2.8 \cong 2 \times \sqrt{2}$, etc.

Consider now the case of finite conjugates, for example, a slide projector. In this case, we expect the magnification to enter in the calculation of E (compare with the depth of focus discussion in Section 5.1.3). Equation (5.45) holds in this case as well, and we may substitute for $\Delta\Phi$ from Eq. (5.44) to obtain (with $\tau = n = 1$ for simplicity)

$$E = \frac{\pi^2 L H^2}{\pi \eta'^2}$$

or, using $H = \sin \alpha\eta$

$$E = \frac{\pi L \sin^2 \alpha}{M^2} \tag{5.49}$$

where M is the paraxial magnification. Thus for a slide projector the screen illuminance is determined by the source luminance, the aperture of the projection lens, and the slide magnification.

5.3.6. Radiance (Luminance) of an Aerial Image

Neglecting transmission losses by the lens, the radiance of an aerial image is equal to that of the object. This law is a direct consequence of the optical invariant: If the available flux is spread over a large image (η' large), then u' must be small (to maintain $n'u'\eta'$ constant); that is, the flux is distributed over a proportionately smaller solid angle, thus keeping the radiance constant. We may state this theorem as follows: For a system in air, and in the absence of absorption and vignetting, radiance is conserved.

The optical invariant applied to the object/image and pupil conjugates gives:

$$nu\eta = -nh_e\bar{u} = -n'h'_e\bar{u}' = n'u'\eta'$$

Upon squaring and multiplying by π^2, we have

$$n^2\pi u^2\pi\eta^2 = n^2\pi h_e^2\pi\bar{u}^2 = n'^2\pi h_e'^2\pi\bar{u}'^2 = n'^2\pi u'^2\pi\eta'^2 \tag{5.50}$$

If we consider all the angles as small, then πu^2 is the solid angle subtended at the object by the pupil, $\pi\bar{u}^2$ is the solid angle subtended at the pupil by the object, etc., while $\pi\eta^2$ is the object area, πh_e^2 the entrance pupil area, etc. Call $\Delta\Phi$ the flux emitted by the object that is captured by the entrance pupil; this flux will be conserved in the absence of absorption and vignetting. Therefore, by inverting Eq. (5.50) and multiplying by $\Delta\Phi$, we obtain

$$\frac{1}{n^2} L = \frac{1}{n^2} L_e = \frac{1}{n'^2} L_{e'} = \frac{1}{n'^2} L' \tag{5.51}$$

where L is the object radiance, L_e ($L_{e'}$) the entrance (exit) pupil radiance (considered as an object), and L' the image radiance. We may rewrite Eq. (5.51) for the object and image only as

$$L' = \left(\frac{n'}{n}\right)^2 L \tag{5.52}$$

If the system is in air (or if simply $n = n'$), then $L = L'$ and the radiance is conserved. Our simplified derivation assumed small aperture and field, but the theorem can be extended to larger values.

5.3.7. Photometry of Illumination Systems

A good example is the slide projector, which comprises two lenses with very different functions. The first lens (condenser) is used to illuminate the slide, while the second lens images the slide on the screen (Figure 5.28). The slide is located at the exit pupil of the condenser, while the source is imaged at the entrance pupil of the projection lens.

Since H determines the amount of light captured by a lens, we normally want the projection lens and condenser to have $H_p \cong H_c$. If $H_c > H_p$, then some light from the condenser is lost, either the source image is larger than the entrance pupil of the projection lens, or the illuminating beam exceeds the slide diameter. On the other hand, if $H_p > H_c$, then either the edges of the slide are not illuminated or the full aperture of the projection lens is not used (like stopping it down). Of course, any of these effects may be tolerated to some extent, but the total system H will be the smaller of the two. Normally, the design goal would be to have H_c a little larger than H_p.

Example 5.6. It is required to design a slide projector for 35 mm slides according to the following brief specification: (1) screen size 2.56 m × 1.76 m, at a distance of 7.76 m from the projector; (2) screen illuminance (with no

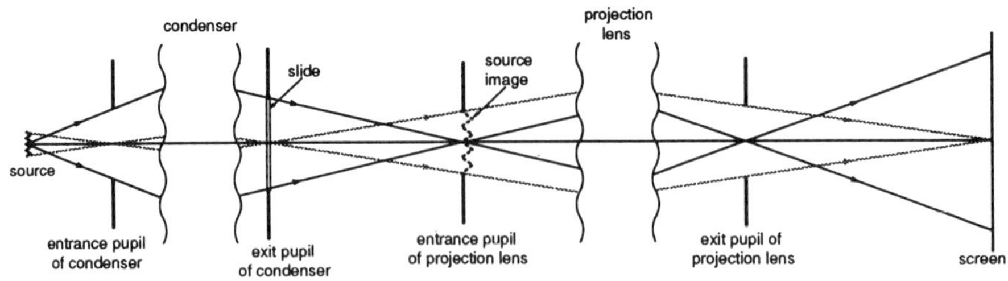

Figure 5.28. Schematic of a slide projector.

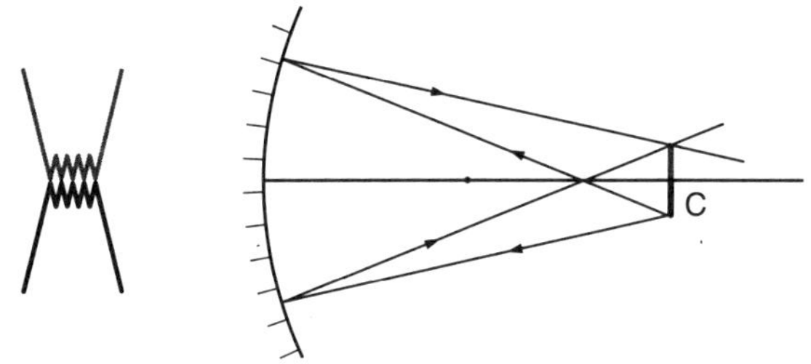

Figure 5.29. Source–mirror arrangement for a slide projector.

slide) to be 100–150 lux; (3) slide size: 34 mm × 22 mm. (At this point, we will attack only the photometric aspects of the problem.)

The flux collected by the condenser is [Eq. (5.44)] $\Delta\Phi = \pi^2 LH^2$, where L is the circular Lambertian source luminance. We expect "from previous experience" to be using a tungsten halogen lamp. These usually have a rectangular, flat, dense filament with an aspect ratio of roughly 2:1. When this filament is located near the center of curvature of a concave mirror, an "effective" square source is formed (Figure 5.29). Because Eq. (5.44) refers to a circular source, and assuming that the real source dimension implicit in H is the semidiagonal, a factor of $2/\pi$ must be used to correct for this. In addition, there will be losses in the system due to scattering, absorption, unwanted reflections, etc. It might be estimated that 40% of the light will be lost this way. Thus Eq. (5.44) becomes:

$$\Delta\Phi = 1.2\pi LH^2 \tag{5.53}$$

For tungsten halogen lamps, L is an insensitive function of wattage; a typical figure would be $L = 30\ \text{cd/mm}^2$. From the given screen size and the required illuminance, we find the necessary flux on the screen as $\Delta\Phi = 700$ lm. Substituting these two values into Eq. (5.53), we find $H = 2.5$ mm; so considering source imagery,

$$H = \sin\alpha_L \eta_L = 2.5\ \text{mm} \tag{5.54}$$

where the subscript L refers to "lamp," η_L is the effective source semidiagonal, and α_L is the condenser collection semiangle. Table 5.4 compares the relevant parameters for five candidate lamps. Lamp 1 does not output enough lumens,

TABLE 5.4.
Parameters for five lamps

Watts	Filament size (mm²)	η_L	Required $\sin\alpha_L$	α_L
50	3.1 × 1.6	2.2	—	—
100	4.4 × 2.2	3.1	0.806	53.8
150	5.4 × 2.7	3.8	0.657	41.1
250	7.0 × 3.5	5.0	0.500	30.0
500	10.0 × 5.0	7.1	0.352	20.6

but any of the other four lamps may be used. The extremes are a low-wattage lamp with a very large collection angle condenser and a high-wattage lamp with a low collection angle. The condenser design for the 500 W lamp will be the simplest, while the 100 W design would be very difficult. However, the 500 W lamp would require a large lamphouse with a powerful cooling fan to dissipate the excess heat. So we decide on the 250 W lamp as the best compromise; this would have a total collection angle of 60° (π steradians).

We may now turn our attention to the projection lens. For a distant image, the image-space semifield angle $\bar{\alpha}'$ is related to the object size η by

$$F = \frac{\eta}{\tan \bar{\alpha}'} \tag{5.55}$$

From the specification, $\bar{\alpha}' = 0.2049$ and $\eta = 20.5$ mm (the slide semidiagonal). Thus $F = 100$ mm.

The optical invariant gives $H = u\eta = 2.5$ mm at the slide, where u is the PMR angle variable. Therefore, $u = -0.122$, corresponding to an F number $(1/2u)$ of $F/4.1$. This is a reasonable number, and we will be satisfied with the design up to this point.

5.3.8. Off-Axis Illumination

The image illuminance varies across the field of view in proportion with the fourth power of the cosine of the field angle. In Figure 5.30, consider the object to be an extended Lambertian source. The flux collected by the entrance pupil from an extra-axial source element dA at a field angle $\bar{\theta}$ is given by

$$d\Phi = L\, dA \cos \bar{\theta} \frac{S \cos \bar{\theta}}{\bar{R}^2/\cos^2 \bar{\theta}}$$

from the definition of luminance, where S is the area of the pupil normal to the optical axis. Therefore,

$$d\Phi = d\Phi_0 \cos^4 \bar{\theta}$$

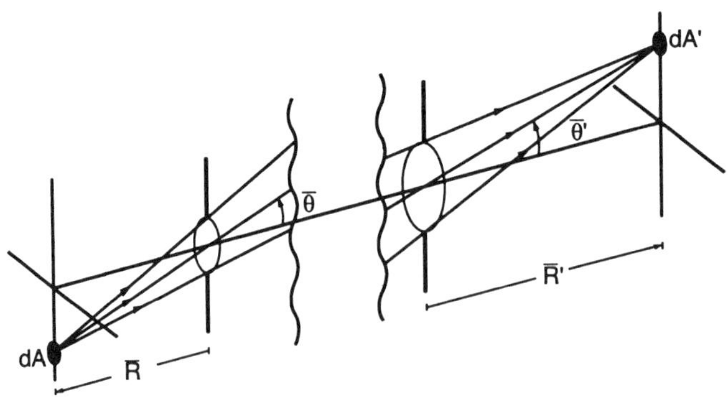

Figure 5.30. Off-axis illumination for an extended image.

where

$$d\Phi_0 = \frac{L\,dA\,S}{\bar{R}^2}$$

the flux collected from the axial source element. The flux $d\Phi$ will be distributed over an area dA' in the image plane, where $dA' = M^2\,dA$. Thus the illuminance at the extra-axial image point will be $E' = d\Phi/dA'$, or

$$E'(\bar{\theta}) = E'_0 \cos^4 \bar{\theta} \tag{5.56}$$

where $E'_0 = LS/(M\bar{R})^2$. The validity of this law assumes that the magnification M is independent of $\bar{\theta}$; that is, there is no distortion. Sometimes, distortion is introduced deliberately in order to improve the evenness of the illumination.* The $\cos^4$ variation is subjectively difficult to notice when observing a busy scene (such as a typical photograph), but could be noticeable if the object had little structure.

5.3.9. Illuminance from a Large Source

So far we have considered only the case of an infinitesimal source, noting that the formulae apply with a good approximation to finite but small sources. But what happens if the source size is large? In Figure 5.31, we have a circular Lambertian source of radius R, and we wish to determine the illuminance (irradiance) on a receiving surface dA perpendicular to the axis, a distance z away.

First we find the flux emitted into the elemental solid angle $d\Omega$ subtended at the source element dS by the area dA. The solid angle $d\Omega$ is

$$d\Omega = (\text{area})/(\text{squared radius}) = \frac{dA \cos\theta}{(z/\cos\theta)^2} = \frac{dA}{z^2}\cos^3\theta$$

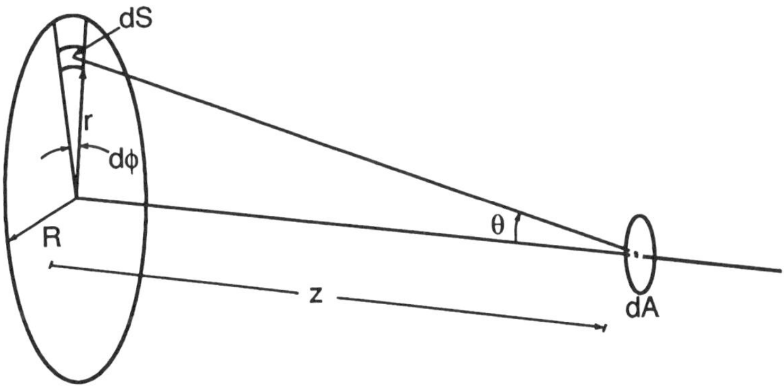

Figure 5.31. Illuminance from a large source.

---

* R. Kingslake, "Illumination in optical images," in *Applied Optics and Optical Engineering*, Vol. II, R. Kingslake, Ed., Academic Press, 1965.

From the definition of luminance, the flux at dA due to dS is

$$d^2\Phi = L\,dS \cos\theta\, d\Omega = L\,dS \cos\theta \frac{dA}{z^2} \cos^3\theta$$

Therefore, the illuminance

$$\frac{d^2\Phi}{dA} = \frac{L}{z^2} \cos^4\theta\, dS = \frac{L}{z^2} \cos^4\theta\, r\, dr\, d\phi$$

where $dS = r\,dr\,d\phi$ in polar coordinates. Now $r = z\tan\theta$ and $dr = (z/\cos^2\theta)\,d\theta$; so

$$\frac{d^2\Phi}{dA} = L\cos^2\theta \tan\theta\, d\theta\, d\phi = \tfrac{1}{2}\sin 2\theta\, d\theta\, d\phi$$

The total illuminance is then

$$E = \int \left(\frac{d^2\Phi}{dA}\right) = \frac{1}{2}\int_0^{\theta_0}\int_0^{2\pi} L\sin 2\theta\, d\theta\, d\phi$$

Performing the integration, we have

$$E = \pi L \sin^2\theta_0 \tag{5.57}$$

where θ_0 the angle subtended *by the source* at the element dA. Notice the similarity between this equation and Eq. (5.40) giving the flux emitted into a cone by a small Lambertian source. The same result is obtained if we swap the roles of source and receiving area!

5.3.10. Luminance of a Distant Source

For a distant source, such as the sun or sky, the definition of luminance as candelas per square meter breaks down (where is the square meter measured?). We need then an alternative definition of luminance in this case. There are only two measurable quantities associated with a distant source such as the sun: the illuminance it produces on a surface (δE) and the solid angle ($\delta\Omega$) it subtends, as seen from the Earth. Therefore, the definition of radiance of a distant source must contain only these two quantities:

$$L = \frac{\delta E}{\delta\Omega} \tag{5.58}$$

This definition can be justified in light of Eq. (5.57): The illuminance produced by the source is proportional to the luminance of the source and the solid angle $\delta\Omega = \pi\sin^2\theta_0$.

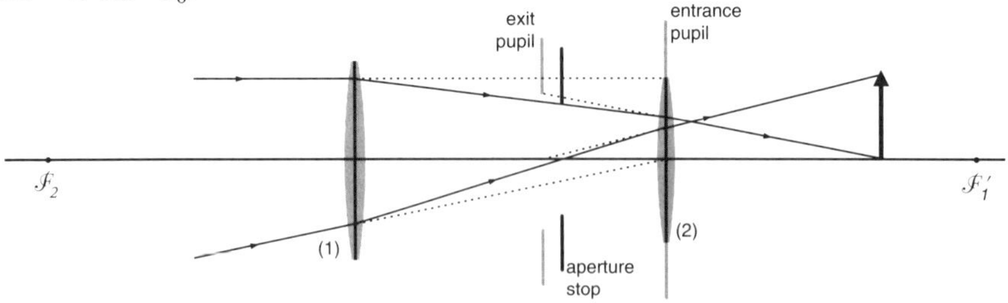

Figure 5.32. Solution to the problem of Figure 5.14.

PROBLEMS

1. Determine the position and size of the exit pupil of the two systems shown in Figure 5.33. The stop (and entrance pupil) has a radius r. The lenses are considered thin and large enough to avoid vignetting.

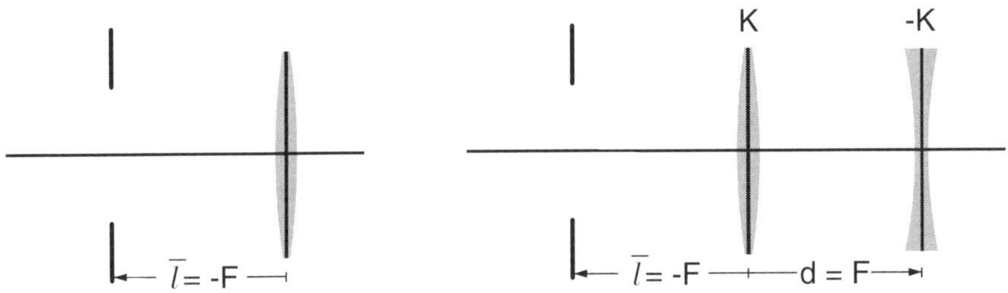

Figure 5.33

2. A thin positive lens of focal length F and diameter D has an iris diaphragm located at $-F/2$ in front of it. The diameter of the iris is $D/2$. Determine the exit pupil for the following two cases: (a) object at infinity, (b) object at $-3F/4$ in front of the lens.

3. Figure 5.34 shows a telephoto lens with an object at infinity. Determine graphically the location and size of the exit pupil.

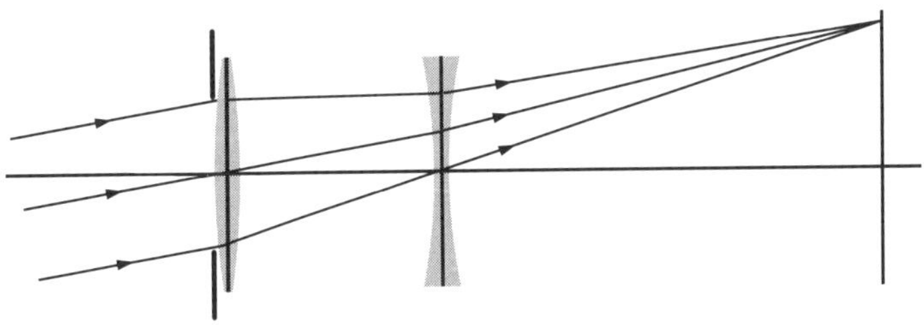

Figure 5.34

4. A camera lens, working at a magnification of $-1/10$, can tolerate a longitudinal displacement of the film of up to ± 1 mm before unacceptable image blurring occurs. What is the depth of field? If the maximum tolerable image blur is given as a blur disk diameter of 0.2 mm at the film, what working relative aperture of the camera lens does this imply?

5. Calculate the hyperfocal distance for an $F/2.8$ and an $F/11$ lens, both used with film that can resolve 100 lines/mm. Take $F = 50$ mm.

6. Refract a paraxial ray through a single surface of radius 100 mm, with n and n' equal to 1 and 1.5, respectively. The object point is located 150 mm to the left of the surface. Associate the paraxial ray height and angle variables of the ray with a finite ray of aperture angle 30°. Consider three forms of association: (a) $u = \sin \alpha$, (b) $u = \tan \alpha$, (c) $u = \alpha$ (in radians). In each case, calculate h_1, u', and l' and compare the values obtained in the three different cases. (Note: Revisit this problem after having studied the finite ray-tracing equations of Appendix 3, in order to compare these results with the real ray also.)

7. If Kepler had the money, what could he have done to "improve the intensity" of his 20× telescope? What do you think might have been the problem?

8. Calculate the optical invariant for a photographic objective with $F = 50$ mm, $\mathcal{N} = 2$, field of view $\pm 26.6°$.

9. Given the system shown in Figure 5.35 and the condition that the entrance pupil is at infinity and has a known angular size, draw and label the PMR and the PPR in the object and image spaces. (Note: A system with a pupil at infinity is called *telecentric*; see also Section 6.3.3.)

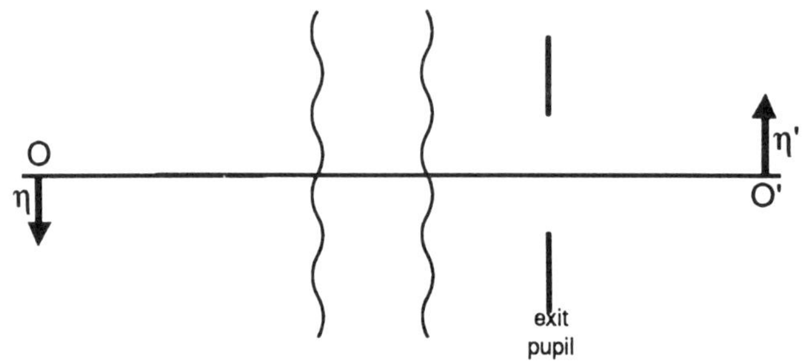

Figure 5.35

10. A certain optical instrument produces an image that is just external to it and is 100 mm in diameter. The exit pupil of the instrument is 500 mm inside it, and the image-forming beam has an aperture of $F/20$. It is required to photograph this image with a large instant camera borrowed from another piece of equipment. This has an $F/10$ objective lens, with an EFL of 250 mm, which is fixed focus at an object distance of 500 mm. At this conjugate the camera magnification is -1. The negative size of the camera is 100 mm^2. Neither the camera lens nor its back may be separated from its body. The aperture stop of the objective is in contact with its outside surface. What problems will be met attempting to take the required photograph, and how can these be overcome? Auxiliary equipment at your disposal consists of a set of large equiconvex lenses (150 mm in diameter)

with EFLs of 200, 250, 350, and 500 mm, a set of small equiconvex lenses (50 mm diameter) with EFLs of 100, 150, and 200 mm, and various pieces of cardboard and cardboard tubing.

11. The lens system shown in Figure 5.36 (a thin-lens version of the Petzval projection lens) has an infinite object, a focal length of 100 mm, a relative aperture of $F/2.5$, and a field of view of $\pm 11.3°$. You are also given $F_1 = F_2 = 150$ mm. Calculate the following quantities: h_1, H, u'_k, η'_k, $\bar{h}_1$, h_2, $\bar{h}_2$.

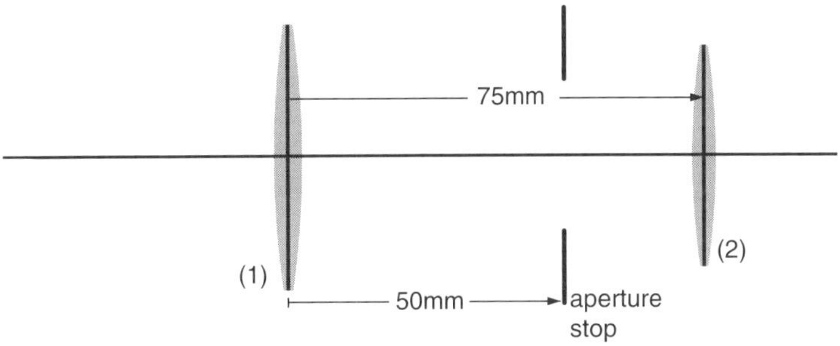

Figure 5.36

12. A positive thin lens ($F_1 = 10$ cm) is followed by a negative thin lens ($F_2 = -5$ cm). The distance between the two lenses is 2 cm. The first lens has a diameter of 1.5 cm, and the second lens has a diameter of 1 cm. For an object at infinity, determine the following: position and size of the entrance and exit pupils, position of the field stop (or nearest equivalent), and the paraxial angle variable $\bar{u}_1$ (semifield of view).

13. Derive the illuminance at noon from the known luminance of the sun. Assume that the sun subtends 0.5°. Also determine the luminance of an "average" scene, considered as a Lambertian reflector.

14. You (an optics expert) are in a darkened auditorium and you are taking photographs of the slides projected on the screen in front of you (thus joining the ranks of many annoying conference participants). The illuminance on the screen is m lux. Assume that the screen is a Lambertian diffuser of reflectance r, that your camera lens has a transmittance τ, and that the illuminance needed to expose the film properly is E (lux). (a) After exiting the room, the person next to you confesses little knowledge of optics, and asks you how you determined the required $F\#$. Explain. (b) Of course, you really have an automatic camera. But now you must explain to the persistent stranger how the exposure meter of your camera might determine the $F\#$. Assume that the camera is equipped with a photodetector of known characteristics, placed at the same conjugate location as the film (through the use of a retractable mirror).

15. For a slide projector, if the screen illuminance is 1000 lux, what kind of lamp is required? Assume a slide magnification of $50\times$ and an $F/2.8$ projection lens.

16. You are given the approximate expression

$$\text{Exposure} = 0.8/\text{ISO number (lux seconds)}$$

Determine the F number and the exposure for photographing a scene under overcast conditions, using an ISO-100 film and a camera objective with a transmission of 0.8.

17. Assuming that the illuminance at sunset is 2000 lux, determine the exposure for photographing the sun with ISO-25 film. You may consider that the sun is a uniform disk subtending 0.5°, and that your camera objective, including filter, has a transmission of 0.3. If the resulting number is unusual, why might that be so?

18. You want to illuminate an optical fiber of core radius $r = 60\ \mu\text{m}$, and acceptance angle (or NA) $\alpha = \pm 11^\circ$, using a tungsten–halogen source with a filament area of $5 \times 4\ \text{mm}^2$ and a lens of NA $= 0.3$. At what magnification should the source be imaged onto the fiber face to maximize the light into the fiber? (Hint: There are more than one solution; think of what determines the amount of light accepted by the fiber.)

19. We wish to locate the focal plane of a lens by focusing on an object whose distance from the lens is about 1000 focal lengths. Calculate the error in locating the focal plane. Suppose that the lens $F\#$ is 11, and that we can resolve about 100 lines/mm in the image plane. Compare the depth of focus with the focus error. Determine the focal length for which the depth of focus becomes approximately equal with the focus error.

20. Prove Eq. (5.19). [Hint: One possible way is to apply the general conjugate formula (4.22) to the pupil conjugates.]

21. In a Cassegrain-type mirror, the convex secondary mirror causes a magnification of m and places the final image at the apex of the primary (mirror distance d). If the entrance pupil is at the primary, determine the location of the exit pupil and the pupil magnification.

Gaussian Optics of Optical Instruments and Components*

6

"... the perfecting of [optical instruments] is the most likely way to afford us the greatest help for the Detection of the Nature of Bodies."

Robert Hooke, in *The Method of Improving Natural Philosophy*, ca. 1705 A.D.

6.1. THE TELESCOPE

6.1.1. Visual Telescopes

In this section, we concentrate on telescopes, whose function is to produce a magnified virtual or infinite image of a distant object, normally used for visual observation. Varieties of these visual telescopes are found as opera glasses, binoculars, periscopes, rifle scopes, and surveying instruments.

The basic telescope configuration comprises two lenses separated by the sum of their focal lengths; this system has been already encountered several times in previous chapters. It has zero power, or infinite focal length. For this reason, telescopes are also referred to as *afocal systems*. This name persists despite the fact that in practice departure from the afocal condition is the rule rather than the exception. One reason for this persistence is found in the long-held incorrect belief that the relaxed eye accommodates at infinity; nowadays, we know that the preferred distance for accommodation (i.e., the distance that the eye prefers to focus) is at around 1 m. Another reason is that the afocal condition makes it easier to analyze the basic characteristics of the telescope.

In Figure 6.1 we show three types of telescopes. The first type, employing two positive lenses, produces an inverted image. By using a third lens, typically

* Sections 6.1 and 6.2 are essential complements of the previous theory, even for the beginner. The remaining sections of this chapter are optional; they are not a prerequisite for the following chapters.

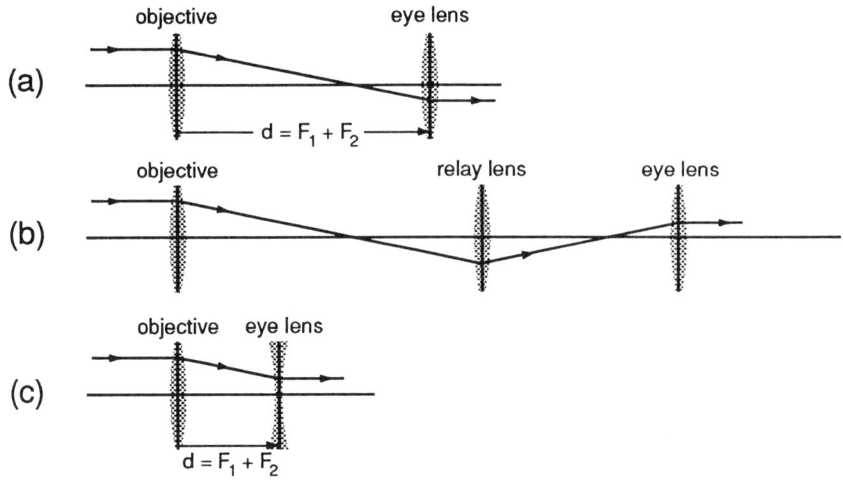

Figure 6.1. Types of afocal refractive telescopes.

at symmetric conjugates ($M = -1$), as in Figure 6.1b, we can obtain an erect image. The intermediate lens is called a *relay lens*; the telescope length is increased by the throw for this lens ($4F$). If the second lens is negative, we have the *Galilean telescope*, which produces an erect image (Figure 6.1c).

If both object and image are at infinity, it makes sense to definc the *angular magnification* as the angular image size divided by the angular object size. In our usual notation,

$$M_{\text{ang}} = \frac{\bar{u}'}{\bar{u}} = \frac{\tan \bar{\alpha}'}{\tan \bar{\alpha}} \tag{6.1}$$

Consider now the inverting (also called *Keplerian*) telescope of Figure 6.2, with the entrance pupil at the front lens (objective). The paraxial marginal and pupil rays are shown. The following relation can be proven by the reader as an exercise:

$$M_{\text{ang}} = -\frac{F_1}{F_2} = \frac{h_e}{h_e'} \tag{6.2}$$

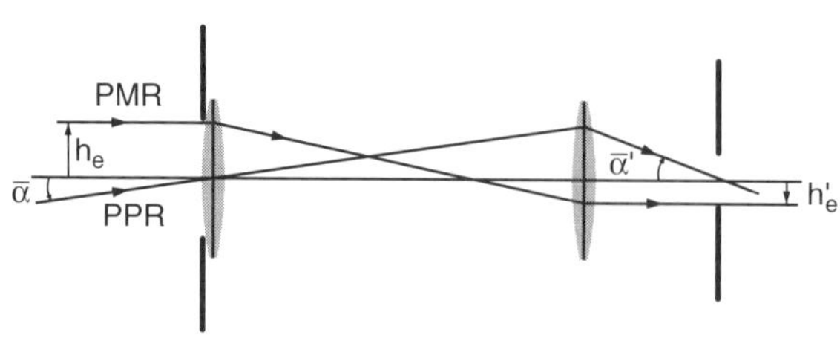

Figure 6.2. Paraxial marginal and pupil rays for the Keplerian telescope.

where h_e and h'_e are the PMR height at the entrance and exit pupil, respectively. The linear magnification of the telescope was shown in Chapter 4 to be constant, and equal to $-F_2/F_1$. Therefore,

$$M_{\text{ang}} = \frac{1}{M_{\text{lin}}} \tag{6.3}$$

Equations (6.1)–(6.3) hold for any type of afocal system, not just the inverting telescope. The entrance pupil of an inverting telescope is at the objective lens. The exit pupil is then the image of the objective through the second lens. The exit pupil must be placed at least 7 mm away from the lens to provide adequate *eye relief.* A Galilean telescope has no obvious aperture stop, and the exit pupil is normally defined by the observer's own eye pupil.

The Galilean telescope (also known as "opera glasses" or, nowadays, as "sports glasses") is compact and presents an erect image. It is used in applications where small size and light weight are important, and high magnification is not required. Ordinary binoculars are Keplerian telescopes with no relay lens; a prism system is used to provide the observer with an erect image (see Section 6.5). Binoculars are specified by a product (e.g., 7 × 50). This means that the angular magnification is 7 and the diameter of the entrance pupil (objective) is 50 mm. From this specification then we can find the size of the exit pupil (50 mm/7).

For an inverting telescope as in Figure 6.3, the size of the field of view is limited by the size of the eyelens L_2. A *field lens* placed at or near the focal point of the objective increases the field of view and allows a more reasonable size for the eyelens. Because of its position, it has no effect on magnification. Any surface imperfection or dust on the field lens will appear superposed on the final image; therefore, it is desirable to move the field lens a small distance away from the intermediate image. The combination of field lens and eyelens is called an *eyepiece.* In systems where an intermediate image is formed, a *field stop* may be used; it has the effect of providing a sharp border around the field and reducing stray light.

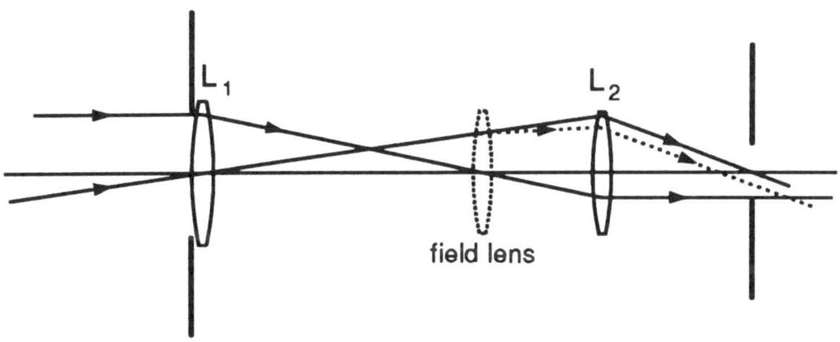

Figure 6.3. Action of a field lens: By bending the pupil ray, the field lens allows smaller size for the eye lens and/or an increased field of view. The field lens also relays the pupil to a different position without affecting the image location or magnification.

Finally, it is worth repeating that afocal systems cannot be described in terms of principal planes, although they can still be reduced to a set of three parameters (magnification, and the location of a pair of conjugates) for the purpose of determining the image location, as in section 4.4.4. Even if the system is not strictly afocal, the magnification is better than the power as a parameter for describing the telescope. For example, the reader can verify that a defocused Keplerian telescope with $d > (F_1 + F_2)$ has negative power, but it presents a *converging* bundle of rays to the eye.

6.1.2. Astronomical Telescopes and Resolution*

Modern astronomical telescopes are not afocal; they are best thought of as very large cameras; that is, they only have an objective but no eyepiece. The eyepiece is only needed for visual observations, requiring a virtual image; in modern astronomy virtually all observations are recorded on film or electronic cameras; so a real image must be formed. The objective is a mirror rather than lens, because large mirrors are easier to produce than large lenses, and they do not suffer from chromatic aberration (see next chapter). An example of an astronomical telescope objective is the Cassegrain mirror discussed in Chapter 4. As with the camera example, the larger the focal length of the objective, the larger the magnification. For two nearby stars separated by an angle $\Delta\phi$, the distance at the focal plane of the two corresponding images will be $F\,\Delta\phi$.

The resolution and size of an astronomical telescope are critical considerations. Because of diffraction, the image of a point is not a point but has a certain minimum size. We borrow this result from wave optics, that the minimum distance r for resolving two points after diffraction spread by the lens (mirror) aperture is

$$r = \frac{0.61\lambda}{\text{NA}} \tag{6.4}$$

where $\text{NA} = \sin\alpha'$, the numerical aperture of the objective ($n' = 1$). For two nearby stars to be just resolvable, we set their separation $F\,\Delta\phi = r$.† Using Eq. (6.4) and writing the NA as $D/2F$, we have

$$\Delta\phi = \frac{1.22\lambda}{D} \tag{6.5}$$

where D is the diameter of the objective. Thus the larger the diameter, the higher the resolution. For example, taking $D = 1$ m and $\lambda = 0.5$ μm, we obtain $\Delta\phi = 6.1 \times 10^{-7}$ rad or 0.13 arcseconds. Unfortunately, that kind of resolution

* For an excellent review of astronomical telescopes, see Daniel J. Schroeder's *Astronomical Optics*, Academic Press, 1987.

† The reader should clearly understand here that the concept of the resolution limit depends on several parameters, and can differ from the values given by Eq. (6.4) or (6.5). Even in the absence of aberrations, the true resolution will depend on the ability of the detector to detect small variations in luminance, as well as the object being imaged. Eqs. (6.4) and (6.5) are traditionally used for an object comprising two nearby points.

is not possible for ground-based telescopes because atmospheric turbulence degrades the image, and the resolution is normally not better than that of a 10 cm diameter mirror; the prime reason for the large size of astronomical telescopes is not to achieve higher resolution, but to collect a lot of light and thus detect faint stars. But a space-based telescope, such as the Hubble, is not subject to these limitations; an increase in size also means an increase in resolution, at least in principle. The performance of space-based telescopes (as unfortunate experience has proven) will then be limited by residual aberrations, if the mirror shape is not exactly right.

6.1.3. Information Capacity of an Optical System

The information capacity can be defined as the maximum number of resolvable spots in the image. If r is the resolution limit and η' the radius of the image, the number of resolvable spots in the image is proportional to $(\eta'/r)^2$, assuming no aberrations. From the formula for the resolution limit (6.4), we see that

$$(\eta'/r)^2 \propto (\mathrm{NA})^2 \eta'^2 = H^2 \tag{6.6}$$

Thus *the information capacity of an optical system is proportional to the square of the optical invariant*. So once again, we have the opportunity to appreciate the fundamental significance of the optical invariant.

If the system has aberrations (which is practically unavoidable), then the spot size is larger than that predicted by Eq. (6.4). In that case H^2 represents the upper limit in information capacity that is achievable by any (incoherently illuminated) system performing the given imaging task.

6.1.4. Laser Beam Expander

Lasers emit approximately collimated beams of relatively small diameter, of the order of 1 mm (a notable exception is semiconductor laser diodes, which emit highly diverging beams). A laser beam expander is a device that expands and recollimates the laser beam. As Figure 6.1 suggests, this can be achieved with a reversed telescope, with the light entering through the short focal length lens, and exiting through the objective. Both an inverting and a Galilean telescope can be (and are) used for this purpose. The inverting telescope allows a pinhole or *spatial filter*, to be placed at the intermediate focus. If the pinhole is of the right size, it will allow only the central lobe of the point image to pass while stopping all light from stray reflections and scatter, thus resulting in a "cleaner" output beam. High-power lasers focused near a pinhole may produce air breakdown, as well as enough heat to melt the pinhole material, in which case only the Galilean beam expander is possible.

A typical laser beam may have a divergence of about 1 mrad; this is small but not negligible. Can we use a beam expander to improve the collimation of the beam? The general answer is yes, as we can see from simple diffraction considerations [or Eqs. (6.4) and (6.5)]: The larger aperture of the exit beam will allow smaller divergence $\Delta\theta$, if diffraction is the only cause of divergence. However, the behavior of laser beams is somewhat different from the nominally uniform beans we have encountered so far. With uniform beams, we can generally apply the methods of geometrical optics until we get very near the focus, and then apply diffraction theory, as in the example of astronomical telescopes. But with laser beams, the effects of diffraction must be taken into account at the beginning: There are several cases where the geometrical optics predictions are not adequate, even far from the region of focus. For this reason, we examine the problem of the laser beam expander as well as the general characteristics of laser beams in more detail in Appendix 2.

6.2. THE MICROSCOPE

There is a traditional distinction between the "simple" and the "compound" microscope. The simple microscope is a single lens magnifier, discussed in Chapter 3*; In this section, we are concerned only with the compound microscope; it consists of an objective, which forms a magnified real image of a nearby object, and a magnifier (eyepiece), which presents that image to the eye.

6.2.1. Magnifying Power and Resolution

The *tube length* of a microscope is the distance between the second focal point of the objective and the first focal point of the eyepiece. This is a standard length (160 mm). The reason for standardizing the tube length is so that the magnifying power of the microscope can be determined unambiguously from the focal lengths of the two lenses. We use the term *magnifying power* rather than magnification. This is because the magnifying power of the microscope is the product of the linear magnification of the objective and the visual magnification of the eyepiece.

In Figure 6.4, let the eyepiece be adjusted to form an image at infinity (i.e., intermediate image formed at its front focal point). The magnifying power of the microscope is

$$M = M_o M_v = M_o \frac{250\ \text{mm}}{F_e}$$

where M_o is the magnification of the objective and M_v the visual magnification of the eyepiece, taken from Eq. (3.34). Also,

$$l' = F_o(1 - M_o) \rightarrow M_o = \frac{F_o - l'}{F_o} = -\frac{160\ \text{mm}}{F_o}$$

* A highly entertaining account of the simple microscope is to be found in Brian J. Ford's *Single Lens: The Story of the Simple Microscope*, Harper and Row, 1985.

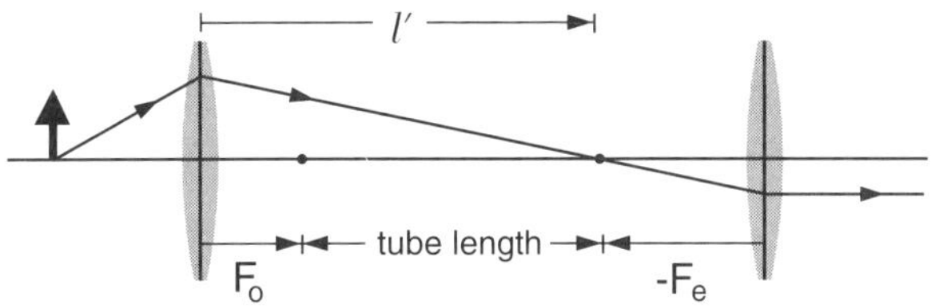

Figure 6.4. Microscope schematic.

since $l' - F_o$ is the tube length. Therefore, in terms of absolute value, the magnifying power of the microscope is

$$|M| = \frac{160\ \text{mm}}{F_o} \frac{250\ \text{mm}}{F_e} \tag{6.7}$$

From geometrical optics alone, there is no limit to the magnification that can be achieved with a microscope, since F_o and F_e can be made arbitrarily small in principle. There is, however, a practical limit, placed by the resolution of the eye and by diffraction. Diffraction theory dictates that there will be a minimum spread to the image of a point, given by Eq. (6.4), with $\text{NA} = n \sin \alpha$. Note that the NA of the objective can be defined in the object space because the image distance is large compared with the focal length (large magnification); correction for the finite conjugate condition is unnecessary. There are then three ways to increase resolution (i.e., reduce r): decrease λ, increase α, or increase n. The first way is employed in ultraviolet or electron microscopes, but it is obviously not very useful for visual microscopes. The last way is used in so-called *oil-immersion objectives*, where the space between the lens and the object is filled with a transparent oil. Finally, the acceptance angle α can be increased at the expense of a more complicated design.

Suppose now that we wish to resolve 1 μm size detail, that is, $r \leq 1\ \mu\text{m}$. Equation (6.4) gives $\text{NA} \geq 0.3$, for $\lambda \cong 0.5\ \mu\text{m}$. The spot separation at the intermediate image plane is $rM_o = 160r/F_o(\text{mm})$. The angular separation of the spots in the eye space must be larger than the resolution of the eye, which in angular terms is about 1 arcminute for high-contrast objects. Since most objects seen through a microscope tend to be low contrast, we may take the eye resolution to be about 3 arcminutes or 0.9×10^{-3} rad (see Section 6.4.6). The angle subtended at the eye by the two resolved spots is $rM_o/F_e = 160r/F_oF_e \geq 0.9 \times 10^{-3}$, from which

$$F_oF_e \leq 1.8 \times 10^5 r \quad (r \text{ in mm}) \tag{6.8}$$

or

$$F_oF_e \leq 180\ \text{mm}^2$$

This condition can be satisfied with a variety of objective and eyepiece focal lengths. For example, we can choose $F_o \cong 9$ mm and $F_e \cong 20$ mm, giving a magnifying power of 222 from Eq. (6.7).

Suppose now that the only available 9 mm objective has $\text{NA} = 0.15$, or, equivalently, that the NA is limited by an inadequate illumination system that does not fill the aperture of the objective. In this case, the objective will only resolve $r \cong 2$ μm from Eq. (6.4) and Eq. (6.8) shows that we can now use an objective (or eyepiece) with twice the focal length (and half the magnification). In other words, a magnifying power of 111 is all that is needed in order to take full advantage of the resolution of the objective. If we were to use the same two lenses as before, we would obtain a larger image, but we would not resolve any additional detail within that image. In this example (for NA = 0.15), the magnifying power of 111 is the *useful magnification*, while any magnification above this limit is referred to as *empty magnification*.

About the highest NA that can be achieved with an oil-immersion objective is 1.2–1.3 ($r \cong 0.25$ μm). If this separation is to subtend 3 arcminutes in the eye space, we have, from Eq. (6.8), $F_o F_e = 45$ mm^2 or a useful magnification of about 900. Since the resolution of the eye depends on the object contrast (Section 6.4), there is no fixed value for the maximum useful magnification, but a magnification above 900 is probably empty for visual observations. For example, if we use a 100× objective and a 10× eyepiece, we obtain a magnifying power of 1000; replacing the 10× eyepiece with a 20× one would likely be not useful.

Although the concept of empty magnification is more often associated with visual observations under a microscope, it is applicable to any optical system. If the final detector is not the eye but some form of a camera, the maximum useful magnification may be found with similar reasoning. The key considerations are that the objective must produce a resolved image in the intermediate space, and that the second (eyepiece) lens must provide sufficient angular magnification to satisfy the resolution of the detector.

Microscope objectives are specified by their magnification and NA. Table 6.1 gives values for some typical objectives. The field of view of an objective decreases with the magnification. A 4× objective may have a total field of view of 3–4 mm, while a 60× or a 100× objective will have only a small fraction of a millimeter.

6.2.2. Microscope Illumination Systems

The illumination system is a very important part of the microscope. There are two basic modes of operation, *bright field* and *dark field*. In bright-field

TABLE 6.1.
Typical microscope objectives

Magnification	Focal length (mm)	NA
4	30	0.10
10	16	0.25
20	8	0.40
60	3	0.85
100	1.8	1.3 (oil imm.)

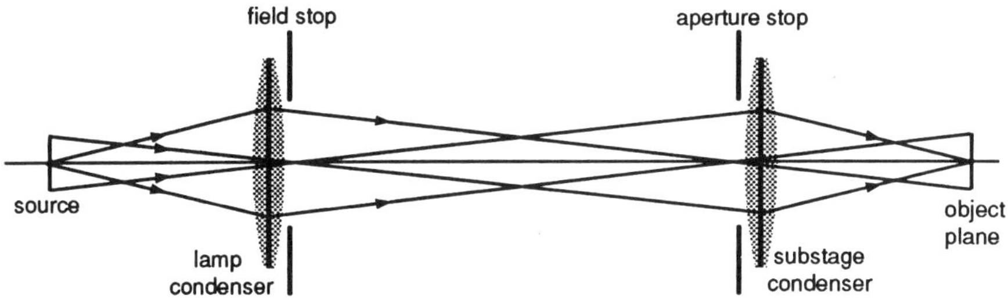

Figure 6.5. Schematic of the Köhler illumination system.

modern microscopes employ the so-called *Köhler illumination* (Figure 6.5) for nominally transparent objects. This system is similar to the illumination system for the slide projector, examined in the previous chapter. There are two lens groupings, the source condenser, near the source, and the substage condenser, near the sample. The essential design of the system has the source imaged at the aperture stop of the substage condenser, and the field stop, which is located near the source condenser, imaged on the sample. The field stop helps by excluding light that originates far from the detail of interest, which can then result in increased contrast through stray light reduction. Also, because the source is not imaged near the object, the illumination of the object can be uniform, independent of the uniformity of the source itself. As might be expected, the aperture stop controls the illumination of the object but has no effect on the field size. The illumination control is achieved by reducing the stop size until it becomes smaller than the source image. If the stop is larger than the source image, then it is the source itself (or its image) that becomes the stop.*

Dark-field illumination (Figure 6.6) is useful in increasing the contrast of certain small details, thus rendering them visible. It works by allowing only the light scattered or diffracted from an otherwise invisible detail to enter the objective, while light from the background is suppressed.

Both bright-field and dark-field illumination modes are used for viewing a large field at the same time. A different and very important way is to illuminate only a point at a time and scan across the sample in order to acquire a large image. For example, in the *confocal microscope* (Figure 6.7), the illumination is provided by a high-NA objective, similar to the one used for collecting the light. The critical difference between this and an ordinary microscope is that the object can be probed in depth as well as extent, because the confocal arrangement interrogates only one point (or very small volume) within the sample. This is accomplished by the combined action of the pinhole, which

* Perhaps the single most useful geometrical optics experiment that the student can perform is to attempt to set up a Köhler illumination system, as shown in Figure 6.5. All that is needed is two lenses, a source, two variable-aperture stops, a screen to observe the image of the field stop, and a piece of white paper to trace the beam and observe the source image. If two simple thin lenses are used, the best location for the stops is close to the corresponding lenses. By deviating from this condition and bringing the two stops closer together, one can appreciate why both the field and aperture stops must be properly placed if they are to function independently.

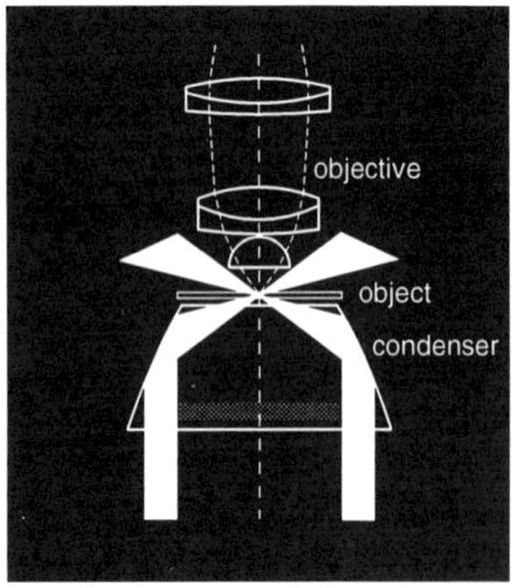

Figure 6.6. Dark-field illumination. (Reproduced from: L. C. Martin, *Technical Optics*, Vol. II, Pitman Publishing, 1961.)

rejects light from outside the focal spot, and the high NA of the objectives, which spreads the light rapidly, thus averaging out the effect of the sample away from the focal spot. Confocal systems can operate in both transmission and reflection modes, the latter being the simpler of the two as it uses the same objective for illumination and imaging.

Example 6.1. Microscope illumination systems are designed to cope with a variety of objectives. This means that they must provide a large illumination cone to fill the NA of the high-power objectives, as well as illuminate an area that is at least as large as the field of view of the lowest-power objective. Suppose that our two extreme cases are a high-power objective with an NA of 0.8 and a low-power one with a semifield of view of 3 mm (radius). If a single illumination system is to be used for both cases, its optical invariant must satisfy

$$H = n' \sin \alpha' \eta' \geq 0.8 \times 3 \text{ mm} = 2.4 \text{ mm}$$

Suppose that the lamp available has a filament of 2 mm radius (largest inscribed circle). For the object space of the condenser, we have $H = 2 \sin \alpha$ (mm) $= 2.4$ mm. The required collection angle in the object space would be given by $\sin \alpha = 2.4/2 = 1.2$, which is impossible. Therefore, we need a larger filament. Professor H. H. Hopkins used this example to illustrate the importance of choosing the correct correspondence between paraxial and finite quantities, explained in Section 5.2: if we had chosen $u = \tan \alpha = 1.2$ (instead of $u = \sin \alpha$), the problem would have appeared possible.

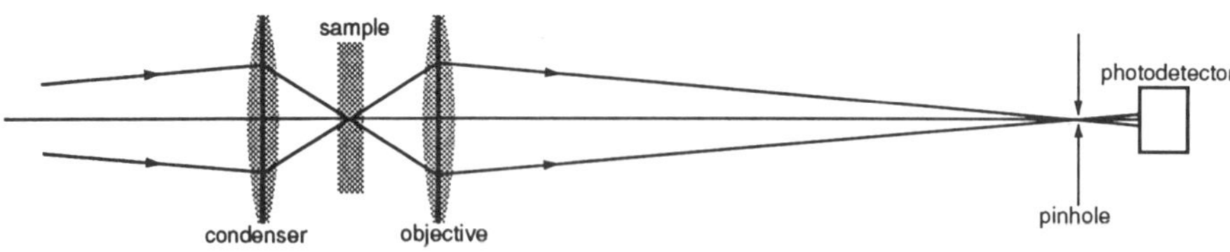

Figure 6.7. Schematic of the confocal microscope in transmission mode.

6.3. PROJECTION SYSTEMS

As we have already examined the slide projector in some detail in Section 5.3, we use this section primarily to give an idea of the variety of projection systems and the different solutions needed for the illumination assemblies. The typical projection system has the task of forming a normally distant real image of an illuminated object. There are two very different cases: (1) The object is opaque and diffusing (e.g., a book page), and (2) the object has variations in transparency (e.g., a slide). The projection system for opaque objects (historically called an *episcope*) is of little interest from the Gaussian viewpoint, because it is reduced to merely a single lens imaging an effectively self-luminous object. The second case is of more interest because the illumination system must be considered together with the projection system, as we saw for the slide projector.

6.3.1. The Overhead Projector

Examples of projection systems for transparencies are the photographic enlarger, the slide projector, and the overhead projector. The illumination system of the photographic enlarger is the least challenging, because the required image illuminance is typically low, and the magnification not very large (less than $10\times$); therefore, the condenser is not required to collect a lot of light, and a single-element lens often suffices. The calculations needed to specify such a system are essentially identical with those performed for the slide projector.

The overhead projector presents some additional interest, mostly because of the large object size ($\cong 30$ cm in diameter). A schematic of a typical overhead projector is shown in Figure 6.8. Again, the source is imaged by the condenser at the entrance pupil of the projection lens, and the object is placed at the exit pupil of the condenser. If the latter is a single lens, then the object must be in contact with or close to it. In order to maintain a reasonable overall system length, the focal length of the condenser must be short; so let the condenser be used at the minimum throw condition ($l' = -l = 2F$, $T = 4F$). A reasonable value for the lamp-to-projection-lens distance is 80 cm, requiring a focal length of 20 cm. For a thin equiconvex lens of $n \cong 1.5$, the required radius of curvature for the two surfaces is $R = 20$ cm. But such a lens cannot have a clear aperture of 30 cm (almost a sphere!). The solution is to use a *Fresnel lens*, named after its inventor and shown in Figure 6.9. This type of lens can have both large aperture and short focal length with minimum weight and bulk. A lens of 30 cm diameter and 20 cm focal length is possible, thus satisfying the requirements of the above design.

Fresnel lenses are normally made out of moulded plastic; for this reason, the parent surface does not have to be spherical, but can take an *aspheric* shape, which is very useful in reducing the large aberrations that result from the extreme aperture of the lens. However, the use of Fresnel lenses is limited to illumination systems; in an imaging system the scatter and glare produced by the flat facets would be unacceptable.

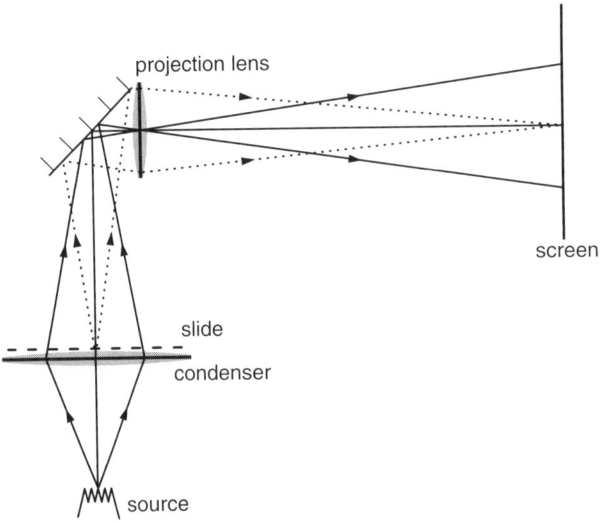

Figure 6.8. Schematic of the overhead projector.

Figure 6.9. A Fresnel lens is obtained by "removing" the plane-parallel part indicated inside the ordinary lens and replacing it with a thin flat part. Since the removed part has no power, the power of the Fresnel lens is the same as that of the bulk equivalent.

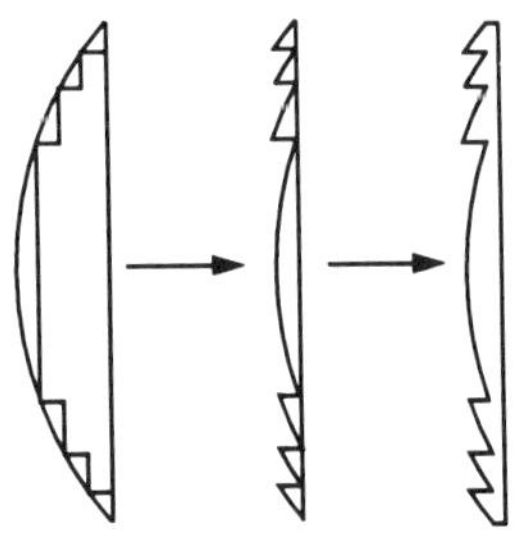

6.3.2. Aspherics in Illumination Systems

Aspheric shapes are very often used if a large collection angle from the source is needed. The most common example is the *parabolic reflector*, used in searchlights, automobile headlights, etc. A spherical reflector, with the source at its focus, would suffer from severe aberration at large aperture (see Figure 2.9); that is, the output beam would not be well collimated. A parabolic reflector, on the other hand, has the property that all the reflected rays will emerge parallel if the object is a point source placed exactly at the focus (Figure 6.10a). If both conjugates are finite, then the mirror shape must be elliptical (Figure 6.10b).

Perfect imaging for an axial point object is also possible for refractive surfaces. In Figure 6.11a, consider the problem of producing a parallel beam from a diverging

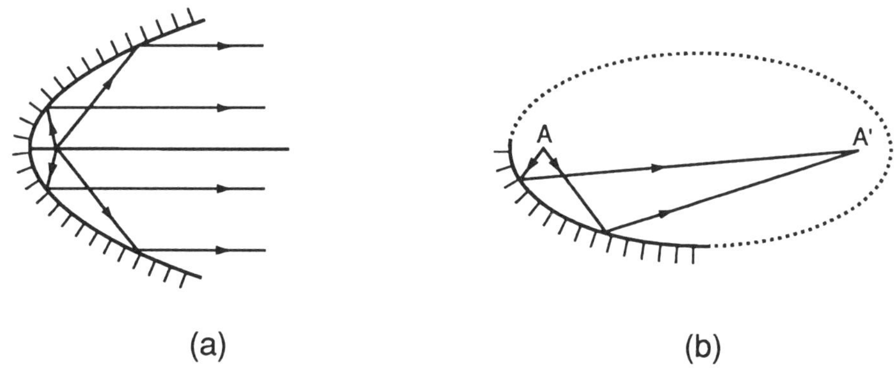

Figure 6.10. A parabolic (a) and an ellipsoidal (b) reflector.

one, after a single refraction. The path lengths between two wavefronts must be equal for all rays; so

$$n_1(AB) + n_2(BC) = n_1(AO) = n_1 l \quad \text{(constant)}$$

The coordinates of the various point are A: $(0, l)$, B: (x, z), C: $(x, 0)$. Thus

$$n_1[x^2 + (z - l)^2]^{1/2} + n_2 z = n_1 l$$

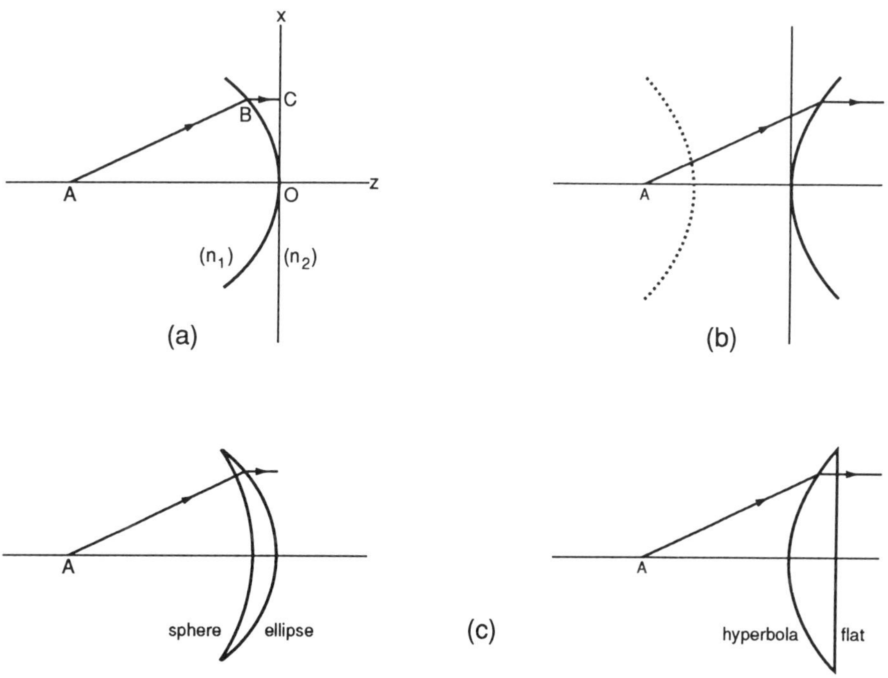

Figure 6.11. Perfect axial imagery by a single surface and a single lens.

This equation can be rewritten as

$$x^2 + z^2\left(1 - \frac{n_2^2}{n_1^2}\right) - 2zl\left(1 - \frac{n_2}{n_1}\right) = 0$$

This is the equation of a conic section with the apex at O. If $n_2 < n_1$, the curve is an ellipse. If $n_2 > n_1$, the curve is a hyperbola, with point A at the focus of the other branch (Figure 6.11b). If $n_2 = -n_1$ (reflection), we have a parabola.

Complete lenses can be constructed for both the elliptical and the hyperbolic solution: In the first case, we include a spherical surface centered at A, and in the second case a plane surface (Figure 6.11c). In both cases, the additional surfaces produce no ray deviation and hence do not have an effect on the image of point A. For finite conjugates, the required shape is a fourth degree curve, called the *Cartesian oval*. In practice, the aspheric often has a general polynomial form that derives from additional considerations. For example, notice the large angle of incidence of the ray on the hyperbolic surface of Figure 6.11c. Such an angle would lead to a large amount of light being reflected instead of transmitted at that interface. Typically, the source is placed on the flat side of the aspheric condenser, with the second (aspheric) surface being used to correct the aberration arising from the first.

6.3.3. Other Projection Systems

The illumination systems described previously are essentially similar to the Köhler microscope illumination. An alternative system, called *critical* or *Abbe* illumination, has the source imaged directly on the object (Figure 6.12). There are two conditions for this arrangement to be practical: (1) The object must not be large, so that it can be covered fully by the image of the source, and (2) the image of the source must be uniform, since any nonuniformities will appear superposed on the final image. This type of illumination is used in motion picture projection, where the film size is small and the arc sources used are reasonably uniform.

The so-called engineer's or contour projector uses the arrangement shown in Figure 6.13. This type of projection system is used to magnify machine tools, so that accurate measurements of their dimensions can be made. The magnification must then be known accurately. For this reason, the aperture stop is placed at the focal plane of the projection lens, making the PPR parallel to the axis. The entrance pupil is then at infinity, and the system is called *telecentric*. This system is tolerant to axial displacements of the object (defocus): Such a

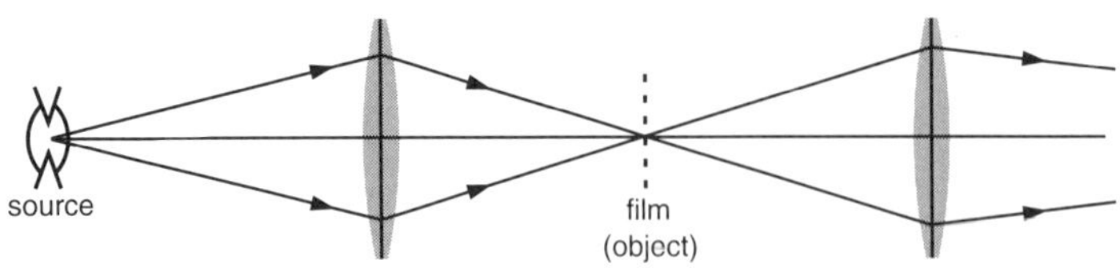

Figure 6.12. The Abbe or critical illumination (compare with Figure 6.5).

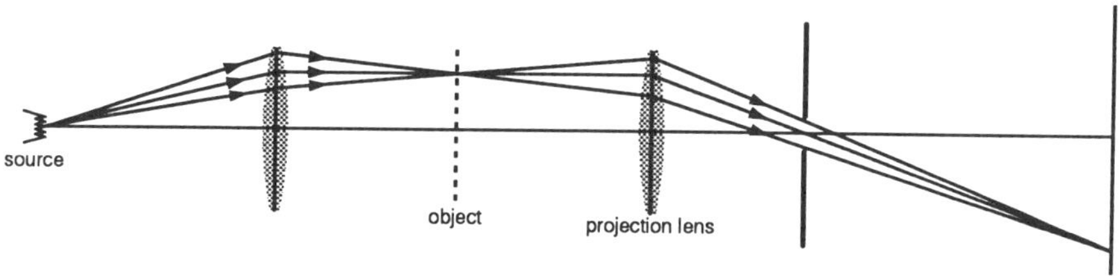

Figure 6.13. The contour projector. The aperture stop is at the rear focal point of the projection lens. Since the PPR is parallel to the axis in the object space of the projection lens, the measured size at the receiving (image) plane is not affected by movement of the object. Such movement will produce blur at the receiving plane, but no overall displacement.

displacement causes an image blur, but does not affect the image size, as can be seen from the figure (compare also with the telescope Example 5.5).

6.4. THE EYE

We restrict our discussion of the eye to geometrical optics and a few basic properties.

6.4.1. Basic Anatomy

A schematic section of an eye is shown in Figure 6.14. The fundamental parts are:

The *sclera*, which is the hard outer part that gives the eyeball its shape.

The *cornea*, the transparent part of the sclera through which light passes. It does approximately two-thirds of the total refraction of the eye. The cornea contains no blood vessels, as they would interfere with its transparency;

Figure 6.14. Schematic horizontal section of the right eyeball. (Reproduced with permission from W. J. Smith, *Modern Optical Engineering*, 2nd ed. McGraw-Hill, 1990. Figure 5.1, p. 120.)

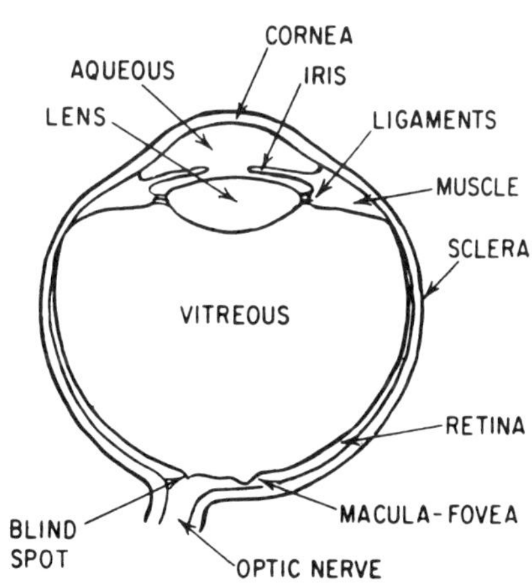

therefore, it is isolated from the immune system, which makes corneal transplants relatively easy. The aqueous humour provides nutrition.

The *vitreous* and *aqueous* humours are transparent fluids that fill the eyeball.

The *lens*, which contributes the remaining one-third or so of the refraction. The shape of the lens, and hence its power, is altered by the ciliary muscles. The resulting change in the eye's focal length is called *accommodation* and is necessary if an object is to be brought into sharp focus irrespective of its position. The power of the lens is between 20 and 30 dioptres, depending on the state of accommodation.

The *iris*, which controls the pupil diameter. The pupil varies in diameter from about 1.5 to 8 mm, depending on illumination conditions.

The *retina* is the mosaic of photoreceptors, blood vessels, and nerve cells that cover the internal surface of the eye. Somewhat surprisingly, the photoreceptors are located behind the nerve and blood cell layer relative to incoming light. There are two types of photoreceptors, rods and cones, named after their shape. The seven million or so cones are mostly concentrated in the *fovea*, which is the part of the retina where vision is sharpest and which subtends about 1.3 degrees of visual angle. They are responsible for *photopic* or daytime vision and the color sensation. There are three types of cones, red, green, and blue sensitive. At low light levels the cones are inactive; *scotopic* or night time vision is provided by the $\sim$130 million rods, which are completely absent from the fovea. In bright light conditions rods are saturated; so they provide no useful signal. The overall spectral sensitivity of the eye has already been discussed in the previous chapter, but we may add here that the photopic and scotopic spectral sensitivities are somewhat different, with the peak of the scotopic sensitivity shifted toward the blue by about 50 nm.

6.4.2. Geometrical Parameters

While eyes vary between individuals, there is an obvious need for defining a standard eye. Various geometrical models of the eye exist that provide different degrees of approximation to the performance of real eyes. From the Gaussian viewpoint, the widely used Gullstrand simplified schematic eye suffices. This is shown in Figure 6.15, in accomodated and unaccommodated version. The "unaccommodated" condition is when we look at an object at infinity (minimum power). It is interesting to note that the front surface of the lens is the one primarily affected by accommodation. Also notice that since the object- and image-space refractive indices are different, the nodal and principal points do not coincide.

6.4.3. Scene Luminance and Retinal Illuminance

The pupil diameter is affected primarily, but not exclusively, by the scene luminance. Table 6.2 gives a rough idea of the dependence.

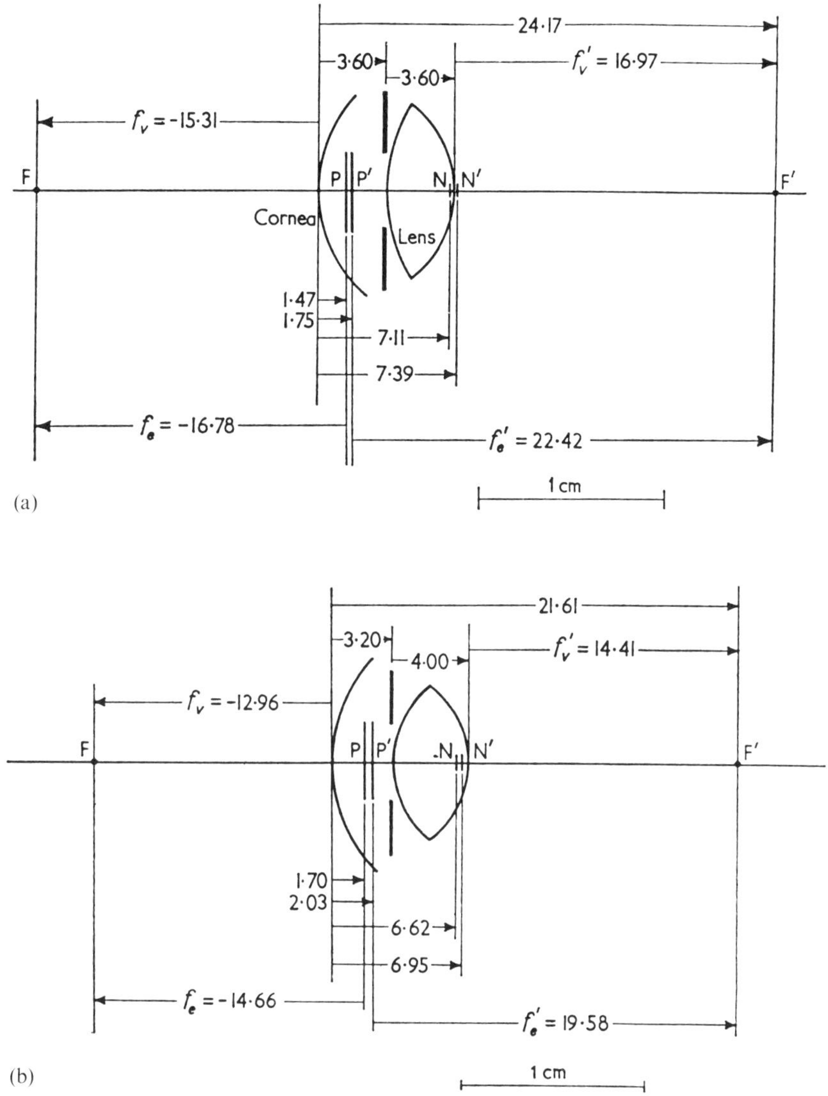

Figure 6.15. The Gullstrand simplified schematic eye (a) unaccommodated, (b) accommodated. F: front focal point, F': rear focal point, f_e: first focal length, f_e': second focal length, f_v: front focal length, f_v': back focal length, P,P': principal points, N,N': nodal points. All distances in millimeters. (Reproduced by permission from H. Davson, Ed., *The Eye*, Vol. 4, Academic Press, 1962, p. 104.)

We now wish to compare the apparent luminance of an object seen through an optical instrument such as a telescope with the luminance of the same object seen with naked eye. We assume that the eye pupil is placed at the exit pupil of the instrument (this type of pupil-matching arrangement is often referred to as the *Maxwellian view*). If the unaided eye is viewing a distant extended object

TABLE 6.2.
Pupil diameter versus scene luminance

Luminance (cd/m²)	Diameter (mm)
10^4	2
10^2	3
10	4
10^{-3}	7

that subtends a small angle 2β, then the flux entering the eye pupil is proportional to $r^2\beta^2$, where r is the pupil radius. This flux is spread over a retinal area that is approximately equal to β^2F^2, where F the focal length of the eye, producing a retinal illuminance (flux/area) that is proportional to r^2/F^2. Suppose now that the eye sees the same object through a telescope of (angular) magnification M, the exit pupil of which is greater than or equal to the eye pupil. The flux at the eye pupil is then proportional to $\tau r^2\beta'^2$, where $\beta'^2 = M^2\beta^2$, and τ the transmission factor of the optics. This flux will be distributed over a larger retinal area, given by β'^2F^2, and the retinal illuminance will be proportional to $\tau r^2/F^2$. Therefore, neglecting transmission losses, the apparent luminance of an extended distant object seen through a telescope is the same as the apparent luminance of the object seen through the naked eye in this case. Another way of seeing this is to consider the image of the eye pupil through the instrument: The telescope effectively magnifies the eye pupil, thus accepting more light from the object. However, the available light is spread over a proportionately larger angle because of the angular magnification, and the two effects cancel out.*

If the instrumental exit pupil is smaller than the eye pupil, then the flux at the eye pupil is $h_e'^2\,\beta'^2$, where h_e' is the exit pupil radius of the instrument. This is smaller than before by a factor $h_e'^2/r^2$, but the illuminated retinal area is still β'^2F^2. Therefore, if the instrumental exit pupil is smaller than the eye pupil, the apparent luminance of the object seen through the telescope will be less than that seen with the naked eye by a factor equal to the transmission times the ratio of the instrumental pupil area divided by the eye pupil area. Since the eye pupil is large in the dark, night vision instruments are designed with a large exit pupil so as to maximize the apparent object luminance.

Finally, let us make the same comparison for viewing point objects, such as stars. For a point object, the relevant quantity is the luminous intensity, given as flux per unit solid angle. The apparent luminous intensity of a star seen with the naked eye will then be proportional to the flux collected by the eye, which in

* This discussion is considerably simplified in assuming that the retinal illuminance and the apparent luminance have a simple relation, and in ignoring the so-called *Stiles–Crawford effect*, according to which rays striking the retina at different angles of incidence produce different sensation of brightness. The relation between retinal illuminance and apparent luminance is rather complex and depends on the object viewed; but we can ignore this complication because we are only interested in a comparison between aided and unaided vision.

turn is proportional to the pupil area, or r^2. The flux collected by the telescope will be proportional to the area of the entrance pupil, or h_e^2. If all this flux enters the eye pupil, then the apparent luminous intensity of the star seen through the telescope will be given by τM^2, where M^2 is the ratio of entrance to exit (eye) pupil area. Thus, neglecting absorption, the apparent luminous intensity of a point object is enhanced by a factor of M^2 by using the telescope.

6.4.4. Refractive Effects and Accommodation

The normal or *emmetropic* eye can bring an object located anywhere between infinity and 25 cm away into clear focus. (The 25 cm limit is somewhat arbitrary; young eyes can accommodate closer.) The near-sighted or *myopic* eye can accommodate closer than 25 cm but cannot accommodate for distant objects. The opposite effect is called *hyperopia* or hypermetropia (see also the discussion in Chapter 2). A negative lens is required to correct myopia, and a positive lens is needed for hyperopia. The eye may also be *astigmatic*, in which case the cornea is not rotationally symmetric, but has different power, depending on the meridional azimuth; this is corrected with cylindrical lenses, discussed in Section 6.7.

The refractive state of the eye is measured in dioptres. The sign of the effect is that of the lens required to correct it; thus -2 D of myopia means that a lens with power of -2 D is required in order to restore the eye to the normal accommodative range of ∞ to -25 cm. Accommodation is similarly measured in dioptres, with the same sign convention: If the eye is viewing an object 1 m away, we say that -1 D of accommodation is required.

There are two important effects associated with accommodation: the first one is the gradual loss of accommodation with age, or *presbyopia*. The lens loses its flexibility and gradually assumes its unnaccommodated shape. This leads to a recession of the near point.

The second effect is that the relaxed eye does not accommodate at infinity, as was assumed traditionally, but at a distance of about 1 m. This can be verified in the absence of a stimulus, in which case it is given the names "empty-" or "dark-field" myopia. A similar effect, called "instrument myopia," has been demonstrated in observations using visual instruments, such as telescopes: The preferred distance for accommodation is also at about -1 m. The state of accommodation of the relaxed eye is called the *resting state*. For a given stimulus vergence, the accommodative response of the eye is a compromise between the resting state and the stimulus vergence; the response is perfect only when the stimulus vergence coincides with the resting state. In other words, the eye does not necessarily accommodate in such a way as to provide the best possible image, but only enough to provide an acceptable image.

6.4.5. Resolution and Acuity

The resolving power of the eye is measured in various different ways. Optometrists use *Snellen letter* charts, and use the name *visual acuity* to describe resolution. Visual acuity is quoted as a fraction, where the numerator is the distance of the chart (normally 6 m or 20 ft) and the denominator is the distance at which the smallest correctly read chart line subtends 5 arcminutes. Snellen letters are such that the detail (e.g., the distance between two horizontal bars of the letter E) is one-fifth of the letter size. A 6/6 acuity means that the eye can resolve detail of 1 arcminute.

Visual acuity varies drastically across the retina, having a sharp maximum at the fovea. It also depends on the pupil size; at very small pupil sizes, diffraction spread degrades acuity, while at larger pupil diameters the degradation is caused by aberrations. Optimum acuity is for a pupil diameter of 2–3 mm.

6.4.6. Contrast Sensitivity

A very important characteristic of most imaging systems, including the eye, is that the resolving power depends on the contrast or modulation* of the target, defined as

$$\mathcal{M} = \frac{L_{\max} - L_{\min}}{L_{\max} + L_{\min}} \tag{6.9}$$

where $L_{\max}$ and $L_{\min}$ are the maximum and minimum luminance of the target, respectively. Maximum resolving power is only obtained for high-contrast targets. To measure the dependence of resolution on target modulation, use is made of sinusoidal gratings, which are sinusoidal luminance distributions such as shown in Figure 6.16. Sinusoidal gratings are chosen because, according to a theorem by Fourier, any luminance distribution can be expressed as a sum (or integral) of sinusoidal gratings of different periods, orientations, and phases. The inverse of the grating period is called *spatial frequency*. An observer is presented with gratings of various spatial frequencies; for each grating, the observer determines the minimum modulation necessary for the grating to be just visible, which is called the *threshold modulation* (contrast). The inverse of threshold modulation as a function of the grating frequency is called the *contrast sensitivity function* (Figure 6.17). It expresses the fact that the contrast needed to see a high-frequency grating (small detail) is higher than that needed for lower frequencies. However, at very low frequencies the sensitivity drops again; optimum sensitivity is in the region 3–5 cycles/degree. All optical systems can be similarly characterized by a curve showing the degradation of the contrast of a sinusoidal grating when it is imaged by the system; such a curve is called the *modulation transfer function* (MTF). It plays a key role in the assessment of image quality.

---

* Formally, there is a difference between contrast and modulation: The contrast is more appropriately defined as $(L_{\text{target}} - L_{\text{background}})/L_{\text{background}}$. The two terms, however, are often used interchangeably. Since there is a one-to-one correspondence between the two, care is needed only when one is dealing with numerical values.

Figure 6.16. A sinusoidal luminance distribution (grating).

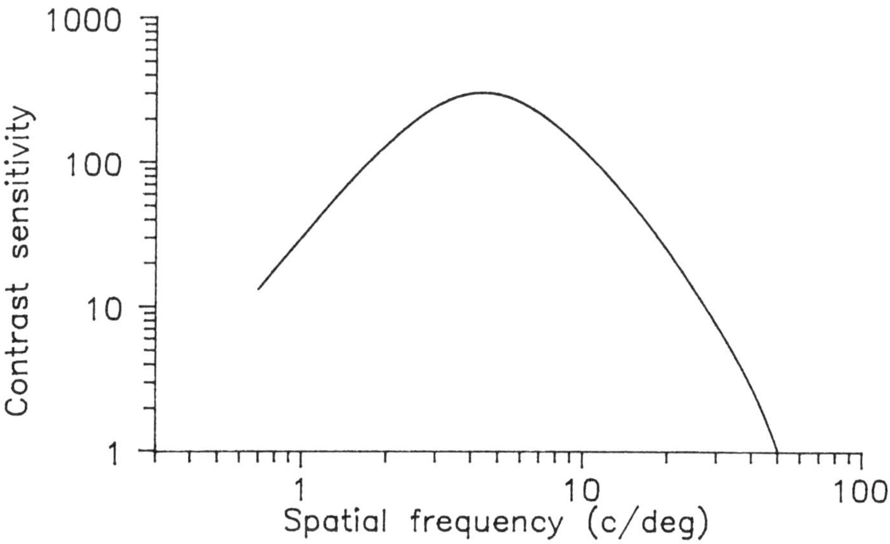

Figure 6.17. Typical contrast sensitivity curve for the human eye.

6.5. REFLECTING PRISMS

Reflecting prisms are very frequently used in optical instruments; so some understanding of their function is necessary. More often than not, reflecting prisms utilize total internal reflection (TIR). Prisms are used in preference to mirrors in many cases because of: (1) mechanical stability and simplicity of mounting, (2) high reflectivity (TIR is 100% efficient over a broad spectral range), and because (3) mirror coatings can deteriorate with age, whereas an optical surface that is protected from dust and scratches will retain its high reflectivity indefinitely.

For light to be totally internally reflected, the angle of incidence must be larger than the critical angle, $\theta_c = \sin^{-1}(1/n) = 41.8°$ for $n \cong 1.5$. This restricts the acceptance angle of the prism (NA of the beam), as well as the overall angle of incidence. If necessary, prism faces are coated with a reflective coating to remove these limitations while maintaining the first advantage.

6.5.1. Geometrical Aspects of Reflection

When you look in the mirror, you do not see yourself, at least not as others see you. If you part your hair to the right, your mirror image parts it to the left. Likewise, you can not shake hands with your mirror image; we call such an image *enantiomorphic* from the Greek εναντι = counter or opposite, and μορφη = face or appearance. However, your image through a lens parts its hair the same way as you: Such an image is called *congruent*. The right and left hands are enantiomorphic, but the left hand is congruent with the mirror image of the right hand.

To illustrate these concepts in less colloquial terms, let us consider the letter F as an object (Figure 6.18). Any displacement and/or rotation of the object around an axis normal to the paper (or around the optical axis) produces a congruent image. The lens image of Figure 6.19 is congruent (apart from a magnification factor). *The image must always be viewed from a point receiving the light*, in order to maintain a consistent definition. The mirror images of Figures 6.20a and 6.20b cannot be made to match the object by rotation about the optical axis; this holds for plane as well as curved mirrors.*

Figure 6.18. Congruent and enantiomorphic images of a two-dimensional object.

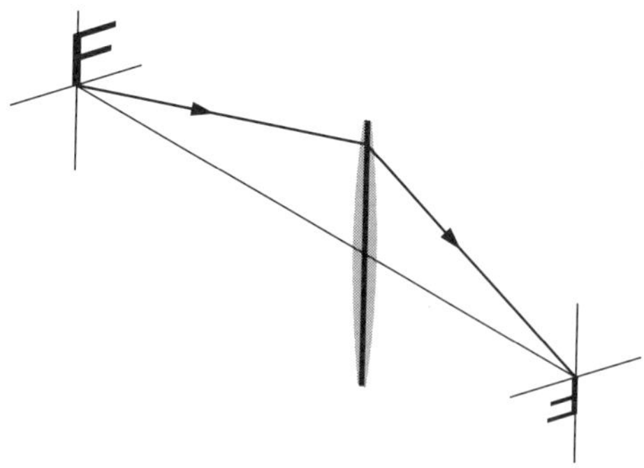

Figure 6.19. The image produced by a lens is congruent.

---

* If the object is restricted to two dimensions, an enantiomorphic image can match the object through rotation about an axis normal to the optic axis. But a three-dimensional object can never match its enantiomorphic image: Think of the right and left hands.

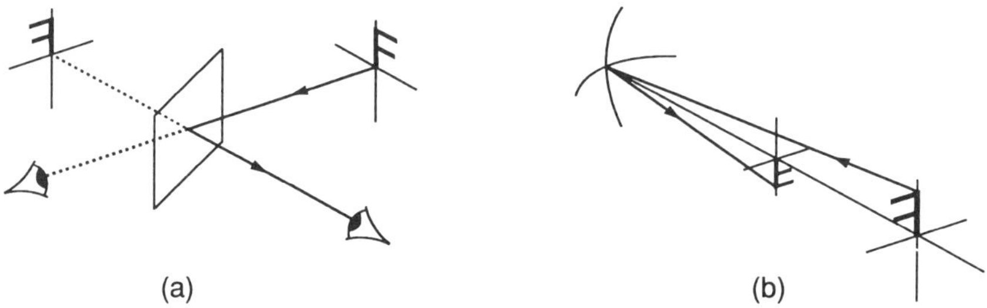

Figure 6.20. Nature of a mirror image: The image is enantiomorphic if seen from the direction receiving the light.

Two reflections produce a congruent image; therefore, to find the nature of the image in a complex system, we have only to count reflections. An odd number of reflections will produce an enantiomorphic image; an even number will produce a congruent image.

6.5.2. Tunnel Diagram. Effect of a Plane Block of Glass

Let us consider the right-angle prism, as in Figure 6.21. It produces an axis deviation of 90°. The axis deviation is ignored by our previous convention for ray tracing through many surfaces. This is not a problem because neither the magnification, nor the image distance from the last refracting surface, is affected (as measured along the beam axis). Thus if we ignore the reflection, the effect of the prism is the same as that of a block of glass that has a thickness equal to the length of the axial ray segment inside the prism. This "unfolding" of the prism, as indicated in Figure 6.21, is referred to as a *tunnel diagram.*

A plane parallel block of glass of thickness t causes an image shift calculated as follows (Figure 6.22). The paraxial refraction formula is

$$\frac{n'}{l'} - \frac{n}{l} = (n' - n)c = 0 \rightarrow l' = l\left(\frac{n'}{n}\right)$$

Figure 6.21. A right-angle prism and the corresponding tunnel diagram, shown in dashed lines. The tunnel diagram is obtained by "unfolding" the prism around the reflecting surface(s). In this way, the equivalent block thickness of the prism is immediately seen, and so is the relative orientation of the entrance and exit faces. If these are not parallel, the prism will act dispersively.

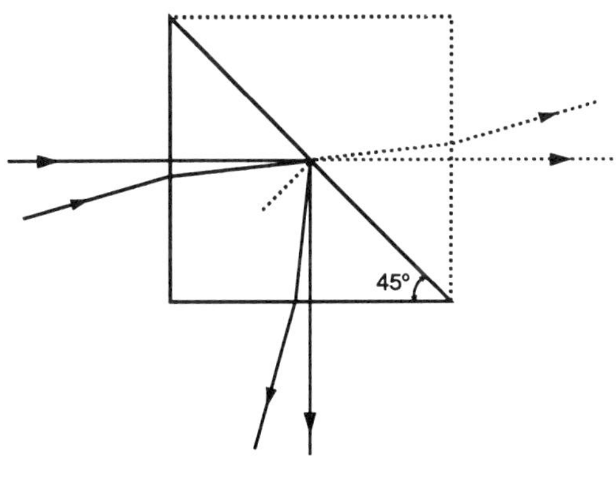

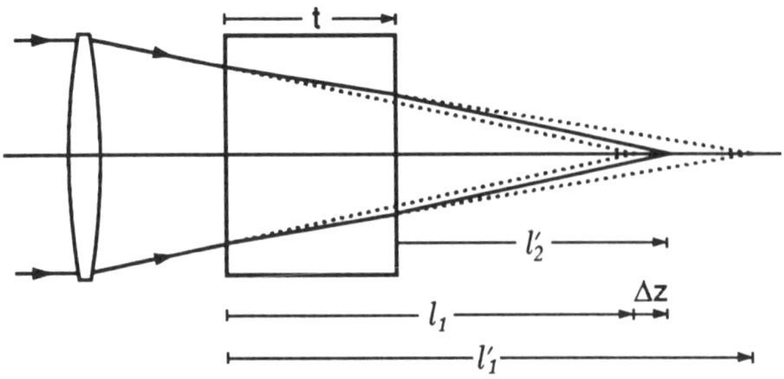

Figure 6.22. Image displacement by a plane-parallel block of glass.

Note that the magnification $M = nl'/n'l = 1$ for a plane surface. For the first surface, $n = 1$, $n' = \mu$, so $l_1' = \mu l_1$ and for the second surface $n = \mu$, $n' = 1$, $l_2 = l_1' - t$; so $l_2' = l_2/\mu = (l_1' - t)/\mu$. By eliminating l_1', we get $l_2' = (\mu l_1 - t)/\mu = l_1 - t/\mu$. The image shift is

$$\Delta z = t + l_2' - l_1 = t - t/\mu$$

so

$$\Delta z = t\left(1 - \frac{1}{\mu}\right) \tag{6.10}$$

which is independent of l_1, the original image position. For $\mu = 1.5$, $\Delta z = t/3$.

The second effect of the glass block is to introduce aberrations, which are discussed in some detail in subsequent chapters. At this point we may mention that it is advantageous to have the optical axis perpendicular to the entrance and exit faces of the prism. Under this condition, the prism can be used in a converging or diverging beam. But when the entrance and exit faces are oblique to the optical axis, large amounts of aberrations can be introduced. This situation is almost entirely avoided, except in the case where the beam is essentially collimated.

6.5.3. Common Prism Types

The *right-angle prism* is the simplest and most common. It produces a deviation of the optical axis by 90° and an enantiomorphic image.

The *penta prism* (Figure 6.23) also gives a 90° axis deviation but with a congruent image. The two reflecting surfaces must be coated as the angle of incidence is less than θ_c. The ray deviation is constant (90°) if the apex angle is 45°, independent of the initial angle of incidence (see Section 2.1.1).

The *roof prism* or *Amici prism* (Figure 6.24) also deviates the axis by 90° after two reflections. The effect of the roof is to split an incident plane wavefront in two sections, which would result in a double image unless the roof angle is 90° (within seconds). Also the roof must have a razor-sharp edge; otherwise, a line will appear in the middle. The image is congruent.

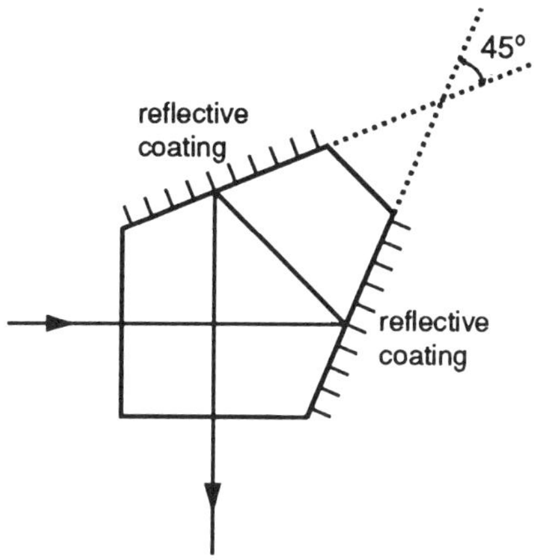

Figure 6.23. The Penta prism.

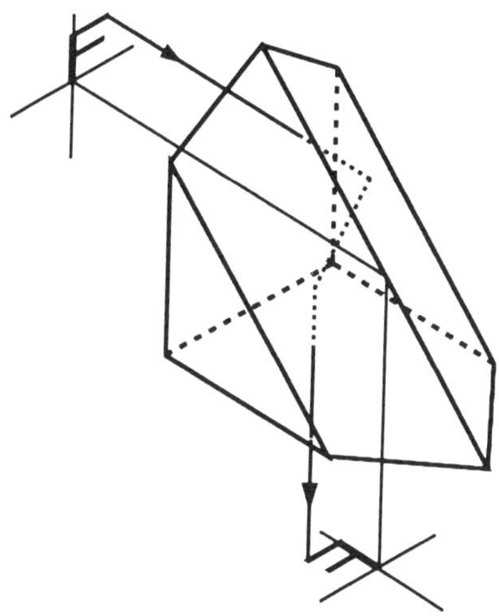

Figure 6.24. The roof or Amici prism.

The *Dove prism* (Figure 6.25) has its input and output faces oblique relative to the axis; so it can only be used in collimated light. It is used as an image-rotating device: When the prism is rotated by an angle ϕ, the image rotates by 2ϕ.

The *K prism* (Figure 6.26) can also be used for image rotation. Any prism with an odd number of reflections, and with input and output axes parallel, will rotate the image at twice the speed of its own rotation. A very interesting property of the K prism is that, with the appropriate refractive index, it can be inserted into a beam without producing an image shift. This is because the image shift can be compensated by the folding of the axis inside the prism.

The *Pechan prism* (Figure 6.27) is an image-rotating prism that can be made shorter than the Dove prism. Also, the incident and emergent axes are normal to the prism face; so it can be used in converging or diverging beams.

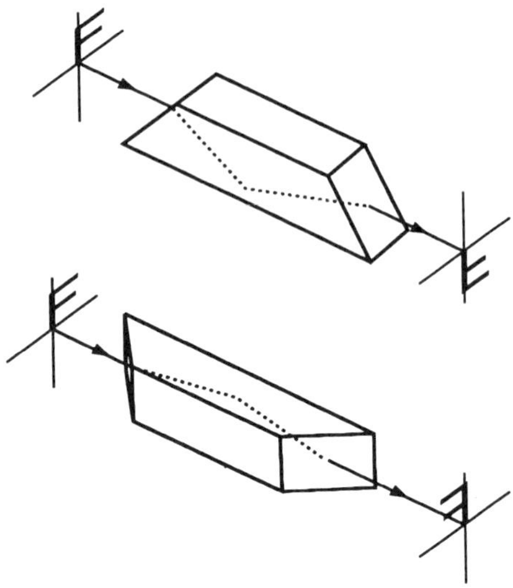

Figure 6.25. The Dove prism. Notice that this is essentially a right-angle prism with the top cut off. The entrance and exit facets are at 45° with respect to the reflective surface. Also notice that the prism rotated by 90° and the image by 180°.

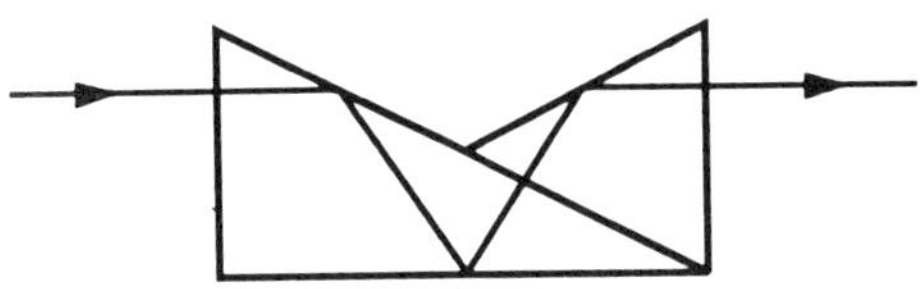

Figure 6.26. The K prism.

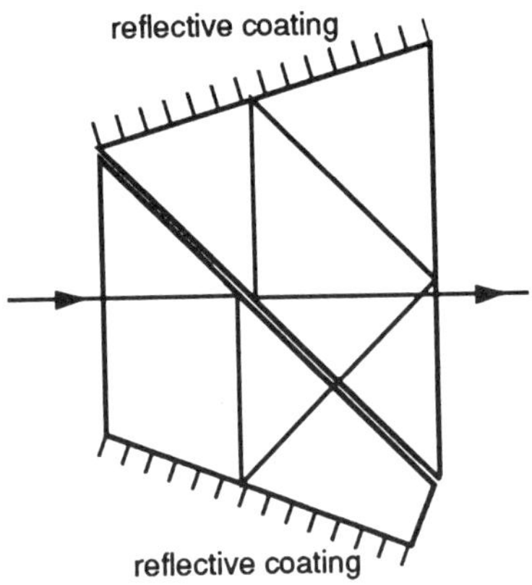

Figure 6.27. The Pechan prism.

The *corner cube* (Figure 6.28a) is a corner of a cube cut normal to the cube diagonal. It is a retroreflector; that is, a ray incident on it will emerge parallel to itself, independent of the angle of incidence. This prism is not normally used in imaging systems, but it provides a very useful way of reversing a beam without having to deal with the strict alignment tolerance that would be required of a single plane mirror. If you look into a corner cube, you will always see your own eye! A two-dimensional version of a retroreflector is the right-angle prism, used as shown in Figure 6.28b. This produces a

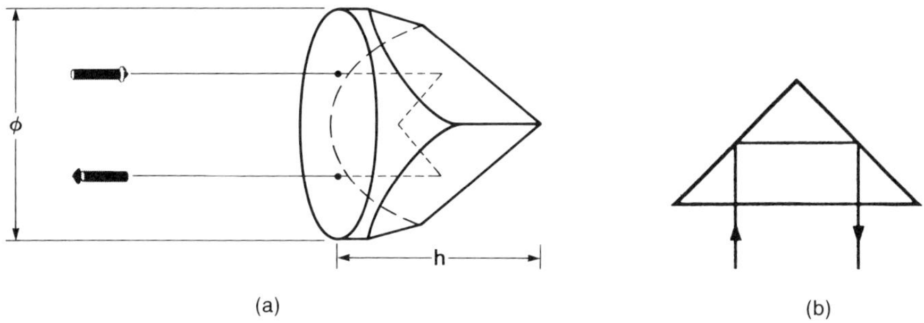

Figure 6.28. (a) The corner cube (courtesy of Melles Griot). (b) The right-angle prism used as a retroreflector.

congruent image; so it is the prism of choice for shaking hands with one's image.

There are many other kinds of prisms, some finding more general use and some made for specialized applications.* The important thing to remember is that whenever you want an image inversion, rotation, translation, or an optical axis deviation or displacement, there is a prism that will do the job.

6.5.4. Some Instrumental Applications of Reflecting Prisms

Ordinary binoculars are inverting telescopes; in order to present an erect image, a combination of two right-angle prisms is used, called a *Porro prism.* There are four reflections to provide a congruent image, which is inverted in both the x and y direction (Figure 6.29).

The optical system schematic of a single-lens-reflex (SLR) camera is shown in Figure 6.30. The lens produces an inverted image that becomes enantiomorphic after reflection on the mirror. This image is projected on a ground glass screen, and is magnified by the viewfinder lens. In order to present the eye with a congruent and erect image, three more reflections are needed. This is accomplished by using a roof-penta prism, which is a penta prism with one of the two reflecting faces replaced by a roof (which accounts for the characteristic bulge on top of SLR cameras).

An interesting device, called a *panoramic telescope*, is shown in Figure 6.31. Using the properties of the prisms described and the name of the device as a guide, the reader can attempt to determine the function of the various prisms shown. (Hint: The "head prism" rotates through 360° in azimuth; what happens to the image orientation when the head prism points backwards?)

---

* See also "Optical Design," Military Handbook 141, and L. Levi, *Applied Optics: A Guide to Optical System Design*, John Wiley and Sons, Inc., New York, 1968.

Figure 6.29. The Porro prism. The finished form of the prism (bottom) has rounded edges to save weight and a ridge to cut down stray light. (Reproduced with permission from W. J. Smith, *Modern Optical Engineering*, 2nd ed., McGraw-Hill, 1990.

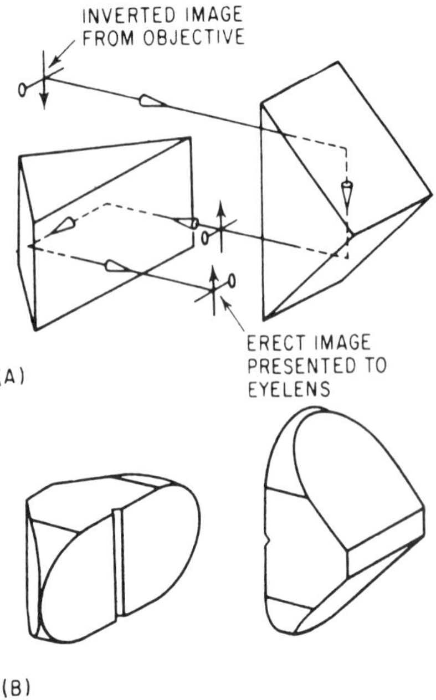

Figure 6.30. A single-lens-reflex (SLR) camera schematic.

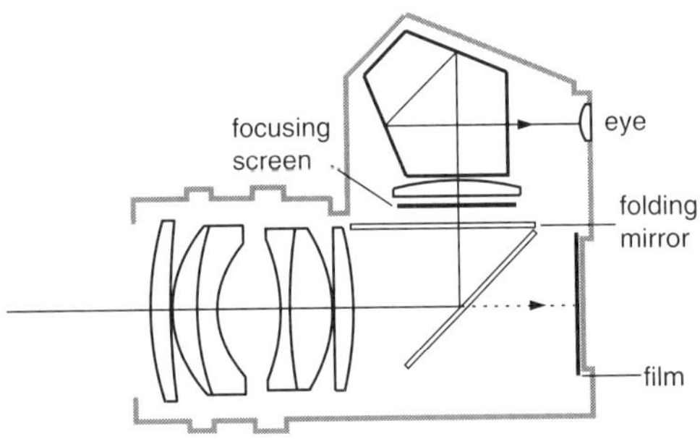

Figure 6.31. A panoramic telescope. (Adapted from T. B. Patrick, "Military optical instruments," in *Applied Optics and Optical Engineering*, Vol. V, R. Kingslake, Ed., Academic Press, 1969.)

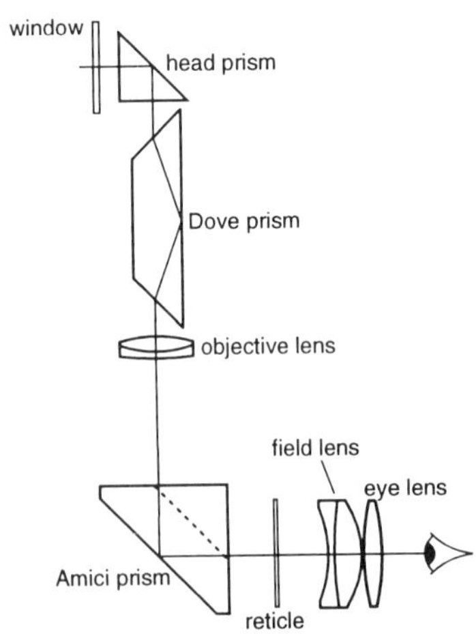

6.6. CYLINDRICAL AND ANAMORPHIC OPTICS

For systems with spherical surfaces, such as examined so far, the image of a square (perpendicular to the axis) is another square, at least paraxially. In anamorphic systems, the image of a square is a rectangle. The simplest anamorphic system is a wedge (transmission) prism, which expands or contracts a parallel beam along one direction only (see Section 2.1, Problem 2.7). The next simplest system is a cylindrical lens (Figure 6.32), which focuses the light along one direction only. A single cylindrical lens has power along one meridian only, but a general anamorphic system (such as a pair of crossed cylindrical lenses) has power along both meridians. Accordingly, we expect two different magnifications along the two meridians, M_x and M_y. We then define the *anamorphic factor* as

$$A = \frac{M_x}{M_y} \tag{6.11}$$

For a single cylindrical lens we take the meridian perpendicular to the cylinder axis (along which there is power) to be the x axis.

6.6.1. Image of a Point through a Cylindrical Lens

Figure 6.33 shows the geometry; we consider the lens as thin. Along the x meridian, the drawing looks the same as for a spherical lens; along the y meridian, the lens acts as a thin glass plate. Therefore, as the figure indicates, the image of a point source is a line, parallel to the cylinder axis. The length of the line can be found as follows. If σ_y is the length of the cylindrical lens and η'_y the length of the line, Figure 6.33 gives

$$\frac{\eta'_y}{\sigma_y} = \frac{l' - l}{-l}$$

or

$$\eta'_y = (1 - M_x)\sigma_y \tag{6.12}$$

where $M_x = l'/l$, the magnification along the x meridian.

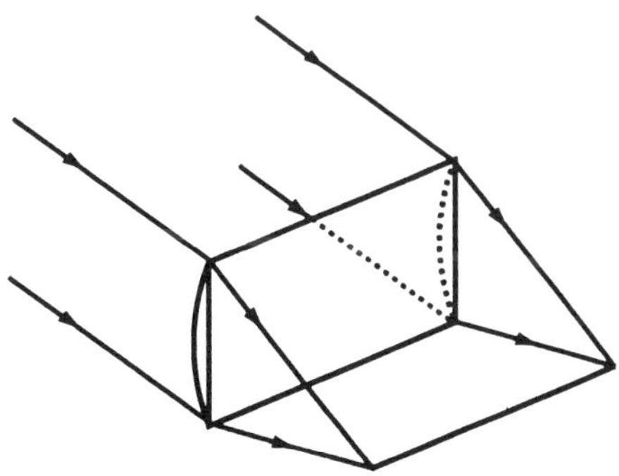

Figure 6.32. A cylindrical lens.

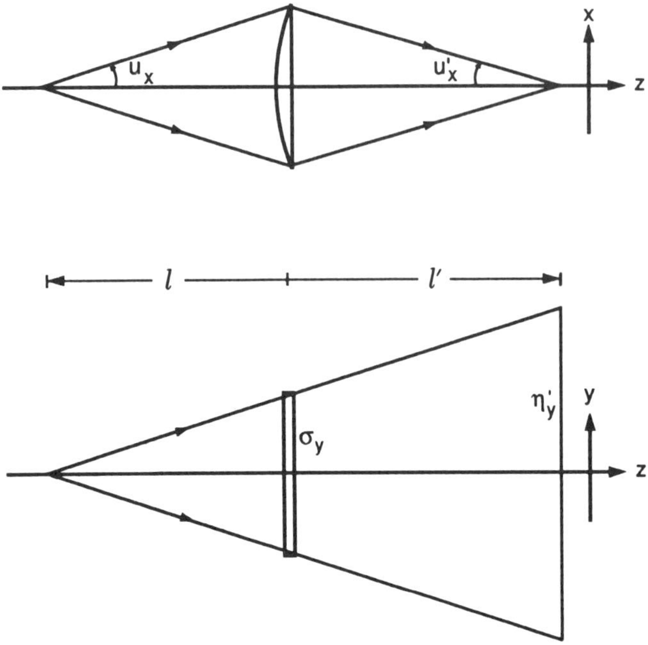

Figure 6.33. Geometry for imaging a point through a cylindrical lens.

6.6.2. Image Illuminance through a Cylindrical Lens

Let us assume that the object is a Lambertian emitter, in the shape of a small (strictly infinitesimal) square of side η, as in Figure 6.34. The flux collected by the lens is

$$\delta\Phi = L\eta^2\,\delta\Omega \cong \frac{L\eta^2\sigma_x\sigma_y}{l^2}$$

where $\sigma_x\sigma_y$ is the area of the lens. We have assumed that the solid angle $\delta\Omega$ is small, for simplicity. The flux $\delta\Phi$ is spread over an image area $\delta A'$, given by

$$\delta A' = \eta'_x\eta'_y = \eta M_x(1 - M_x)\sigma_y$$

by virtue of Eq. (6.12), for $\eta \ll \eta'_y$. Therefore, if τ is the transmission factor of the lens, the image illuminance is

$$E' = \tau\frac{\delta\Phi}{\delta A'} = \frac{\tau L\eta\sigma_x}{M_x(1 - M_x)l^2}$$

Using $M_x = l'/l$, we obtain

$$E' = \frac{\tau L\eta\sigma_x}{l'(l - l')} = -\frac{\tau L\eta\sigma_x}{l'T}$$

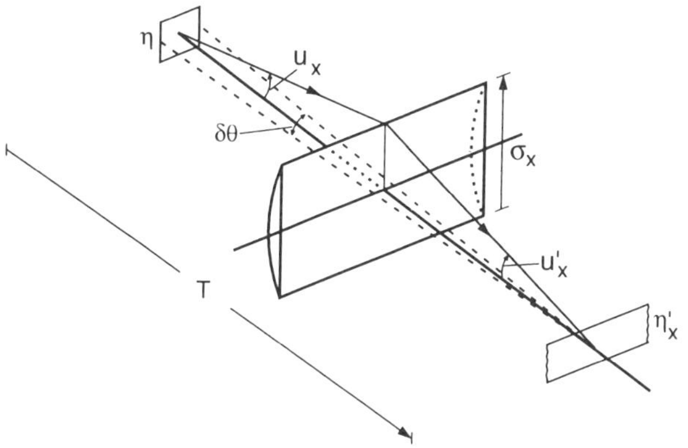

Figure 6.34. Geometry for imaging a small square source through a cylindrical lens.

where T is the object to image throw. Further, $-\sigma_x/l' = 2u'_x$, and $\eta/T = \delta\theta$, the angular subtense of the source at the image plane. Therefore,

$$E' = 2\tau L u'_x \, \delta\theta \tag{6.13}$$

Notice that u'_x is similar to the numerical aperture of spherical lenses when the object is at infinity. Thus Eq. (6.13) is consistent with the equations derived for spherical lenses: The numerical aperture enters along the x meridian, where there is imaging, and the angular subtense of the source enters along the y meridian, where there is no imaging [compare with Eqs. (5.47) and (5.57)].

Example 6.2. Laser diode beam conditioning. The light-emitting facet of a laser diode has typical dimensions of a few micrometers in the vertical direction by tens of micrometers in the horizontal direction. Diffraction causes the output beam to spread more rapidly along the vertical direction. If a collimated beam is desired, we may place the diode at the focal point of a (spherical) lens; the resulting beam will be collimated, but the profile of the beam will be elliptical rather than circular, with the major axis of the ellipse along the vertical direction. If a circular profile is desired, an anamorphic system must be used to compress the beam along the vertical direction (or expand it along the horizontal). There are two options: We may either use a prism system as in Figure 6.35a, or a cylindrical telescope, as in

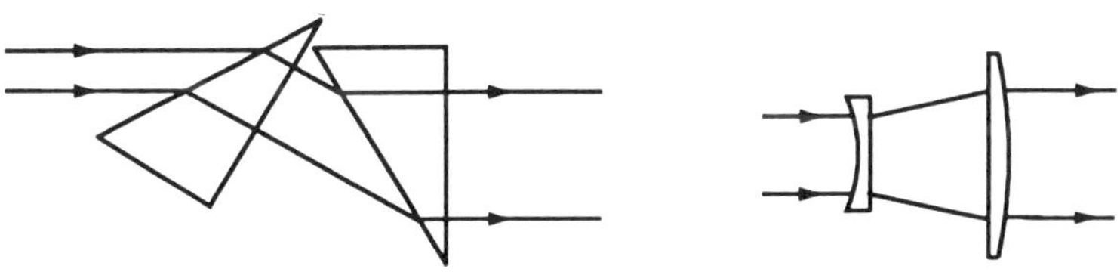

Figure 6.35. Two anamorphic systems that can be used to change the cross section of a beam from elliptical to circular.

Figure 6.35b. The cylindrical telescope may be of the Galilean or the Keplerian type.

Note that in both the preceding cases, the anamorphic system was placed in a collimated beam. If it was placed in a converging or diverging beam, intolerable amounts of aberration would result. Also note that some laser diode designs suffer from inherent astigmatism; for example, the focus for the x meridian might be 5 μm inside the facet, but the focus for the y meridian might be 30 μm inside the facet. Since an anamorphic system will have in general different foci for the x and y meridians, it may be used to correct the astigmatism. A tilted block of glass is often used to perform this function.

6.7. GRADIENT INDEX OPTICS

The refractive index of a gradient index material is not constant but varies in a continuous fashion according to some law. In recent years, there has been a great increase in the use of gradient index optics, both for focusing or collimating laser beams and for incoherent imaging applications. The most widespread gradient index optic is the so-called GRIN rod or GRIN lens. It is a cylindrical rod, typically $\cong 2$ mm in diameter and $\cong 5$ mm in length, in which the refractive index decreases radially with the distance from the axis.

There are several other types of gradient index components; the general treatment necessary to cover all possible cases is outside our present scope, and can be found in several books and articles, listed as references. We will restrict our attention to the paraxial properties of gradient index materials with cylindrical symmetry or a plane slab.

6.7.1. Snell's Law

Consider first a stratified medium (slab) as in Figure 6.36, where the refractive index is maximum in the middle and decreases toward the sides in a stepwise fashion. The same diagram would represent a meridional section of a cylinder (one containing the optical axis). Two possible ray paths are shown: Both rays

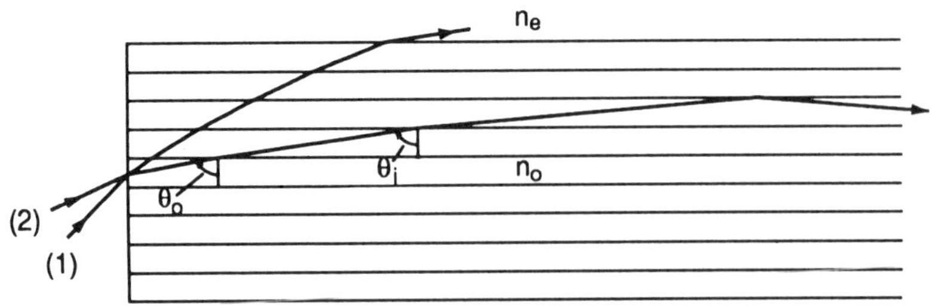

Figure 6.36. A stratified medium with varying refractive index.

tend toward larger angles of incidence, but ray 2 is totally internally reflected, while ray 1 escapes. After crossing the axis, the angle of incidence for ray 2 will again increase until it is totally internally reflected on the bottom side. So this ray will follow a periodic path.

Whether or not a ray follows a periodic path depends on the launching point and the launching angle. We can predict this as follows. Call θ_i the angle of incidence of the ray at the ith interface, and n_i the refractive index of the corresponding layer. Snell's law gives:

$$n_i \sin \theta_i = n_{i+1} \sin \theta_{i+1}$$

If n_e is the refractive index of the surrounding medium, then the ray will escape if

$$\theta_k < \sin^{-1}\left(\frac{n_e}{n_k}\right) \qquad \text{or} \qquad n_k \sin \theta_k < n_e$$

If n_0 and θ_0 are the index and angle of incidence in the layer that the ray is launched, Snell's law gives

$$n_0 \sin \theta_0 < n_e$$

It is convenient to replace the sine of the angle of incidence θ with the cosine of $(\pi/2 - \theta)$. This is the z-direction cosine of the ray N. Thus the condition for a ray to escape the slab is:

$$N_0 < \frac{n_e}{n_0} \tag{6.14}$$

where N_0 is the z-direction cosine of the ray inside the first (launching) layer.

We can now extend these results to media with continuous refractive index variation, by assuming that the number of layers becomes infinite while the refractive index step becomes infinitesimal. Thus we can write Snell's law in a gradient index medium as:

$$\boxed{nN = b} \tag{6.15}$$

where $b = n_0 N_0$, a constant that depends only on the launching conditions.

6.7.2. The Parabolic Index Profile

The periodic path that is followed by ray 2 raises an interesting possibility: If all rays launched at a single point could be made to follow a periodic path with period independent of the launching angle, then the medium would act as a lens, by focusing all the rays periodically as shown in Figure 6.37. The simplest possible periodic path is a sinusoid of the form

$$x = x_0 \sin(\alpha z) \tag{6.16}$$

where α is independent of the launching angle (N_0). We can then ask the

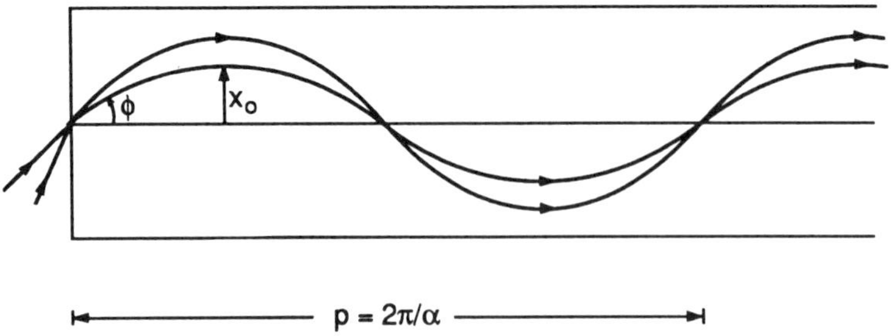

Figure 6.37. A medium with continuous refractive index variation, in which the rays follow a smooth oscillatory path.

question: Is there a refractive index profile $n = n(x)$ that would result in ray paths of the form of Eq. (6.16)?

To answer this question, let ds be the path-length element measured along the ray, then Eq. (6.15) can be written as

$$n\frac{dz}{ds} = b = n_0 N_0 \qquad \text{or} \qquad n\frac{dz}{dx}\frac{dx}{ds} = b \tag{6.17}$$

We assume for simplicity that incoming light is focused at the input face, $x = z = 0$, so the constant $n_0 = n(0)$. If the ray is restricted in the x–z plane, then $dx/ds \equiv L = (1 - N^2)^{1/2} = (1 - b^2/n^2)^{1/2}$. Substituting into Eq. (6.17) and squaring, we obtain

$$\left(\frac{dz}{dx}\right)^2 (n^2 - b^2) = b^2$$

From Eq. (6.16) we have

$$\frac{dz}{dx} = \frac{1}{\alpha}(x_0^2 - x^2)^{-1/2}$$

Upon eliminating dz/dx from the last two equations and solving for n, we find:

$$n = b(1 + \alpha^2 x_0^2 - \alpha^2 x^2)^{1/2} = n_0 N_0 (1 + \alpha^2 x_0^2 - \alpha^2 x^2)^{1/2} \tag{6.18}$$

This index profile depends on the ray parameters x_0 and N_0. However, for a sinusoidal curve, the amplitude and the slope at the origin are related. Let us call $\phi = \cos^{-1} N_0$. The slope at the origin is:

$$\tan\phi = \left.\frac{dx}{dz}\right|_0 = \alpha x_0$$

and since

$$\cos\phi = (1 + \tan^2\phi)^{-1/2}$$

we have

$$N_0 = (1 + \alpha^2 x_0^2)^{-1/2} \tag{6.19}$$

Substituting for N_0 in Eq. (6.18), using the approximation $(1+x)^{\pm 1/2} \cong 1 \pm x/2$, and carrying out the multiplication, we obtain

$$n = n_0\left(1 - \frac{\alpha^2}{2}x^2 - \frac{\alpha^4}{4}x_0^4 + \frac{\alpha^4}{4}x^2x_0^2\right) \tag{6.20}$$

which upon neglecting terms of fourth power in x becomes

$$n(x) = n_0\left(1 - \frac{\alpha^2}{2}x^2\right) \tag{6.21}$$

Thus the desired index profile is of the second degree in x, and is called a *parabolic profile*. Notice that this type of profile is not an exact solution to the problem of periodic focusing, because the more precise form (6.20) contains terms in x_0. This shows that the period of the ray path depends on x_0, or equivalently, N_0. Periodic focusing is obtained only for small x_0 or large N_0 (small launching angle), which is similar to the usual paraxial approximation.*

6.7.3. Paraxial Ray Tracing for Gradient Index Media

We now wish to determine the equations for tracing a ray inside a medium with a parabolic index profile. We already know from Eq. (6.16) that the ray will follow an oscillatory sinusoidal path. The most general form of a sinusoidal oscillation is

$$x = c_1 \cos(\alpha z) + c_2 \sin(\alpha z) \tag{6.22}$$

where c_1 and c_2 are constants to be determined by the initial (launching) conditions. If we launch at $z = 0$, we have $c_1 = x(0)$. Also,

$$\left.\frac{dx}{ds}\right|_0 \equiv L_0 = \left.\frac{dx}{dz}\frac{dz}{ds}\right|_0 = \alpha c_2 N_0 \rightarrow c_2 = \frac{L_0}{\alpha N_0}$$

To return to our usual notation, notice that $L_0 = \sin\phi = u_0$, the paraxial angle variable. Also paraxially, $N_0 \rightarrow 1$, and we can replace x with the paraxial height variable h, so $c_1 = x(0) \equiv h_0$. Equation (6.22) then becomes

$$\boxed{h = h_0 \cos(\alpha z) + \frac{u_0}{\alpha}\sin(\alpha z)} \tag{6.23}$$

This is the equivalent of the paraxial transfer formula. To find the paraxial refraction formula we note that $u = dh/dz$; so from Eq. (6.23),

$$\boxed{u = -\alpha h_0 \sin(\alpha z) + u_0 \cos(\alpha z)} \tag{6.24}$$

---

*Exact periodic focusing of meridional rays can be achieved with a profile of the form $n = C\,\mathrm{sech}(Dx) \cong C(1 - D^2x^2/2)$, called a *hyperbolic profile*. In this case, the ray paths are periodic with a period that depends only on the constant D and not on the ray parameters. However, the ray paths are not exactly sinusoidal. If the variation of n is not strong, the parabolic and hyperbolic profiles are similar.

6.7.4. Gaussian Constants of GRIN Rods

We can characterize a GRIN rod through its power and principal planes, just like any lens of nonzero power; it can then be incorporated easily into Gaussian design calculations. To determine the power, we apply Eqn. (6.24) for $z = d$ and take a ray parallel to the axis in the object space, $u_0 = 0$. Then

$$u = -\alpha h_0 \sin(\alpha d)$$

If the rod is in air, the emergent ray will have an angle $u' = n(h)u = n_0 u$ paraxially. From the definition of power [Eq. (4.13)], we then have

$$K = n_0 \alpha \sin(\alpha d) \tag{6.25}$$

From Eq. (4.24), the principal point P' is located at $\delta' = (h_0 - h)/u'$ with $h = h_0 \cos(\alpha d)$, from Eq. (6.23). Therefore,

$$\delta' = \frac{\cos(\alpha d) - 1}{n_0 \alpha \sin(\alpha d)} \tag{6.26}$$

$$\delta = -\delta' \tag{6.27}$$

by symmetry. An additional useful parameter is the working distance or back focal length of the lens. We have $F_b = F + \delta'$; therefore,

$$F_b = \frac{\cot(\alpha d)}{n_0 \alpha} \tag{6.28}$$

The length corresponding to one full period of the ray path is called the *pitch*, $p = 2\pi/\alpha$, of the lens. Thus a $0.5p$ lens images a point on its input face to a point on its output face. A $0.25p$ lens will produce a collimated beam if a point source is focused at its input face, or will focus a parallel beam to a point on the output face (Figure 6.38). Other fractions such as $0.29p$ can be used to give a focus that is located a short distance away from the input or output face.

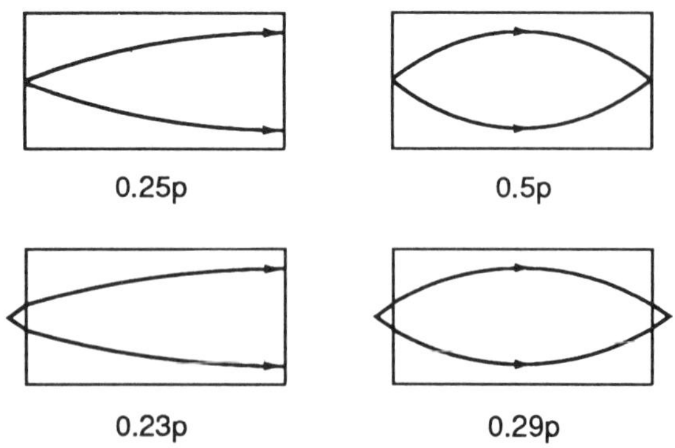

Figure 6.38. Various types of GRIN rod lenses.

6.7.5. The Optical Invariant

The optical invariant H can be defined for GRIN rods in exactly the same way as for uniform index media. First we note that the ray-tracing equations for the PPR are simply

$$\begin{aligned} \bar{h} &= \bar{h}_0 \cos(\alpha z) + \frac{\bar{u}_0}{\alpha} \sin(\alpha z) \\ \bar{u} &= -\alpha \bar{h}_0 \sin(\alpha z) + \bar{u}_0 \cos(\alpha z) \end{aligned} \tag{6.29}$$

Let us now form the expression $H = n_0(\bar{h}u - h\bar{u})$. Substitution from Eqs. (6.29) (6.23), and (6.24) yields "in a few lines" $H = n_0[\bar{h}_0 u_0 - h_0 \bar{u}_0]$, which is a constant. Therefore, H is an invariant.

6.8. DIFFRACTIVE OPTICS

As the name suggests, the operation of diffractive optical elements is understood through diffraction theory, which is outside the scope of this book; for this reason, several formulas are given in this section without proof, and the entire treatment is qualitative. However, diffractive optics are increasingly used in optical instrumentation; even from the viewpoint of geometrical optics we need to understand some of their properties so as to be able to incorporate them in a first-order Gaussian design.

There are several different types of diffractive optical elements. Although diffraction is the common thread, important physical details associated with the various types are sufficiently different to make a common treatment difficult. We therefore examine separately three basic types of diffractive elements, starting from the oldest.

6.8.1. The Diffraction Grating*

The diffraction grating came to be well known as a result of the work of Joseph Fraunhofer, in the early part of the nineteenth century. Fraunhofer used gratings to measure the absorption lines in the solar spectrum, which now bear his name. In the latter part of the nineteenth century and the beginning of the twentieth century, H. A. Rowland, R. W. Wood, and A. A. Michelson made significant contributions to the understanding and the production of gratings. There are two basic types of diffraction gratings, ruled and holographic; more details of the latter type are found in Section 6.8.2. A ruled grating is made by ruling on a glass or metal substrate thousands of parallel lines with a diamond tip; a typical line spacing is on the order of a micrometer. The ruled grating is then used as a master from which replicas are made. We distinguish between

---

* For a full discussion, see M. C. Hutley, *Diffraction Gratings*, Academic Press, 1982.

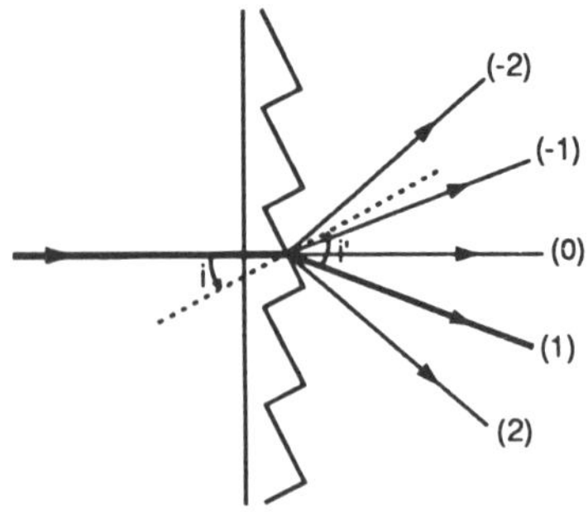

Figure 6.39. A typical transmission grating. The diffracted light is distributed among many orders, but with the appropriate groove profile, most of the light can be made to diffract along a single order. For the grating shown, the profile must be such that Snell's law ($n \sin i = \sin i'$) is satisfied for the first order.

transmission and reflection gratings. The profile of a typical transmission grating is shown in Figure 6.39. A reflection grating may be ruled directly on a reflective material or may be coated on a glass substrate.

When light is incident on a grating such as that of Figure 6.39, some part is transmitted along the original direction of propagation; this is the undiffracted, or zero order, which is normally not useful. Part of the light is diffracted along other directions. The permitted directions of diffraction are easily found by considering the light beams arising from each individual groove; light will be diffracted along the directions of constructive interference of all these beams. For this to happen, the path difference between the different beams must be a whole number of wavelengths. As Figure 6.40 shows, the path difference Δp between two successive beams is given by

$$\Delta p = |d \sin I| + |d \sin \theta|$$

where I is the angle of incidence and θ the angle of diffraction, respectively. If we apply our usual sign convention, the angle I is negative as shown, while θ is positive. Therefore,

$$\Delta p = d(\sin \theta - \sin I)$$

This path difference must equal a whole number of wavelengths for constructive interference; so we have

$$d(\sin \theta - \sin I) = m\lambda \tag{6.30}$$

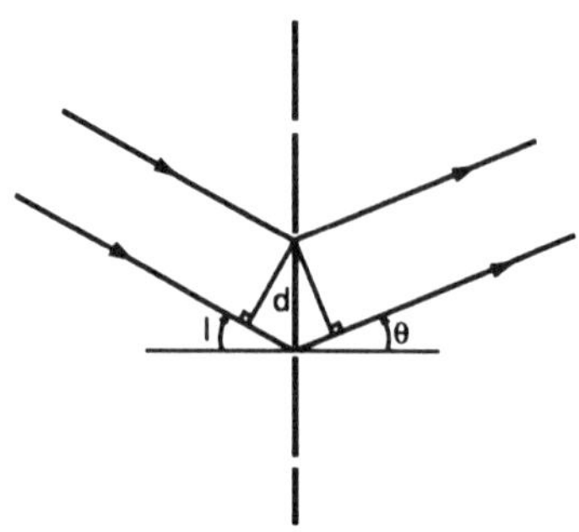

Figure 6.40. A simplified derivation of the grating equation. We consider the grating as an opaque aperture with many parallel slits. The same equation holds for the transparent transmission grating of Figure 6.39.

where m is an integer, called the order of diffraction. This is known as the grating equation: It determines the permissible angles of diffraction for a transmission grating. It can take a more general form if the refractive indices on either side are n and n'. We then have

$$d(n' \sin \theta - n \sin I) = m\lambda \tag{6.31}$$

The equation for a reflection grating can now be obtained by setting $n' = -n$. In air, we have

$$d(\sin \theta + \sin I) = m\lambda \tag{6.32}$$

where a negative sign has been absorbed into m. Since the angle of diffraction depends on the wavelength, incident polychromatic light will be dispersed, that is, separated into its component wavelengths, which will be diffracted at different angles. For this reason, the diffraction grating is used primarily as a spectral analysis tool. In performing such analysis, the available light from the source in a narrow wavelength band may be very little: It is therefore important to concentrate the light into one diffracted order, rather than have it be distributed among all diffracted and undiffracted orders. The *diffraction efficiency* of a grating is given by the amount of light diffracted in the useful (or desired) order, divided by the amount of incident light. Unless specially constructed, a grating will have a diffraction efficiency of only a few percent.

The process by which a grating is made to diffract most of the light in the desired order is called *blazing*, and the grating is then called *blazed*.* Blazing can be understood from an intuitive viewpoint rather simply: Returning to Figure 6.39, it is seen that each facet will refract the light individually, along a direction predicted by Snell's law that depends on the slope of the facet. If this direction of refraction coincides with an allowed direction of diffraction, the diffraction efficiency is maximized. Similarly, for a reflection grating blazing occurs for a facet slope such that the direction of reflection coincides with the direction of diffraction.

If a grating is used in spectral analysis, it is important to know the minimum resolvable separation between two neighboring wavelengths, $\Delta\lambda$. This is characterized by the *chromatic resolving power* of a grating, given through

$$\frac{\lambda}{\Delta\lambda} = mN \tag{6.33}$$

where λ is the average wavelength, m the order, and N the total number of illuminated grooves (if the grating is the aperture stop, then N is the total number of rulings of the grating).

---

* The name derives historically from the subjective impression of "lighting up" or blazing, if a properly made grating is held up against a light source at the proper angle.

6.8.2. Holographic Optical Elements*

The three-dimensional appearance of holographic images is due to the fact that the hologram is a record of both the irradiance and the phase of the light emanating from the object. It is the phase that contains the distance information missing from an ordinary photograph. However, radiation detectors, such as silver halide film, that can be used for recording the hologram, are sensitive only to the irradiance and not the phase of the light incident on them. In order to record phase, we must convert it to an irradiance variation. This is accomplished by interfering the wave emanating from the object with another wave, called the *reference wave*. The hologram then is a record of the interference pattern between the object wave and the reference wave.

Holographic optical elements (or HOEs) may be thought of as holograms that are not meant for visual display (although they can be used as part of a display system). The function of HOEs is akin to that of ordinary optical elements, such as lenses, mirrors, or gratings. The simplest case is that of a plane grating. Physical optics shows that when two (coherent) plane waves interfere, a sinusoidal fringe pattern is formed, the period of which is given by

$$d = \frac{\lambda}{2\sin(\phi/2)} \tag{6.34}$$

where ϕ is the angle between the two waves (wave vectors). Suppose that the interference pattern is recorded on film (Figure 6.41); the film will then resemble a diffraction grating. If this grating (or hologram) is illuminated with one of the two beams used for recording, it will diffract along a direction given by the grating equation (6.30), which for $\theta = -I$ becomes $2d \sin\theta = m\lambda$. Thus the direction of diffraction (for $m = 1$) is $\theta = \sin^{-1}(2d/\lambda) = \phi/2$, that is, along the direction of the other recording beam. We may thus say that we have reconstructed the second beam by illuminating the hologram with only the first beam. Since the hologram does not have 100% diffraction efficiency, there will be, in general, two more beams, corresponding to the zero and -1 orders. The existence and magnitude of higher orders ($|m| > 1$) will depend on how

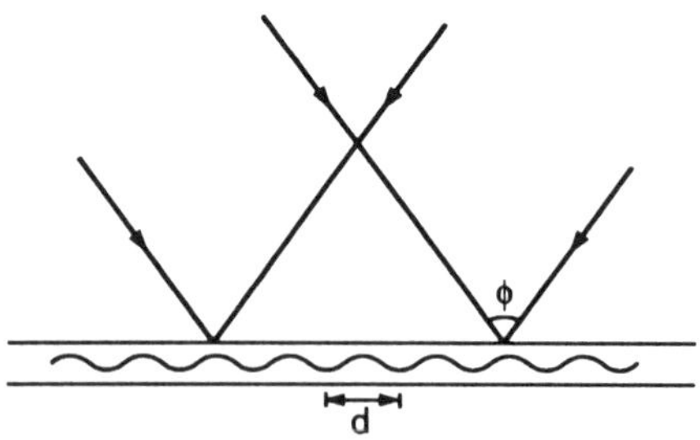

Figure 6.41. A sinusoidal distribution of intensity of period d results from the interference of two coherent beams and is recorded on photosensitive material.

*For an excellent introductory treatment of holography, see M. Françon, *Holography*, Academic Press, 1974. For a more advanced treatment, see P. Hariharan, *Optical Holography*, Cambridge University Press, 1984.

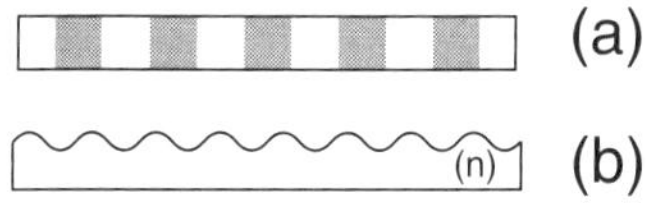

Figure 6.42. Schematic of a phase hologram (a), in which the interference pattern is recorded as a refractive index variation, and of a surface-relief hologram (b).

faithfully the sinusoidal distribution was recorded on the film, that is, on the linearity of the recording. Nonlinear recording will result in higher harmonics.

Holograms and HOEs may be classified according to the recording medium into two types, amplitude and phase. In an amplitude hologram, the interference pattern is recorded as a variation in the real amplitude (or irradiance) transmittance of the material, similar to a photographic negative. A phase hologram records the interference pattern as an optical path variation in an otherwise transparent material. For example, in certain photopolymers, the fringe pattern is recorded as a variation in the polymer concentration, leading to a variation in refractive index. In photoresist systems, the interference fringes are recorded as a variation in the surface height (Figure 6.42). Another classification refers to the thickness of the emulsion: We speak of thick (or volume) holograms and thin holograms. A thick hologram has typically an emulsion thickness of more than about 10 μm; it is characterized by the condition that the incident wave (or ray) must cross several fringe planes before exiting. By contrast, in a thin hologram the ray may encounter only one fringe.

There are limits to the diffraction efficiency that can be achieved with the various types of holograms. Amplitude holograms have typically very low diffraction efficiency; so they are rare. Phase holograms can achieve efficiencies of close to 100%. This is most often accomplished with volume phase holograms, but is also possible with surface-relief-type holograms under certain conditions. The process utilized for increased diffraction efficiency with a volume hologram is called *Bragg diffraction*, and is not too different from the blazing of ruled gratings. In Figure 6.43, we can consider the fringes as partially reflecting mirrors: When the angle of incidence is such that the reflected beams interfere constructively, we have maximum diffraction efficiency. We then say that the Bragg condition is fulfilled, and the corresponding angle of incidence is called the Bragg angle. For the case shown in Figure 6.43, the Bragg condition is $2d \sin \phi = m\lambda$. The Bragg condition provides angular and wavelength selectivity,

Figure 6.43. Bragg diffraction from a thick grating. The reflected beams interfere constructively if the Bragg condition is satisfied (refraction at the boundary of the material is not shown in the figure).

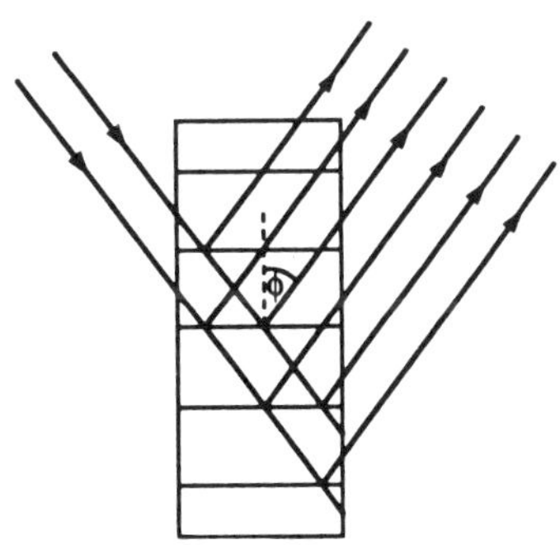

which is advantageous in certain cases. However, because of the strong wavelength dependence, HOEs are normally used with monochromatic or narrow-band light.

The plane holographic grating is the simplest possible HOE. But if the recording waves are not both plane, we can construct HOEs that perform the function of lenses or curved mirrors. For example, suppose we record a hologram using a plane and a spherical wave, as shown in Figure 6.44. If upon reconstruction we illuminate the hologram with the plane wave, it will reconstruct the spherical wave; that is, it will focus the light; this then is a holographic lens. A holographic mirror with $K \neq 0$ is also shown in Figure 6.44. Since such elements are highly dispersive, they may also be used as gratings with nonzero power.

Application: Holographic head-up display. A head-up display (or HUD) is used in aircraft, and even in passenger cars recently. The basic idea of a HUD is to provide a display of vital information (speed, altitude, etc.) that appears projected through

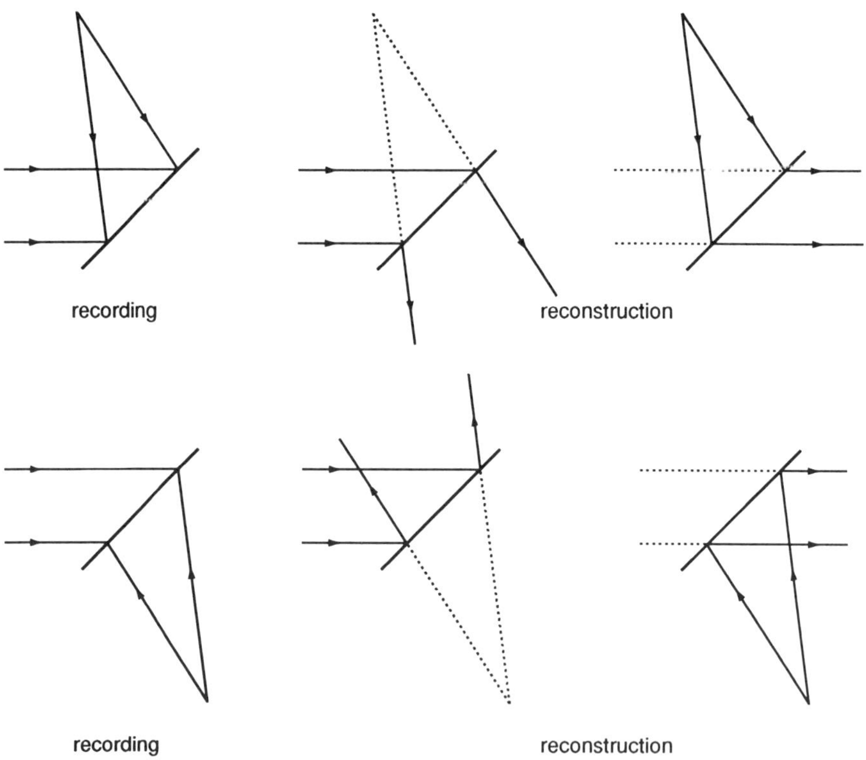

Figure 6.44. Recording and reconstruction of a transmission (top) and a reflection (bottom) holographic optical element. Notice that both HOEs can act also as focusing lenses if the direction of the rays is reversed. In monochromatic light, the HOE acts as a lens. In polychromatic light, it acts as a dispersive grating and lens simultaneously.

the windscreen; the pilot/driver may then look at the displayed information without having to shift her gaze from the sky/road. This function can be (and often is) accomplished by a simple beamsplitter, but an ordinary, spectrally neutral beamsplitter will waste at least half of the light from the display, and will similarly reduce the apparent luminance of outside objects. Furthermore, the beamsplitter will have zero power; so the image must be projected at the appropriate distance in front of the driver by using an auxiliary lens. If a holographic mirror is used in place of the ordinary beamsplitter, the following advantages are possible. (1) Due to the Bragg condition, the mirror is highly wavelength selective: The total band of reflected wavelengths is typically around 20 nm. Other wavelengths are transmitted; so the mirror appears almost invisible, and the attenuation of light from the outside is small. (2) If a narrow-band source is used as the display, most of the source light will be reflected toward the observer's eye. Of course, the above two conditions can be satisfied with an appropriately designed narrow-band reflective coating as well, but the application of such a coating over an extended area of the windscreen can present considerable difficulty. Also, if the holographic mirror is recorded with two spherical beams positioned at different distances, it can be used as a lens at the same time, by projecting a virtual image of the display at the appropriate distance in front of the observer. This can simplify the design of the optical system.

There are also possible disadvantages. The HOE may introduce significant aberrations if proper care is not taken, but may also be correct aberrations from the rest of the optical system. Finally, stray diffraction from localized sources such as street lights is possible.

If the geometry and wavelength of reconstruction match those of recording exactly (and assuming linearity of recording), then the reconstructed beam matches the original object beam exactly. But if the reconstruction conditions differ from the recording conditions, the reconstructed beam will differ from the object beam. The difference can be expressed as a first-order effect in the form of image displacement and a magnification change, and as a higher-order effect in the form of aberrations. We may express the first-order effect as follows. Let us assume that the hologram represents the interference of two beams emanating from points P and Q, at a distance p and q from the holographic plate, respectively (along the normal to the plate). Let the wavelength of recording be λ. Let us further assume that the reconstructing beam comes from a point source that is located a distance q' from the hologram, and has a wavelength λ'. The reconstructed image will then be at a distance p', given by:

$$\frac{1}{p'} = \frac{\lambda'}{\lambda}\left(\frac{1}{p} - \frac{1}{q}\right) + \frac{1}{q'} \tag{6.35}$$

Notice that if $\lambda = \lambda'$ and $q' = q$, then $p' = p$. If we consider the object to be an infinitesimal line element instead of a point, then we can define a magnification as the ratio between the size of the reconstructed image and that of the original object. This magnification is given by

$$M = \frac{\lambda'}{\lambda}\frac{p'}{p} \tag{6.36}$$

which can also be expressed as

$$M = \left(1 - \frac{p}{q} + \frac{\lambda}{\lambda'}\frac{p}{q'}\right)^{-1} \tag{6.37}$$

using Eq. (6.35). Justification for these equations may be found in Françon's *Holography*, or any other text on the topic. It is interesting to notice the similarity between Eq. (6.35) and the thin-lens equation. If we rewrite Eq. (6.35) as

$$\frac{1}{p'} - \frac{1}{q'} = \frac{\lambda'}{\lambda}\left(\frac{1}{p} - \frac{1}{q}\right)$$

we see that the right-hand side may be seen as the power of the holographic lens, since it is the difference in curvature between the incident and emergent wavefronts. If $p = q$, or, if $p = q = \infty$ (plane waves), then the lens has zero power, that is, $p' = q'$. The higher-order effects (aberrations) can be determined by several methods, including a geometrical ray-tracing method in which the HOE is modeled as a thin lens with a very high (almost infinite) refractive index.*

The focusing action of a HOE can be understood through the idea of the *Fresnel zone plate*. Referring to Figure 6.45, let the plane of an aperture be at S, and let A and A' be two points on either side of the aperture, with a point source located at A. Consider now the optical path between A and A' via some point P on the plane S, $[APA']$. Following our usual sign convention, we let S be the origin of coordinates, so $(AS) = -l$, $(SA') = l'$, $(SP) = r$. Considering the medium to be air for simplicity, we may write

$$[APA'] = |s| + |s'| = (l^2 + r^2)^{1/2} + (l'^2 + r^2)^{1/2} \cong l' - l + \frac{r^2}{2}\left(\frac{1}{l'} - \frac{1}{l}\right) \tag{6.38}$$

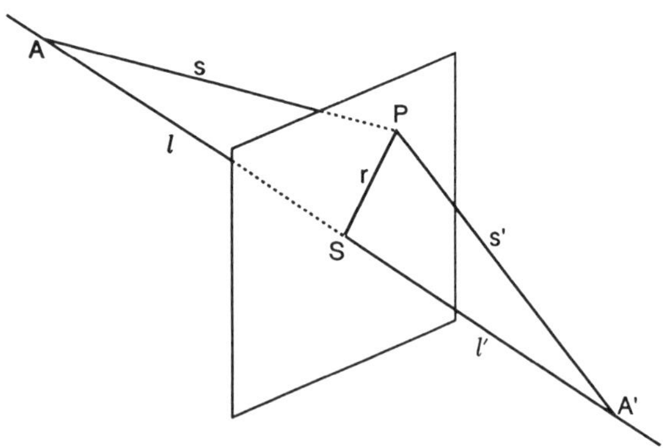

Figure 6.45. Geometry of Fresnel diffraction.

* See W. C. Sweatt: "Describing holographic optical elements as lenses," *J. Opt. Soc. Am.* **67**, 803–8 (1977), and "Mathematical equivalence between a holographic optical element and an ultra-high index lens," *J. Opt. Soc. Am.* **69**, 486 (1979).

where $-l = |l|$. The approximation is valid on the assumption that $|l|, |l'| \gg |r|$. Suppose now that the aperture is divided into concentric regions, defined so that $|s| + |s'|$ differs by $\lambda/2$ from one boundary to the next. These regions are called *Fresnel zones*. Their radius can be calculated from the definition, as follows:

$$|s_{m+1}| + |s'_{m+1}| = |s_m| + |s'_m| + \lambda/2$$

Using Eq. (6.38) and setting

$$\frac{1}{f} \equiv \frac{1}{l'} - \frac{1}{l} \tag{6.39}$$

we obtain

$$r_{m+1}^2 - r_m^2 = f\lambda \tag{6.40}$$

from which $r_1 = \sqrt{f\lambda}, r_2 = \sqrt{2f\lambda}, \ldots, r_n = \sqrt{nf\lambda}$. Notice that the area of a zone $\pi(r_{m+1}^2 - r_m^2) = \pi r_1^2$ so all zones have the same area. Call U the amplitude of the wave arriving at A'. Since the mean phase changes by π from one zone to the next, we can write

$$U \cong |U_1| - |U_2| + |U_3| - \cdots \pm |U_n| \tag{6.41}$$

where U_m is the amplitude due to the mth Fresnel zone, and where the sign of the last term depends on whether n is odd or even. Suppose now that we block the alternate (even) zones, as in Figure 6.46; the resultant device is called a *Fresnel zone plate*. Since all the negative terms have been removed, the amplitude (and irradiance) now reaching A' will be greater than before. Therefore, the zone plate has acted as a lens by concentrating the radiation at A'. Equation (6.40) shows that the focal length of the zone plate is $f = r_1^2/\lambda$.

The zone plate is not a very efficient lens. More detailed analysis shows that there are several foci, similar to the diffracted orders from a grating (as well as the undiffracted order). Some improvement in diffraction efficiency is possible if we remove the sharp edges from the transmittance profile, by using a transparency with a transmittance function of the form $1 + C\cos^2(\pi r/r_1)^2$, where r the radial distance and $C < 1$. This function in fact represents the form of the interference fringes obtained when two spherical waves interfere; a (linear)

Figure 6.46. The zone plate.

recording of that interference pattern is a holographic lens (or mirror), and has only the 0 and ± 1 diffracted orders. So the focusing action of the HOE can be understood intuitively from the viewpoint of the Fresnel zone plate: the partial blocking of the alternate Fresnel zones leads to a concentration of energy at the image point.

Application: The kinoform. It is possible to achieve 100% theoretical diffraction efficiency if the zone plate is made to modulate not the amplitude but the phase of the incident wave. A kinoform is a blazed phase hologram, in which the phase modulation has the form of a surface relief profile. The profile of a kinoform lens can look identical (at least schematically) to the Fresnel lens of Figure 6.9. This similarity sometimes leads to the indiscriminate use of the term *Fresnel lens* for both devices. However, a bulk Fresnel lens is not the same as a kinoform. In the former, there is no requirement that the area of the zones be constant. Also, in the kinoform, the maximum surface relief step is limited to 1λ. Finally, refraction suffices to explain the essential idea of a Fresnel lens, whereas diffraction is needed to understand the kinoform.

In order to understand the kinoform operation, let us first introduce the idea of full Fresnel zones. A full Fresnel zone occupies two of the zones of the Fresnel zone plate shown earlier; in other words, the phase between successive full Fresnel zones changes by 2π, rather than π. In that case, Eq. (6.40) becomes

$$r_{m+1}^2 - r_m^2 = 2f\lambda$$

Therefore, the radius of the mth zone will be given by

$$r_m^2 = 2f\lambda + r_{m-1}^2 = 2f\lambda + 2(m-1)f\lambda$$

or

$$r_m^2 = 2mf\lambda \tag{6.42}$$

Notice that this solution is not exact, because of the approximation involved in Eq. (6.38). The assumption that the radius of the mth zone is small compared to the focal length f is equivalent to our usual paraxial approximation. If the zone radii satisfy Eq. (6.42), then the kinoform will have a focal length f. Each full zone is then given an appropriate surface relief profile, so as to maximize the diffraction efficiency. Within the paraxial approximation, the appropriate profile is quadratic within each zone. The blazing property of the quadratic profile can be understood by comparing the kinoform with a conventional refractive lens and its Fresnel counterpart, the profile of which is also quadratic (Figure 6.9). Thus the blazing principle (refraction and diffraction in the same direction) can be satisfied. The maximum step height $d_{\max}$ is set by requiring that the maximum optical path difference introduced by the kinoform be equal to λ. Therefore,

$$d_{\max} = \frac{\lambda}{n-1} \tag{6.43}$$

where n is the refractive index of the kinoform material. While the theoretical diffraction efficiency is 100%, the efficiency achieved in practice depends on the method of manufacturing the kinoform, which introduces imperfections.* Also, in the nonparaxial regime (e.g., F numbers less than about 10) the surface relief profile must differ from the simple quadratic in order to reduce the aberrations.† The advantages of the kinoform arise mostly from the fact that it is a thin element with a profile that can be specified as needed. Those are: large collecting aperture and extra degrees of freedom in correcting aberrations. Practical considerations are the ease of replication and light weight. The disadvantages arise from diffraction itself. They are: (1) diffraction efficiency that is not necessarily 100% in practice, and (2) most important, large dispersion (wavelength dependence): As can be seen from the preceding, the kinoform is a wavelength-specific device, although it may be combined with conventional optics to yield good performance over an extended wavelength range.

6.8.3. Binary Optics

A binary optical element is, in principle, any optical element that uses only discrete values in its important constructional parameters. One of the simplest binary optical elements is the Fresnel zone plate, described in the previous

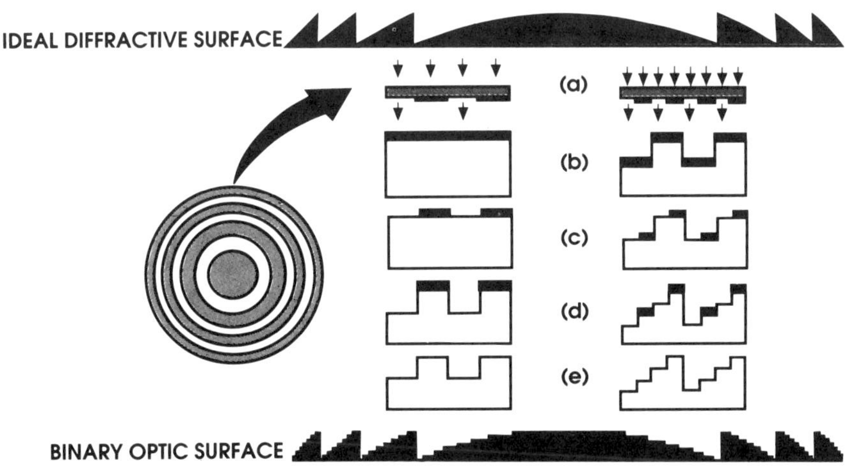

Figure 6.47. Steps in producing a binary optic: (a) Computer-generated microlithography mask is placed over substrate; (b) mask pattern is contact printed into photoresist; (c) exposed photoresist is removed; (d) ion-mill or reactive-ion etch to transfer pattern into substrate; (e) remaining photoresist is removed. Process is repeated to produce multistep pattern. (Reproduced from A. Kathman and E. Johnson, "Binary optics: New diffractive elements for the designer's tool kit," *The Photonics Design and Applications Handbook 1994*, Laurin Publishing Co., Pittsfield, MA).

* The theoretical prediction itself is based on scalar diffraction theory, which is not very accurate as the size of the Fresnel zones becomes small.

† See, for example, D. A. Buralli, G. M. Morris, and J. R. Rogers: "Optical performance of holographic kinoforms," *Applied Optics* **28**, 976–83 (1989).

section: It uses only the transmittance values of zero and one. More often, however, we are interested in binary phase elements, because of their potential for high diffraction efficiency. Figure 6.47 shows the steps involved in the construction of a binary kinoform, as well as the final form of the finished product. The microlithographic techniques that are used in the production of binary optics are substantially the same as those used in the production of integrated circuits (ICs). Therefore, the production of binary optics can draw on a mature technology, which is a major advantage.* The IC production technology affords the optical engineer the ability to create optical elements that perform a desired function by calculating the necessary phase profile and inputting it into the computer. Spherical, aspheric, or irregular shapes can all be produced with equal ease.

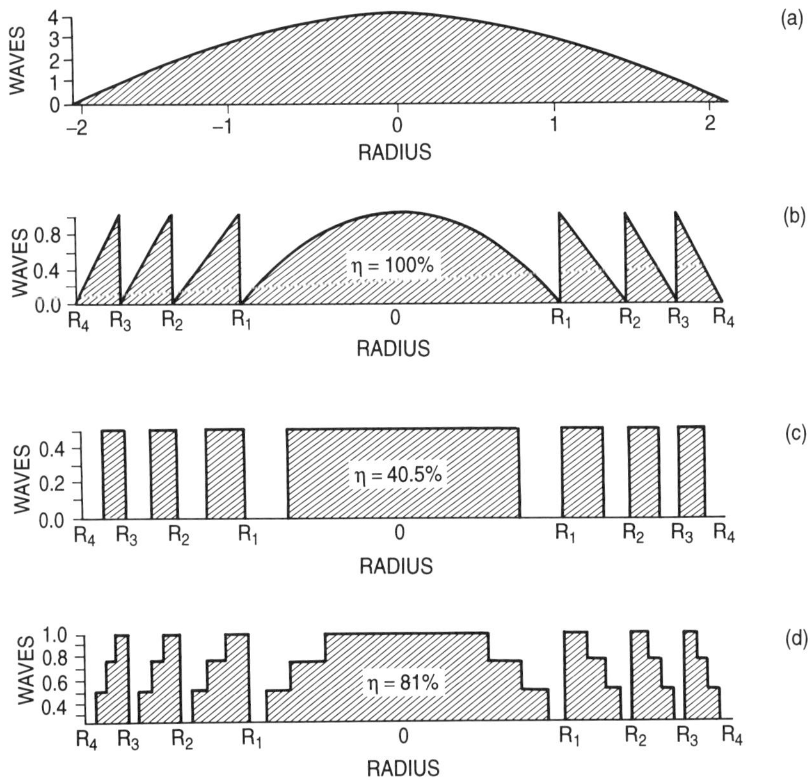

Figure 6.48. Diffractive lens construction: (a) conventional refractive lens, (b) diffractive lens with continuous quadratic blaze profile, (c) phase-reversal (or Wood) lens, (d) four-level approximation to quadratic blaze profile. (Reproduced from: D. Faklis and G. M. Morris, "Taking advantage of diffractive optics to reduce size, weight and costs," *The Photonics Design and Applications Handbook 1994*, Laurin Publishing Co., Pittsfield, MA).

* Diamond point turning has also been discussed as an alternative technique.

As can be seen from Figure 6.47, the binary kinoform attempts to approximate the continuous phase kinoform by having several phase levels. The more the levels, the closer the approximation; so the diffraction efficiency of the binary kinoform begins to approximate that of the continuous-phase kinoform. To understand this, consider the two limiting cases: For a large number of levels, the binary kinoform becomes indistinguishable from the continuous one, and therefore has a high diffraction efficiency. For the minimum number of levels (two), the binary kinoform becomes identical with the Fresnel zone plate, only using a phase instead of amplitude step; there are now several diffracted orders (foci) and the diffraction efficiency is low. This is illustrated in Figure 6.48.

PROBLEMS

1. A Galilean telescope has $F_1 = 6$ cm, $F_2 = -3$ cm and is focused at infinity. The eye pupil is located 1 cm behind the negative lens, and has a diameter of 3 mm. Determine the location and size of the entrance pupil.

2. In the previous problem, assume that the semifield of view is 15° and calculate the minimum aperture of the two lenses.

3. A Keplerian telescope has $F_1 = 8$ cm, $F_2 = 4$ cm, and is focused at infinity. Determine the location of the exit pupil. A relay lens ($F = 3$ cm) is now inserted, operating at symmetric conjugates ($M = -1$). What is the effect of this lens on the pupil location and magnification? Trace the PPR for this system. The field of view is $\pm 5°$, and the exit pupil diameter is 4 mm. Determine H, h_1, and $\bar{u}$.

4. The objective of a visual telescope has a focal length of 5 cm and a diameter of 2 cm. What is the maximum useful magnification? (Assume an eye resolution of 50 cycles/deg.)

5. In a remote sensing experiment, we wish to map the radiance of a distant target, with a resolution of 0.1° (field of view). The photodetector has a diameter of 3 mm. We have at our disposal the following lenses: $K_1 = 0.1\ \text{cm}^{-1}$, $K_2 = 1\ \text{cm}^{-1}$, $K_3 = 0.05\ \text{cm}^{-1}$, $K_4 = -0.5\ \text{cm}^{-1}$, with diameters $D_1 = 4$ cm, $D_2 = 1$ cm, $D_3 = 6$ cm, $D_4 = 1$ cm. Determine and sketch two different systems that maximize the amount of light collected from the 0.1° target spot size. Include any possible stops. An abundant supply of variable diameter iris diaphragms is available. Discuss any relative advantages or disadvantages between the two systems. (Hint: The photodetector can be placed either at the image or at the pupil.)

6. A microscope comprises two thin lenses. Assuming the entrance pupil to be at the objective, determine the location of the exit pupil and the pupil magnification. Assume a 10× eyepiece, a tube length of 160 mm, and an objective focal length of 30 mm.

7. Show that the exit pupil radius of a compound microscope is given by

$$h'_e = \frac{nu(25\ \text{cm})}{M}$$

where nu is the numerical aperture of the objective and M the magnifying power of the microscope (you may assume that the image is at infinity).

8. Determine the diameter of the exit pupil of a compound microscope that has an 8 mm objective of NA = 0.4, a 10× eyepiece, and a tube length of 160 mm.

9. What parameters affect the depth of focus (field) in a contour projector? Discuss the Gaussian design considerations affecting the design of such a projector where the desirable magnification is 10×, and the desirable depth of field is 0.5 cm.

10. A distant object subtends an angle of 2°. Find the size of its image on the retina (use Gullstrand's schematic eye).

11. The apparent brightness of a distant object as seen by the eye appears to be independent of the actual object distance from the eye. Explain.

12. The human eye suffers from longitudinal chromatic aberration, that is, chromatic variation of focus. (This can be easily observed: Try looking at a distant object through a blue filter; the object will appear blurred, because the eye is myopic in the blue.) The difference in power between the red and blue wavelengths is about 1.5 D. Determine the blur spot size on the retina caused by chromatic aberration, assuming that the red wavelength is in focus on the retina. Do this for two pupil sizes: 2 and 5 mm diameter.

13. Draw tunnel diagrams for the Dove prism and the Pechan prism.

14. Suppose that a Dove prism is badly mounted so it is tilted and/or displaced when rotating. What is the effect of this on the image? How can the prism be aligned in its holder?

15. You have two thin lenses of $F = 50$ mm, and an abundant supply of reflecting prisms of all types. Determine optical systems that can produce a real, congruent, and upright image for the following two cases: (a) $T = 400$ mm,

(b) $T = 220$ mm. The tolerance in T is ± 20 mm, and the magnification must be approximately one.

16. Determine the image of a point through a pair of crossed cylindrical lenses.

17. Design a pair of prisms to compress (or expand) the beam cross section by a factor of 2:1. Assume that the refractive index is 1.52 and that the light exists (or enters) normally one face of each prism.

18. A $0.23p$ GRIN rod is 5 mm long and has $n_0 = 1.5$. Determine the Gaussian constants of the rod.

19. A GRIN rod is required to form a real image of an object at $M = -1$. The throw T is 0.8 cm. The refractive index is given by $n = B(1 - Ax^2/2)$, where $B = 1.6$ and $A = 0.1156\ \text{mm}^{-2}$. Determine the length of the rod.

20. A diffraction grating has 1000 lines/mm. How wide must the grating be in order to resolve the longitudinal modes of a HeNe laser ($\lambda = 633$ nm)? The frequency difference between the longitudinal modes is 400 MHz. Do this in the first and second order. Conclusion?

21. A diffraction grating is used to resolve the sodium D lines (589 and 589.6 nm). How many rulings are needed? If the spectrum is formed at the focal plane of a $F = 25$ cm lens, and the total width of the grating is 1 cm, determine the linear separation of the two lines.

22. A transmission grating is to be ruled on glass ($n = 1.5$), with 1000 lines/mm. If the grating is to be blazed in the first order for a wavelength of 550 nm, specify the appropriate groove angle, assuming that the incident wave vector is along the grating normal. Repeat for a reflection grating.

23. Determine the radii of the first four zones of a kinoform that has a focal length of 1 mm at a wavelength of 633 nm. Also determine the maximum step height and draw an approximate diagram of the kinoform in cross section.

24. A hologram is recorded with two plane beams. A point source located 50 cm away from the hologram is used for reconstruction. Determine the image distance from the hologram and the magnification. Repeat if one plane and one spherical beam are used for recording, where the spherical beam is 25 cm away from the hologram (take $\lambda = \lambda'$).

25. A hologram is recorded with two point sources on either side of it, equidistant from the plate. Determine the location of the minima and maxima of irradiance on the plate, using zone-plate-based arguments.

Introduction to Aberrations

7

"... if [rays] flow from a lucid Point, so very remote from the Lens, that before their Incidence they may be accounted parallel, the Focus of the most refrangible Rays shall be nearer to the Lens than the Focus of the least refrangible, by about the 27th or 28th Part of their whole distance from it. And the Diameter of the Circle in the middle Space between those two Foci ... is about the 55th part of the Diameter of the Aperture of the Glass. So that 'tis a wonder, that Telescopes represent Objects so distinctly as they do."

Sir Isaac Newton, *Opticks*

We have already alluded to the fact that rays do not behave according to the paraxial optics predictions, except in the paraxial region. In this chapter, we begin to explore in some detail the reasons for and the consequences of that fact. We also begin to consider the limitations imposed by the properties of glass and other materials from which lenses are made.

7.1. CHROMATIC ABERRATION

In the Gaussian optics discussion of the previous chapters, we neglected the fact that optical materials, such as glass, exhibit dispersion, that is, a variation of the refractive index with wavelength. Since the power of a thin lens depends on the refractive index, we expect that different wavelengths will be focused at different points, as Newton observed in the opening quotation of this chapter. We seek to quantitfy this effect here.

7.1.1. Characterization of Dispersion

The variation of the refractive index of three typical glasses with λ is shown in Figure 7.1. The refractive index decreases with wavelength, except in the regions of the so-called *anomalous dispersion*, which are, however, also characterized by strong absorption. Anomalous dispersion need not concern us, as we are dealing with essentially transparent materials.

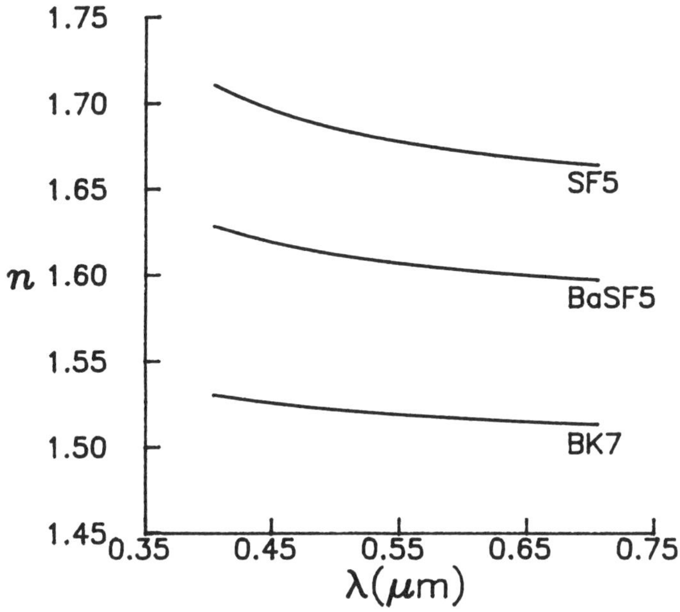

Figure 7.1. Dispersion of three typical glasses.

The refractive index of a material is typically quoted for a few selected wavelengths that correspond to spectral lines associated with specific elements; some of the most common ones are summarized in Table 7.1. The single-letter spectroscopic designation of the line is used as subscript. Thus n_d is the refractive index at the yellow He line, etc.

The term *dispersion* applies not only to the general effect $n = n(\lambda)$, but also to the refractive index difference between two lines [e.g., $n_F - n_C$].

We characterize the effect of dispersion by considering the dispersive properties of a thin prism. Referring to Figure 7.2, recall the thin prism ray deviation formula [Eq. (2.3)]), $\theta = (n - 1)\phi$. Let us choose three commonly used wavelengths, F, d, and C. We then have

$$\theta_F = (n_F - 1)\phi, \qquad \theta_d = (n_d - 1)\phi, \qquad \theta_C = (n_C - 1)\phi$$

The dispersive power of a prism is then given by

$$\frac{\delta\theta}{\theta} = \frac{\theta_F - \theta_C}{\theta_d} = \frac{n_F - n_C}{n_d - 1} \tag{7.1}$$

Notice that the dispersive power does not depend on the prism angle; so it is a property of the glass itself, rather than the prism. The reciprocal of the above ratio is used to characterize the dispersion of an optical

TABLE 7.1.
Typical wavelengths used in the description of dispersion.

Approx. λ (nm)	Designation	Spectral line
656	C	red hydrogen
644	C'	red cadmium
588	d	yellow helium
546	e	green mercury
486	F	blue hydrogen
480	F'	blue cadmium
436	g	blue mercury

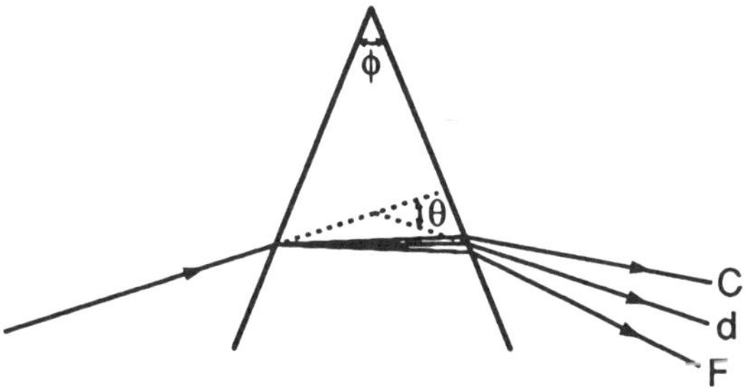

Figure 7.2. The dispersive properties of a thin prism.

material. It is given various names: *constringence*, *Abbe value*, *V number*, or *V value*.

$$V_d = \frac{n_d - 1}{n_F - n_C} \tag{7.2}$$

The three wavelengths used in this definition are the most common for applications in the visible spectrum. A different, although conceptually identical, definition that is sometimes used is

$$V_e = \frac{n_e - 1}{n_{F'} - n_{C'}} \tag{7.3}$$

In general the definition of the V number can be extended to any three wavelengths of interest, with the middle wavelength appearing in the numerator. When the subscript d or e is omitted, $V = V_d$ is normally implied. Typical V numbers are from about 25 to 30, for a highly dispersive glass to about 65 for a less dispersive one. Arbitrary combinations of refractive index and dispersion are not possible. Glass manufacturers provide plots of the available glass types, as an n_d versus V_d diagram, such as shown in Figure 7.3. A shorthand notation, indicated in the following example, is frequently used: glass 517642 has $n_d = 1.517$ and $V_d = 64.2$.* There are different trade names for the various types of glasses: The two most common terms are *crown* glass ($V > 50$ or so) and *flint* glass ($V < 50$).

7.1.2. Chromatic Effects for a Thin Lens

The chromatic variation of power for a thin lens can be found from the thin-lens equation (3.16). For the F wavelength, we have $K_F = (n_F - 1)(c_1 - c_2)$; similar expressions can also be written for the C and d wavelengths. The reader can then easily verify that the chromatic variation of power for a thin lens is given by

$$\frac{K_F - K_C}{K_d} = \frac{n_F - n_C}{n_d - 1} \tag{7.4}$$

or

$$K_F - K_C = \frac{K}{V} \tag{7.5}$$

where K and V are implicitly K_d and V_d. The chromatic variation of power manifests as two effects: chromatic variation of image distance (or focus), and chromatic variation of image height (or magnification). The following alternative names are normally used. The chromatic variation of focus is called *longitudinal chromatic aberration, longitudinal color*, or *axial color*. The chromatic variation of image height is called *transverse* or *lateral chromatic aberration* or *color*.

If l'_F and l'_C are the image distances for the F and C wavelengths, the thin-lens equation is $1/l'_F - 1/l = K_F$, and similarly for K_C. Therefore,

$$\frac{1}{l'_F} - \frac{1}{l'_C} = K_F - K_C$$

or

$$\boxed{\frac{1}{l'_F} - \frac{1}{l'_C} = \frac{K}{V}} \tag{7.6}$$

from Eq. (7.5). This last equation expresses the chromatic variation of image distance. To determine the chromatic variation of image height, let η'_F and η'_C be the image heights for F and C wavelengths, respectively. The magnification formula then gives

$$\eta'_F - \eta'_C = \frac{\eta}{l}(l'_F - l'_C) \tag{7.7}$$

---

* Plastic materials are becoming increasingly popular for less critical applications, due to improvements in manufacturing and moulding techniques. Two typical plastics are acrylic (methyl methacrylate) with $n_d = 1.491$, $V_d = 57.2$, and polystyrene, with $n_d = 1.590$, $V_d = 30.8$.

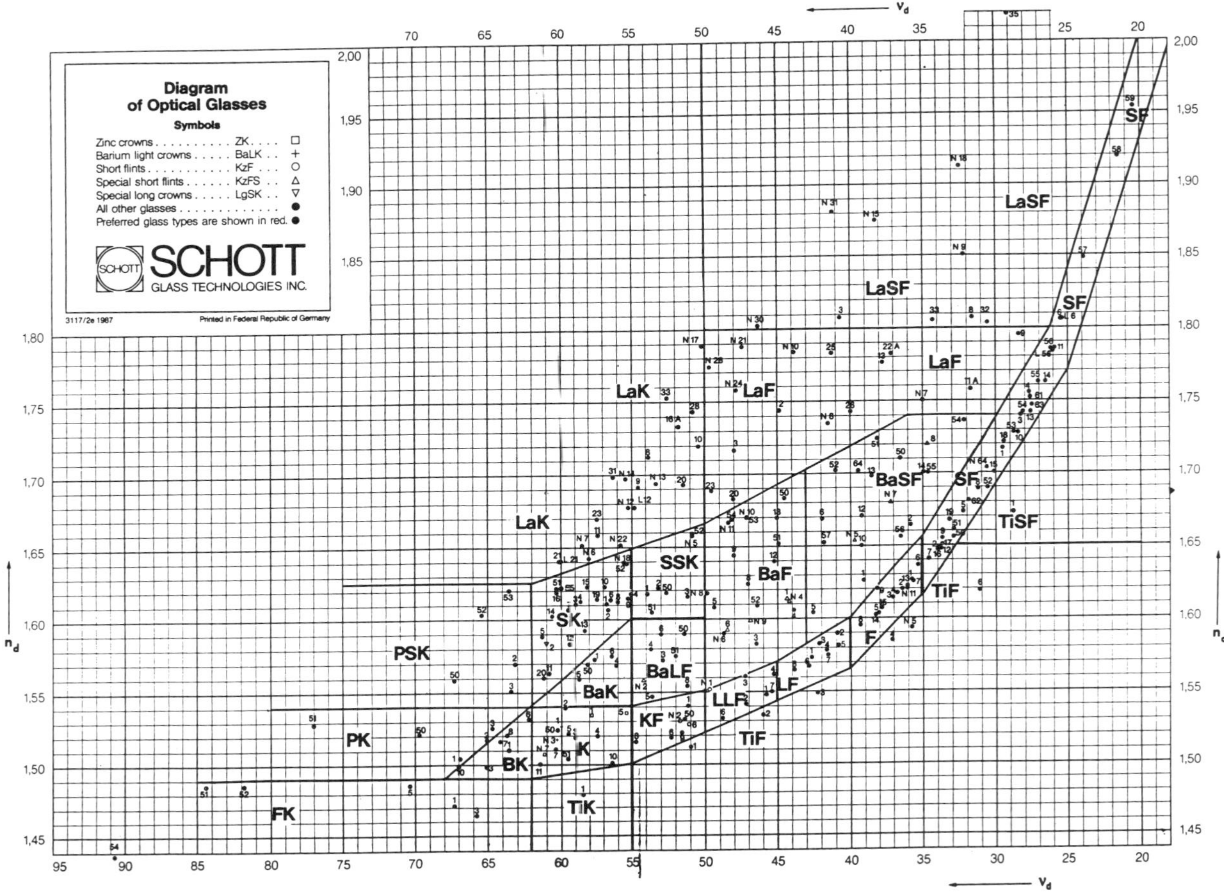

Figure 7.3. The glass chart. Each point on the chart represents a particular type of glass. The designation *K* implies a crown glass, *F* a flint. The other letters in the designation indicate various dopants that determine the refractive properties of the glass. (Courtesy of Schott Glass, Inc.)

Equation (7.6) gives

$$l'_F - l'_C = -\frac{K}{V}\, l'_F l'_C \cong -\frac{l'^2 K}{V}.$$

Substituting into Eq. (7.7) and using $l'/l = \eta/\eta$, we have

$$\eta'_F - \eta'_C = -\eta' l' \frac{K}{V} \tag{7.8}$$

Thus we see that both types of chromatic aberration for a thin lens are determined by the ratio K/V. We note here that Eq. (7.8) is not entirely correct in the sense that it does not express the transverse chromatic aberration alone, but contains a contribution from the longitudinal aberration as well. We show that the correct expression is

$$\boxed{\eta'_F - \eta'_C = -\bar{h} l' \frac{K}{V}} \tag{7.9}$$

where $\bar{h}$ is the PPR height at the lens in our usual notation.

To investigate Eq. (7.8) further, consider a thin lens with a remote stop, as in Figure 7.4. The transfer equation gives

$$\eta' = h + l'\bar{u}'$$

where the subscript d is implied. Substitution into Eq. (7.8) gives

$$\eta'_F - \eta'_C = -l'\bar{h}\frac{K}{V} - l'^2 \bar{u}\frac{K}{V} \tag{7.10}$$

Of the two terms that appear on the right-hand side of Eq. (7.10), only the first one represents transverse chromatic aberration. The second term is due to the existence of longitudinal chromatic aberration, as a result of which the heights η'_F and η'_C that appear in Eqs. (7.7) and (7.10) are measured at different planes. Specifically, the second term describes the difference between the segments (CE) and (FB), which are the heights at which the d ray cuts the F and C focal planes. However, if as is normally the case, the image is received on a single plane, such as our reference plane η_d, then we are only interested in the difference between the segments (DK) and (DH); these

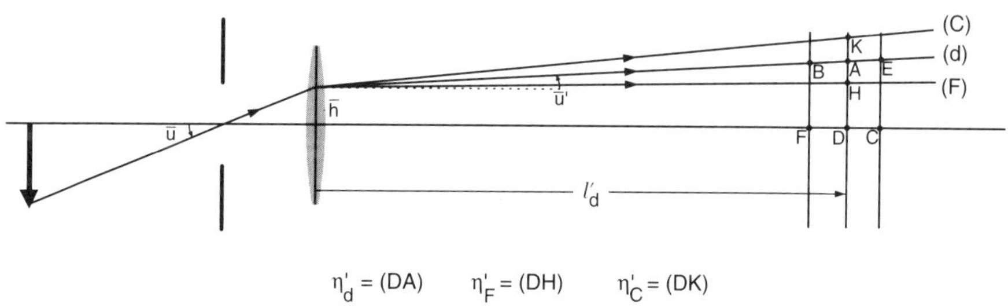

Figure 7.4. Transverse chromatic aberration as a ray effect.

are the intersections of the F and C rays with the d image plane. The following relations now hold:

$$\bar{u}'_F - \bar{u} = -\bar{h}K_F, \qquad \bar{u}'_C - \bar{u} = -\bar{h}K_C$$

so

$$\bar{u}'_F - \bar{u}'_C = -\bar{h}(K_F - K_C) = -\bar{h}\frac{K}{V}$$

Also

$$\bar{u}'_F l'_d = (DH) - \bar{h}, \qquad \bar{u}'_C l'_d = (DK) - \bar{h}$$

Therefore

$$(DH) - (DK) = (\bar{u}'_F - \bar{u}'_C)l'_d$$

Since $(DH) \equiv \eta'_F$, $(DK) \equiv \eta'_C$, and $l'_d \equiv l'$. Eq. (7.9) follows. In that equation, η'_F and η'_C refer to a single image plane.

Notice also the important result, that for $\bar{h} = 0$ (stop at the lens), there is no transverse color. In that case, the pupil ray passes through the middle of the lens and is undeviated; this condition is independent of wavelength. Therefore, all the F, C, and d foci must lie on the same line. At the d image plane, for example, the d rays are sharply focused, while the F and C are not. But the centroids of the ray distributions for all three wavelengths are coincident; hence there is no transverse aberration. It is left up to the reader to verify that under these conditions the second term of Eq. (7.10) describes the difference in image height that is purely due to the longitudinal difference in image location, $l'_F - l'_C$.

7.1.3. The Achromatic Doublet and Related Concepts

Since the chromatic effects for a lens are proportional to K/V, it follows that there will be no chromatic aberration of any kind, if the ratio K/V can be made equal to zero. Of course, this is impossible for a single thin lens; but for two thin lenses in contact, we can make the sum of the two ratios equal to zero. Specifically, we demand that

$$\frac{K_1}{V_1} + \frac{K_2}{V_2} = 0$$

$$K_1 + K_2 = K$$

Solving these two equations simultaneously, we obtain

$$K_1 = \frac{V_1 K}{V_1 - V_2} \qquad \text{and} \qquad K_2 = -\frac{V_2 K}{V_1 - V_2} \tag{7.11}$$

As an example, take $V_1 = 60$ and $V_2 = 30$. Equations (7.11) then give $K_1 = 2K$ and $K_2 = -K$. It is advantageous to have the V values of the two elements differ significantly, otherwise the individual elements must have large powers. For example, for $V_1 = 40$ and $V_2 = 30$, we obtain $K_1 = 4K$ and $K_2 = -3K$ Such powerful elements may be difficult to manufacture if they require extreme curvatures, or may cause substantial monochromatic aberrations, as explained later.

A contacted achromatic doublet is therefore relatively straightforward. But a separated doublet is harder. We will later prove that an achromatic system of two separated components must have each component separately achromatized. But there is one case of a separated doublet that provides good, if not perfect, chromatic correction, without requiring separately achromatic components. That is the *Huygens eyepiece.* It comprises two singlet lenses of the same V value and different powers, whose separation is equal to the mean focal length. Under this condition, we shall prove that the power of the combination is independent of wavelength. We have

$$K = K_1 + K_2 - dK_1K_2 \rightarrow \delta K = \delta K_1 + \delta K_2 - dK_1\,\delta K_2 - dK_2\,\delta K_1 \tag{7.12}$$

by differentiating. The differential power refers to the two wavelengths F and C. Also, from Eq. (7.5), $\delta K_1 = K_1/V$ and $\delta K_2 = K_2/V$. By substituting into Eq. (7.12) and setting $\delta K = 0$, we find

$$d = \frac{1/K_1 + 1/K_2}{2} \tag{7.13}$$

as the desired condition. If the equivalent power is independent of wavelength, the chromatic variation of image height is essentially eliminated. However, the chromatic variation of focus remains: This is because the location of the principal point P' may still vary with λ.

Sometimes it is desirable to have a lens with a V value that cannot be obtained with ordinary glasses. It may be possible to achieve this by designing a contacted doublet with chromatic change in power

$$\delta K = \frac{K_1}{V_1} + \frac{K_2}{V_2}$$

Since $K = K_1 + K_2$ and $V = K/\delta K$, we have a lens with an equivalent V value of

$$V = \frac{K_1 + K_2}{K_1/V_1 + K_2/V_2} \tag{7.14}$$

7.1.4. Secondary Spectrum. Apochromatic Correction

The achromatic condition $K_{\lambda_1} - K_{\lambda_2} = 0$ does not guarantee complete freedom from chromatic effects; the two wavelengths λ_1 and λ_2 are focused at the same spot, but other wavelengths are not. This residual chromatic aberration is called a secondary spectrum.* A secondary spectrum can be essentially

* An interesting bit of history: Glass characteristics were not well understood in Newton's time. He believed that all glasses had the same dispersion, which would make an achromatic doublet impossible. It is for this reason that he developed reflective telescopes, which do not suffer from color effects. In fact, the authority of Newton's opinion delayed the introduction of the achromatic doublet for many years. It was only in the nineteenth century that achromatic doublets were built and used as telescope objectives; the largest one had a diameter of 1 m. However, as manufacturing techniques improved, secondary color became important and limited the performance of the objective. As a result, the use of lenses has been limited to amateur astronomical or terrestrial telescopes.

eliminated by requiring that the lens be achromatized for three wavelengths. Such a lens is called *apochromatic*.

To understand the principle of apochromatic correction, we introduce the *relative partial dispersion*, P_{λ_1, λ_2}, defined as

$$P_{\lambda_1, \lambda_2} = \frac{n_{\lambda_1} - n_{\lambda_2}}{n_F - n_C} \tag{7.15}$$

for example, $P_{Fe} = (n_F - n_e)/(n_F - n_C)$. From this definition it follows that

$$\frac{K}{V_{FC}} P_{Fe} = \frac{K}{V_{Fe}} \tag{7.16}$$

where V_{FC} is the usual V_d and $V_{Fe} = (n_d - 1)/(n_F - n_e)$. Since K/V determines both chromatic aberrations, Eq. (7.16) says that the aberration for the F and e wavelengths equals the aberration for the F and C wavelengths multiplied by the relative partial dispersion P_{Fe}. If the lens is to be achromatized for the F, C, and e wavelengths, we require

$$\sum\left(\frac{K}{V_{FC}}\right) = 0 \rightarrow \frac{K_1}{V_{FC_1}} + \frac{K_2}{V_{FC_2}} = 0 \tag{7.17}$$

$$\sum\left(\frac{K}{V_{Fe}}\right) = 0 \rightarrow \frac{K_1}{V_{FC_1}} P_{Fe_1} + \frac{K_2}{V_{FC_2}} P_{Fe_2} = 0 \tag{7.18}$$

Combining Eqs. (7.17) and (7.18), we have

$$\frac{K_1}{V_{FC_1}} (P_{Fe_1} - P_{Fe_2}) = 0 \tag{7.19}$$

or

$$P_{Fe_1} = P_{Fe_2} \tag{7.20}$$

as the necessary condition. The usual achromatic condition (7.11) gives $K_1 = V_1 K/(V_1 - V_2)$, which, when combined with Eq. (7.19), gives

$$\frac{K(P_1 - P_2)}{V_1 - V_2} = 0 \tag{7.21}$$

after simplifying the notation for P. Neither Eq. (7.20) nor Eq. (7.21) can be easily satisfied. The ratio $(P_1 - P_2)/(V_1 - V_2)$ is essentially constant for all but a few recently developed exotic and expensive glasses. Figure 7.5 is a graph of P versus V for several glasses; we have an essentially linear relationship. That is, with very few exceptions, any two glasses with substantially different V values also have different P values. Apochromatic lenses are therefore rare. In the past, the only way to achieve apochromatism was through the use of crystalline materials, especially fluorite. Apochromatic microscope objectives may still be made with fluorite elements. But crystalline materials are generally not desirable because they do not polish well and they are not chemically stable.

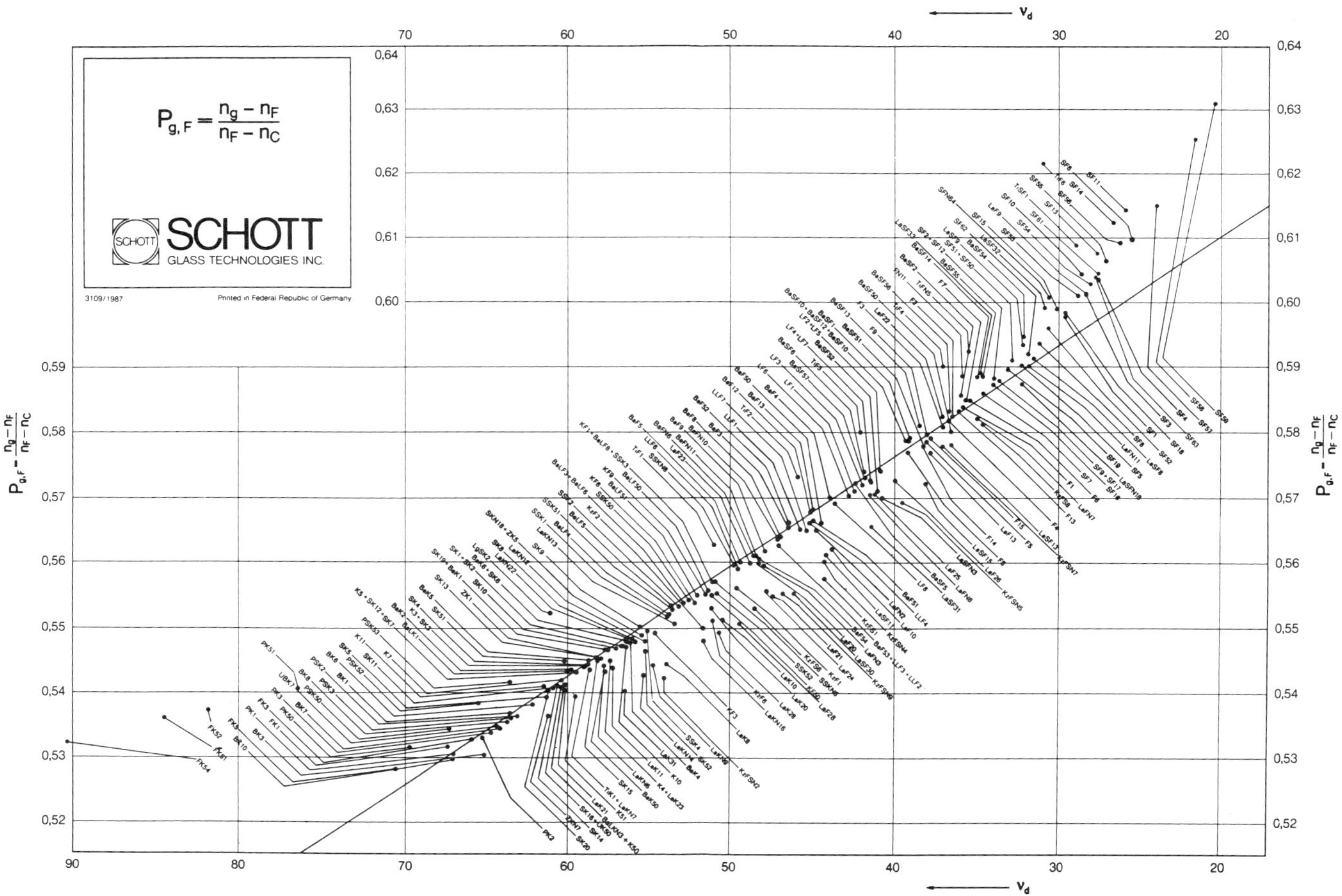

Figure 7.5. Relative partial dispersion versus V value for various glasses. (Courtesy of Schott Glass, Inc.)

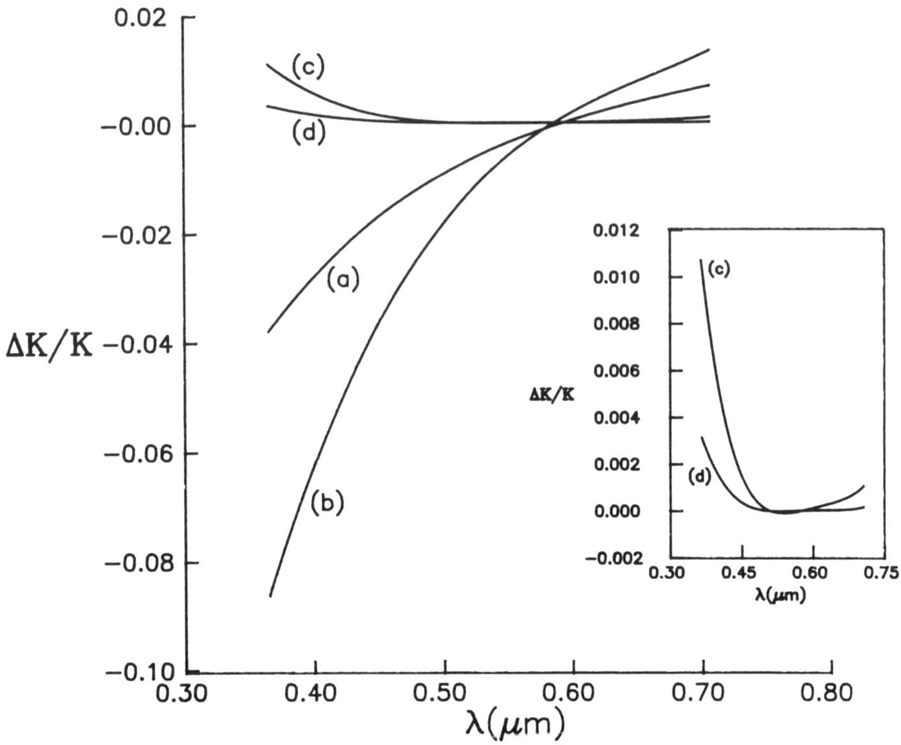

Figure 7.6. Chromatic variation of power for various lenses: (a) crown singlet, (b) flint singlet, (c) achromatic doublet, (d) apochromatic doublet. Inset: expanded view of (c) and (d).

Like achromatism, apochromatism only guarantees freedom from color at three specific wavelengths; so there is still a residual amount of aberration at other wavelengths. This residual, however, is practically zero, at least over the range of the visible spectrum. Figure 7.6 compares the chromatic variation of power between four lenses with the same nominal power: a crown singlet, a flint singlet, an achromatic doublet, and an apochromatic doublet.

7.2. INTRODUCTION TO MONOCHROMATIC ABERRATIONS

7.2.1. The Origin of Monochromatic Aberrations

Monochromatic aberrations are due to the fact that real rays do not refract according to the linear paraxial refraction law ($ni = n'i'$), but according to the nonlinear Snell law ($n \sin i = n' \sin i'$). To understand this statement, refer to the derivation of the single surface conjugate formula in Chapter 3. The usc of the paraxial approximation for all angles (i.e., the linearized approximate form of Snell's law) led to the result that the image distance l' was independent of the ray intersection height h. If, however, we use the real form of Snell's law, then l' becomes a function of h: Rays from the object point no longer meet at the same point in the image space. Thus as soon as the ray angle increases and we move away from the paraxial region (that is, as soon as the approximation $\sin\theta \cong \tan\theta \cong \theta$ is no longer adequate), we encounter monochromatic aberrations.

There are a few special cases—exceptions to the preceding rule, and it is worth listing them here. A single spherical surface can provide aberration-free imaging of an axial point, but this holds only for a specific location (the aplanatic point; see Chapter 8), apart, of course, from the trivial case when the object is at the center of curvature. A single surface can provide perfect imaging of an arbitrary axial point if the shape of the surface is allowed to depart from the sphere* (off-axis points will not be imaged perfectly). This holds for a single lens, if the object and image space refractive indices are the same. However, aspheric surfaces are generally difficult to make and expensive; most optical systems utilize spherical surfaces for that reason. Therefore, a lens system with low aberrations is not normally a collection of low-aberration surfaces; rather it contains surfaces with large aberrations that cancel out in the final image. This then is the art of optical design; how to provide a system with as few (preferably spherical) surfaces as possible that will satisfy the desired image quality specifications.

The basic tool through which we decide whether a system has the desired image quality is *finite ray tracing*. A finite ray is any ray that does not stay infinitesimally close to the axis. The only two exceptions to this definition are the two Gaussian rays, the PMR and the PPR, which are traced according to the laws of paraxial optics. However, the paths of the PMR and the PPR are fictitious. As explained in Chapter 5, they are close to the paths of real rays only in the object and image spaces, assuming that the aberrations are not large and that the correct convention is chosen ($u = \sin\alpha$, $\bar{u} = \tan\bar{\alpha}$). Also, there is no point in tracing more Gaussian rays: Gaussian optics assumes perfect image formation; so no additional information would be obtained. By contrast, each finite ray will, in general, cross the image plane at a different point; so many rays are needed to determine the ray intersection pattern at the image plane. An algorithm for finite ray tracing is given in Appendix 3. Even though such algorithms are incorporated in every optical design program, the student is strongly encouraged to write a computer program using the equations of Appendix 3. This exercise will prove useful in understanding the ray-tracing process.

7.2.2. Wavefront and Ray Aberrations

So far we have discussed aberrations as the failure of rays to meet at the same point. The ray aberration is a measure of this effect: It can be defined as

* See also the discussion in Section 6.3.

the distance between the paraxial image point and the corresponding finite ray point. More precisely, we distinguish two types: longitudinal and transverse ray aberration. The first is the distance from the paraxial image plane at which a finite ray from the axial object point intercepts the axis in the image space. The second is the distance between the extra-axial image point, as determined though paraxial ray tracing, and a corresponding finite ray intersection with the paraxial image plane.

We can have an equivalent description of aberrations through the wavefront shape instead of the rays. Spherical wavefronts give rise to stigmatic (perfect point) images, since the rays, being normal to the wavefront, will all meet at the center. The image may not be perfect either because the focal plane was not chosen to coincide with the center of curvature of the wavefront (e.g., an out-of-focus photograph), or because the wavefront shape is not spherical, so that the rays cannot meet at the same point.

A lens system that does not present aberrations is called *diffraction limited*, as the minimum resolvable spot size is then determined by diffraction. What plays the role of the diffracting aperture in a multielement lens system? One might expect that all of the different lens rims and stops inside the system would contribute to diffraction; so there is not a single diffracting aperture. However, we may recall from diffraction theory that it is the limiting aperture that matters; in the image space, that is the exit pupil.

In modern optical design, the wavefront aberration has emerged as a fundamental concept that unifies the geometrical and diffraction theories of aberrations. The wavefront (or wave) aberration is a measure of the departure of the wavefront shape from the perfect spherical form. It is defined as the difference in optical path length between the wavefront and the reference sphere, measured along the ray. The reference sphere is a sphere centered at the Gaussian image point that passes through the exit pupil point (Figure 7.7). By

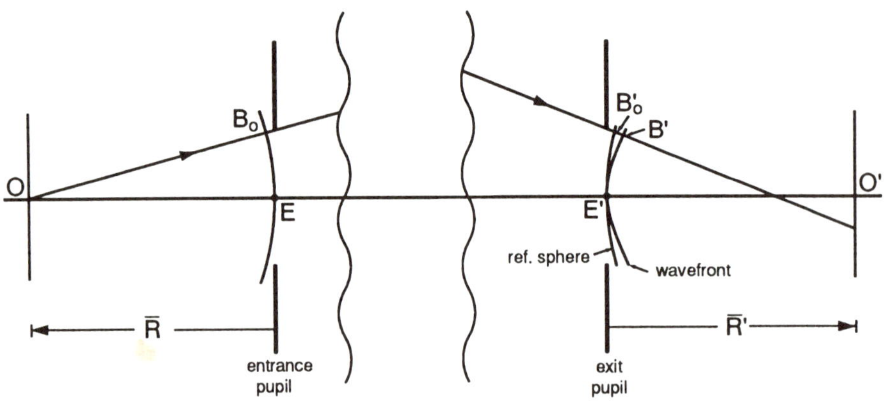

Figure 7.7. Wavefront aberration for a general system. The ray is normal to the wavefront at **B**′. The reference sphere is centered on the Gaussian image point O'. In the absence of aberrations, the ray would also cut the axis at O'.

choosing the reference sphere to pass through the exit pupil point, we can include the effects of diffraction in the calculation, as explained previously. In this book we are not concerned with the diffraction theory of aberrations, but the choice of reference sphere is universal and also convenient.

Referring to Figure 7.7, the optical path $[B_0'B']$ is the wavefront aberration W. It can be computed through finite ray tracing. For example, we have $[OB_0] = [OE]$ and $[OB'] = [OE']$, since the points B_0, E, and B', E, are on the same wavefront. Therefore,

$$[OB'] = [OB_0] + [B_0 \cdots B_0'] + [B_0'B'] = [OE] + [E \cdots E']$$

hence

$$W = [B_0'B'] = [E \cdots E'] - [B_0 \cdots B_0'] \tag{7.22}$$

Thus, in principle, we can ray trace along the two rays shown, and subtract the two optical paths to obtain W. However, this is not an accurate method, since W is of the order of a wavelength, while $[E \cdots E']$ and $[B_0 \cdots B_0']$ can be meters. A better numerical technique that does not involve subtraction of large quantities is given by Welford (1986).

Since the rays are by definition normal to the wavefronts, wavefront and ray aberrations must be related; it is shown in the fine print section below that, constant factors aside, the ray aberration is simply the derivative of the wavefront aberration with respect to the aperture (or exit pupil) coordinates. The exact relation is

$$\delta\xi' = -\frac{\bar{R}'}{n'h_e'}\frac{\partial W}{\partial x'}, \qquad \delta\eta' = -\frac{\bar{R}'}{n'h_e'}\frac{\partial W}{\partial y'}$$

where $\delta\xi'$ and $\delta\eta'$ are the ray aberrations along the ξ and η axes, respectively, h_e' the exit pupil height, and $\bar{R}$ the radius of the reference sphere. The pupil coordinates x' and y' in that expression are normalized with respect to h_e'. The proof is given here.

Referring to Figure 7.8, let the reference sphere, centered at $\bar{Q}'$, pass through E', the exit pupil point. The ray $B_0'B'$ is normal to the wavefront at B', while $\hat{B}_0'\bar{Q}'$ is normal to the reference sphere at B_0'. For small aberration, the angle δu between $B_0'B'$ and $\hat{B}_0'\bar{Q}'$ is small; so we can set

$$W = n'(B_0'B') \cong n'(B_0'B')\cos\delta u = n'(\hat{B}_0'B') \rightarrow (\hat{B}_0'B') = \frac{W}{n'}$$

Also

$$(B'\bar{Q}') = (\hat{B}_0'\bar{Q}') - (\hat{B}_0'B') = \bar{R}' - \frac{W}{n'} \tag{7.23}$$

$$(B'\bar{Q}')^2 = X'^2 + (Y' - \bar{\eta}')^2 + (Z' - R')^2 \tag{7.24}$$

From Eqs. (7.23) and (7.24) we obtain

$$-\left(\bar{R}' - \frac{W}{n'}\right)^2 + X'^2 + (Y' - \bar{\eta}')^2 + (Z' - R')^2 = 0$$

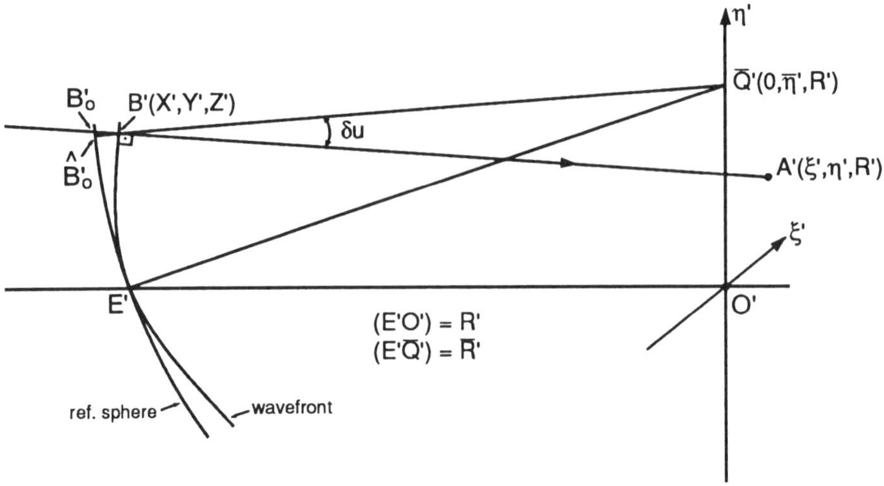

Figure 7.8. Relation between ray and wavefront aberration. The amount of aberration is greatly exaggerated. $\bar{Q}'$ is the Gaussian image point.

The term W^2/n^2 is much smaller than the other terms; so we can ignore it and write

$$F(X',\ Y',\ Z') \equiv X'^2 + (Y' - \bar{\eta})^2 + (Z' - R')^2 - \bar{R}'^2 + \frac{2\bar{R}'W}{n'} = 0 \qquad (7.25)$$

This equation specifies the shape of the geometrical wavefront. The ray is normal to the wavefront; so if (ξ', η', ζ') are the coordinates of a point on the ray, the equations of the normal are

$$\frac{\xi' - X'}{F'_x} = \frac{\eta' - Y'}{F'_y} = \frac{\zeta - Z'}{F'_z} \qquad (7.26)$$

where F'_x, F'_y, F'_z are the partial derivatives of F with respect to X', Y', Z', respectively. From Eq. (7.25) then we have:

$$F'_{x'} = 2X' + \left(\frac{2\bar{R}'}{n'}\right)\frac{\partial W}{\partial X'}$$

$$F'_{y'} = 2(Y' - \bar{\eta}') + \left(\frac{2\bar{R}'}{n'}\right)\frac{\partial W}{\partial Y'}$$

$$F'_{z'} = 2(Z' - R') + \left(\frac{2\bar{R}'}{n'}\right)\frac{\partial W}{\partial Z'} = 2(Z' - R')$$

The last equality follows because W need only be a function of $X'$and Y' to be fully defined; thus we can set $\partial W/\partial Z' = 0$. Substituting the partial derivatives into Eq. (7.26), we have

$$(\xi' - X')\Big/\left(X' + \frac{\bar{R}'}{n'}\frac{\partial W}{\partial X'}\right) = (\eta' - Y')\Big/\left(Y' - \bar{\eta}' + \frac{\bar{R}'}{n'}\frac{\partial W}{\partial Y'}\right) = \frac{\zeta - Z'}{Z' - R'} \qquad (7.27)$$

The ray crosses the image plane at $\zeta = R'$; so the last ratio in Eq. (7.27) equals -1. Thus we have

$$\xi' = -\left(\frac{\bar{R}'}{n'}\right)\frac{\partial W}{\partial X'}, \qquad \eta' = \bar{\eta}' - \left(\frac{\bar{R}'}{n'}\right)\frac{\partial W}{\partial Y'}$$

Now (ξ', η') are the coordinates of the ray intersection with the image plane. Since the Gaussian image point is at $(0, \bar{\eta}', R')$, the x and y components of the transverse ray aberration are given by

$$\begin{aligned} \delta\xi' &= \xi' - 0 = -\left(\frac{\bar{R}'}{n'}\right)\frac{\partial W}{\partial X'} \\ \delta\eta' &= \eta' - \bar{\eta}1 = -\left(\frac{\bar{R}'}{n'}\right)\frac{\partial W}{\partial Y'} \end{aligned} \tag{7.28}$$

which express the desired relation between ray and wavefront aberration. Thus apart from a proportionality factor, the transverse ray aberration is the derivative of the wavefront aberration with respect to the aperture (exit pupil) coordinates of the ray.

7.2.3. Canonical Coordinates

The use of Hopkins' canonical coordinates* allows us to remove the explicit dependence of the ray aberration on $\bar{R}'$. Let us first define the fractional coordinates

$$x' = \frac{X'}{h'_e}, \qquad y' = \frac{Y'}{h'_e} \tag{7.29}$$

where h'_e is the Gaussian exit pupil height. Equations (7.28) then become

$$\delta\xi' = -\left(\frac{\bar{R}'}{n'h'_e}\right)\frac{\partial W}{\partial x'}, \qquad \delta\eta' = -\left(\frac{\bar{R}'}{n'h'_e}\right)\frac{\partial W}{\partial y'} \tag{7.30}$$

We then define reduced image coordinates as the regular coordinates multiplied by the numerical aperture, $n' \sin \alpha' = n'u'$, where u' is the PMR angle, Thus

$$\delta G' \equiv n'u'\ \delta\xi', \qquad \delta H' \equiv n'u'\ \delta\eta' \tag{7.31}$$

For axial conjugates, $u' = -h'_e/\bar{R}'$, so $\delta G' = -n'h'_e\,\delta\xi'/\bar{R}'$, and similarly for $\delta H'$. Thus from Eq. (7.31) we obtain

$$\delta G' = \frac{\partial W}{\partial x'}, \qquad \delta H' = \frac{\partial W}{\partial y'} \tag{7.32}$$

which is the desired formula in normalized form. The set of fractional pupil coordinates and reduced image coordinates is called *canonical coordinates.*

For the general extra-axial image point, a different normalization scheme applies, which is discussed briefly in Section 10.5 (the subject is developed properly in the Hopkins article mentioned earlier). Recall that $u' = \sin \alpha'$ is the

* See H. H. Hopkins, "Canonical and real-space coordinates used in the theory of image formation," in *Applied Optics and Optical Engineering*, Vol. IX, R. Shannon and J. Wyant, Editors, Academic Press, 1983.

sine condition (Chap. 5), which must be satisfied reasonably well for an aberration-corrected system. Thus an additional assumption for the use of canonical coordinates is that the system satisfies the sine condition.

The usefulness of canonical coordinates cannot be fully appreciated here, as they also enter into the diffraction theory of image formation. It should be obvious, however, that they allow the computation to be performed with simpler expressions, it being necessary to return to regular coordinates only at the end of the computation. We also note that by extending the definitions in the object space, a ray that crosses the object and/or entrance pupil at a point with (canonical) coordinates (x,y) will cross the image and/or exit pupil at $x' = x$ and $y' = y$, respectively (assuming a certain degree of aberration correction). Thus canonical coordinates serve to extend the linear rules of paraxial optics to systems of arbitrary field and aperture.

7.3. THE WAVE ABERRATION FUNCTION. CLASSIFICATION OF ABERRATIONS

Our treatment of aberrations revolves around the concept of the wave aberration.* This is somewhat contrary to the traditional way, which relies far more heavily on the ray aberration, and which is still practiced by several optical designers. Our choice is dictated by the fact that the wave aberration is a more fundamental concept than the ray aberration. It not only unifies the geometrical and diffraction theories of image formation, but it also shows that the different aberration types are closely related as terms of the wavefront aberration polynomial.† This fundamental physical insight has acquired additional importance with the proliferation of interferometric techniques and instruments for the testing of lens systems, which allow direct visualization of the wavefront aberration. Also, the wave aberration does not depend on whether the conjugates are finite or at infinity. Finally, whereas it is possible to give general aberration tolerances based on the wave aberration, it is not possible to do so through the ray aberration.

7.3.1. The Wave Aberration Polynomial for Rotationally Symmetric Systems

As we saw in the previous section, the wavefront aberration can be computed through finite ray tracing. But finite ray tracing can only provide a finite number

* See H. H. Hopkins, *The Wave Theory of Aberrations*, Clarendon Press, Oxford, 1950.

† To quote H. H. Hopkins, "...While engaged upon practical lens designing I rapidly came to the conclusion that the former [wave] approach was very much to be preferred. It leads to a simpler picture of what aberrations mean—if the word may be so used. In consulting any ray diagram showing the combined effects of astigmatism and coma, for example, the confusion is striking. In contrast, a few lines suffice to show the shape of the aberrated wave-front with perfect clarity..."

of points on the wavefront (one for every ray traced). To find an analytical expression for the wavefront aberration, we need to perform a two-dimensional polynomial interpolation. As we shall see, this process provides a substantial amount of diagnostic information regarding the causes of image degradation in given system, and their possible cure.

The wavefront aberration is determined (at least conceptually) through the following steps: We first determine the wavefront shape, through finite ray tracing at several points on the wavefront. We then subtract a sphere from that shape (the reference sphere). This now gives the wavefront aberration. Interpolation of a two-dimensional polynomial provides an analytical expression. The wavefront aberration will depend on the pupil point; so we can write $W = W(x,y)$, where (x,y) are the fractional pupil coordinates. However, for an extended object, a different polynomial would be needed in general for every object point (for example, a system may have perfect axial imagery but present significant off-axis aberrations). Thus we must broaden the definition of W to include the object or image coordinates (ξ,η)

$$W = W(x,y,\xi,\eta) \tag{7.33}$$

The polynomial W can contain any combination of the four coordinates and their powers and products, constrained only by the symmetry properties of the system. We will consider here rotationally symmetric systems of smooth surfaces; therefore, only rotationally invariant combinations of the four coordinates can exist. These combinations are:

$$x^2 + y^2, \qquad \xi^2 + \eta^2, \qquad x\xi + y\eta$$

The reasoning for this last statement is provided in fine print at the end of this section. There is no loss of generality if we choose the axes so that $\xi = 0$; so in terms of polar coordinates, $x = r \sin\theta$, $y = r\cos\theta$, $\eta = \tau\cos\phi = \tau$, and $\xi = \tau \sin\phi = 0$. Thus the above three combinations become

$$r^2, \qquad \tau^2, \qquad r\tau\cos\theta$$

Notice that the term $r = (x^2 + y^2)^{1/2}$ is also rotationally invariant. It is excluded on the basis that it would represent a discontinuity of gradient at the origin (cusp), which would have no physical justification. Also the polynomial sought cannot contain a constant term, as such a term would merely represent the difference between two concentric spheres, that is, no aberration. The general term of the polynomial can then be written as

$$(r^2)^i(\tau^2)^j(r\tau\cos\theta)^k, \qquad i,j,k = 0,1,2,\ldots$$

or

$$r^{2i+k}\tau^{2j+k}(\cos\theta)^k = r^m\tau^l(\cos\theta)^n, \qquad \text{where } l = 2j+k,\ m = 2i+k,\ n = k \tag{7.34}$$

We adopt the notation ${}_lW_{mn}$ for the polynomial coefficients. This notation

TABLE 7.2.
Aberration types.

i	*j*	*k*	*l*	*m*	*n*	Term	Name
0	0	1	1	1	1	${}_1W_{11}\tau r \cos\theta$	transverse focal shift
1	0	0	0	2	0	${}_0W_{20}r^2$	longit. focal shift
0	0	2	2	2	2	${}_2W_{22}\tau^2 r^2 \cos^2\theta$	astigmatism
2	0	0	0	4	0	${}_0W_{40}r^4$	spherical aberration
1	0	1	1	3	1	${}_1W_{31}\tau r^3 \cos\theta$	coma
1	1	0	2	2	0	${}_2W_{20}\tau^2 r^2$	field curvature
0	1	1	3	1	1	${}_3W_{11}\tau^3 r \cos\theta$	distortion
						plus higher-order terms	

separates the exponent for the field terms (l) from those of the aperture terms (m, n). The general term of the polynomial is then written as

$$ {}_lW_{mn}\tau^l r^m(\cos\theta)^n \tag{7.35}$$

Note that all the terms ${}_lW_{00}$ do not represent aberrations and are therefore excluded. We then have the breakdown of terms as in Table 7.2.

The wavefront aberration W can then be written as a polynomial containing the following terms:

$$W = {}_1W_{11}\tau r\cos\theta + {}_0W_{20}r^2 + {}_0W_{40}r^4 + {}_1W_{31}\tau r^3\cos\theta + {}_2W_{22}\tau^2 r^2\cos^2\theta$$
$$+ {}_2W_{20}\tau^2 r^2 + {}_3W_{11}\tau^3 r\cos\theta + \text{higher-order terms} \tag{7.36}$$

Note that all the ${}_iW_{jk}$ terms shown have dimensions of length. The τ and r coordinates are normalized so as to be dimensionless.

We shall give a somewhat abbreviated argument for the form of W, based on the symmetry considerations. We have

$$W = W(x,y,\xi,\eta) = W(r,\theta,\tau,\phi)$$

where (x,y) are the pupil coordinates, (ξ,η) are the image coordinates, and (r,θ), (τ,ϕ) are polar versions of the same.

$$x = r\sin\theta, \qquad y = r\cos\theta, \qquad \xi = \tau\sin\phi, \qquad \eta = \tau\cos\phi \tag{7.37}$$

If the axes are rotated by an angle $-\beta$, then the new angles are

$$\theta^* = \theta + \beta \qquad \text{and} \qquad \phi^* = \phi + \beta$$

from which

$$\theta^* - \phi^* = \theta - \phi \tag{7.38}$$

A power series in x, y, ξ, η will involve sines and cosines of θ, ϕ. From Eq. (7.38),

$$\cos(\theta^* - \phi^*) = \cos(\theta - \phi) \tag{7.39}$$

$$\sin(\theta^* - \phi^*) = \sin(\theta - \phi) \tag{7.40}$$

and also

$$\sin^2 \theta^* + \cos^2 \theta^* = \sin^2 \theta + \cos^2 \theta \tag{7.41}$$

$$\sin^2 \phi^* + \cos^2 \phi^* = \sin^2 \phi + \cos^2 \phi \tag{7.42}$$

The last four equations can be seen as equivalent to Eq. (7.38). They also express the only possible combinations of sines and cosines of θ, ϕ, θ^*, and ϕ^* that satisfy Eq. (7.38). From (7.39), we have $\cos\theta \cos\phi + \sin\theta \sin\phi$ as the invariant combination. But $\sin\theta$ and $\cos\theta$ are associated with a factor r [from Eq. (7.37)], while $\sin\phi$ and $\cos\phi$ are associated with a factor τ. So the full invariant combination is

$$\tau r(\cos\theta \cos\phi + \sin\theta \sin\phi) = x\xi + y\eta \tag{7.43}$$

Similarly, Eq. (7.40) gives the following invariant combination

$$y\eta - x\xi \tag{7.44}$$

while Eqs. (7.41) and (7.42) give, respectively,

$$x^2 + y^2 = r^2 \qquad \text{and} \qquad \xi^2 + \eta^2 = \tau^2 \tag{7.45}$$

Thus the only rotationally invariant combinations are Eqs. (7.43)–(7.45); so W must be a function of those parameters and their powers. But there is no loss of generality if we assume $\xi = 0$; that is, the image point is along the η axis. In that case, we see that expressions (7.43) and (7.44) become identical, which implies that they are not independent. Therefore, we finally have

$$W = W(x^2 + y^2, \xi^2 + \eta^2, y\eta) = W(r^2, \tau^2, \tau r \cos\theta) \tag{7.46}$$

7.3.2. Classical Aberration Types

Now we look at each term in more detail. The first two terms represent a shift of focus, and are not aberrations in the sense that they do not imply departure from sphericity. They can be made zero by appropriate choice of focus, which corresponds to a shift of the center of the reference sphere. Specifically:

$_1W_{11}\tau r \cos\theta$ (transverse focal shift). This term implies that the image of an off-axis point is located at some point other than the Gaussian image point. It can be seen also as the difference between two spheres of the same radius but slightly displaced centers. This is explained in more detail in Appendix 4. As we shall see later, this term arises in conjunction with coma, since one of the effects of coma is to shift the optimum image point. Also the chromatic variation of this term is the transverse chromatic aberration, which we discussed earlier from the ray aberration viewpoint.

${}_0W_{20}r^2$ (longitudinal focal shift). This term represents the difference between two spheres of different radii; so it can be eliminated by changing the radius of the reference sphere, which is equivalent to choosing a different focal plane (see also Appendix 4). However, it is often very convenient to treat defocus as an aberration. It also interacts with astigmatism and spherical aberration (see below); that is, when the system suffers from either of these two aberrations, then the optimum image is not found at the Gaussian image plane but at some other plane for which ${}_0W_{20} \neq 0$. The chromatic variation of this term is the longitudinal chromatic aberration, discussed earlier as a ray aberration.

The next three terms astigmatism, spherical aberration, and coma) are called *point-imaging aberrations*, because they degrade the image of a point. This is to distinguish them from the last two (field curvature and distortion), which are called *field aberrations*. In the latter case, the image of a point is still a point, but the image of an extended object suffers from some deterioration. Specifically:

${}_0W_{40}r^4$ (spherical aberration). This is the only aberration that exists on axis; that is, $\tau = 0$ (including higher orders). As can be seen by the r^4 dependence, the departure of the wavefront from sphericity increases very rapidly toward the edges of the aperture. If a system suffers from spherical aberration only, then the image of a point is rotationally symmetric. The wavefront shape that corresponds to spherical aberration is shown in Figure 7.9, where for simplicity

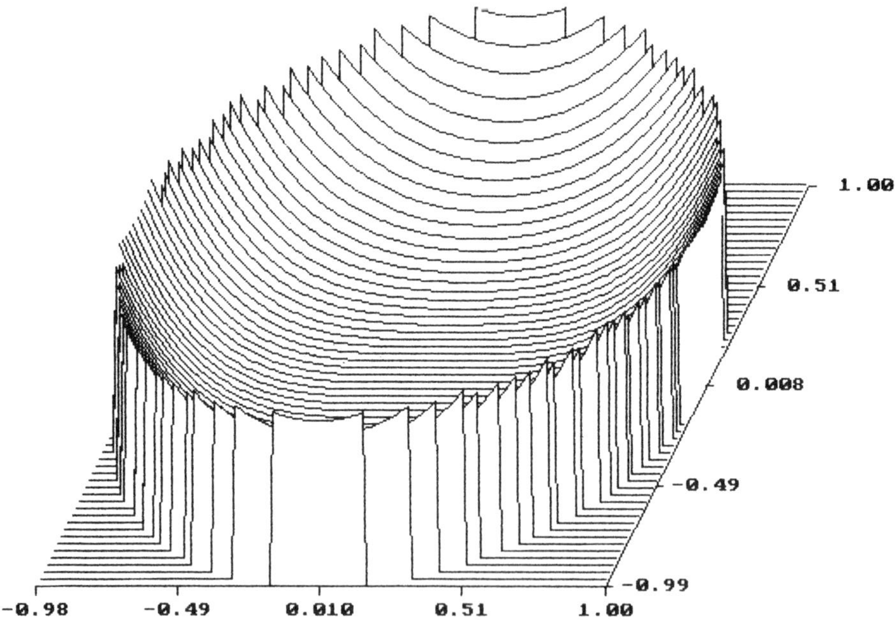

Figure 7.9. Wavefront shape for spherical aberration in "star space" (i.e., with the Gaussian image point at infinity, or flat reference sphere). Wavefront shapes and spot diagrams were generated using the fringe analysis program WISP by Wyko Corp.

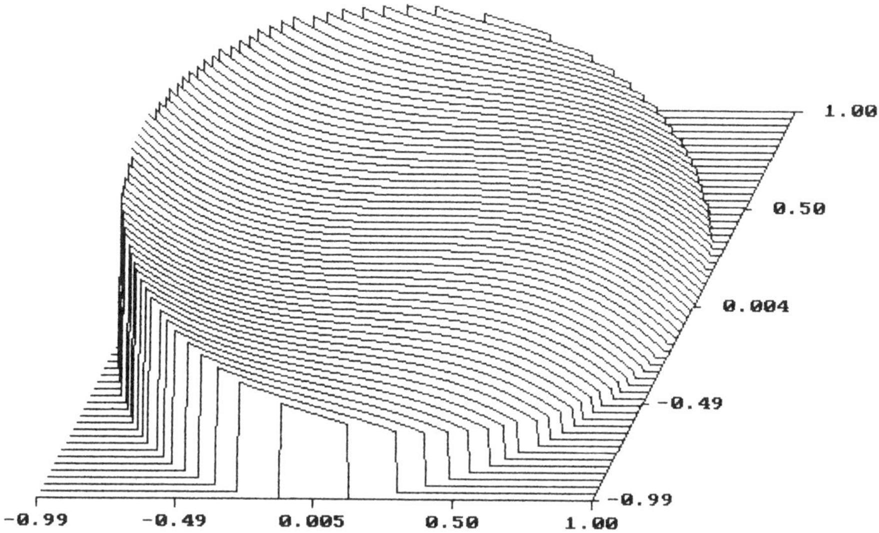

Figure 7.10. Wavefront shape for coma in star space.

the reference sphere is taken with an infinite radius. It can be seen, at least in a qualitative sense, that choosing a reference sphere with shorter radius would produce a better fit to the spherically aberrated wavefront: This is equivalent to saying that the optimum focal plane in the presence of spherical aberration is not the paraxial image plane.

$_1W_{31}\tau r^3 \cos\theta$ **(coma).** The image of a point degraded by coma has a cometlike appearance, hence the name (see also spot diagrams in the next section). It can be seen (Figure 7.10) that the comatic wavefront does not have an overall curvature; hence the best focus is at the paraxial image plane. However, a tilted reference wavefront would produce a better fit; so coma interacts with transverse focal shift.

$_2W_{22}\tau^2 r^2 \cos^2\theta$ **(astigmatism).** This is a cylindrical deformation of the wavefront, and results in two line foci, located at different focal planes. In between the two, the "circle of least confusion" is found. The wavefront shape is shown in Figure 7.11a, with no defocus term, and in Figure 7.11b with the amount of defocus ($_0W_{20}$) needed to bring the circle of least confusion into focus.

In plotting the wavefront shapes (Figures 7.9–7.11), we note that the field dependence is not in fact relevant. Of course, spherical aberration ($_0W_{40}$) is independent of the field. The other two aberrations do depend on the field, but the only thing that changes with τ is the amount of aberration; the basic shape of the wavefront remains the same. But for the two remaining aberrations below,

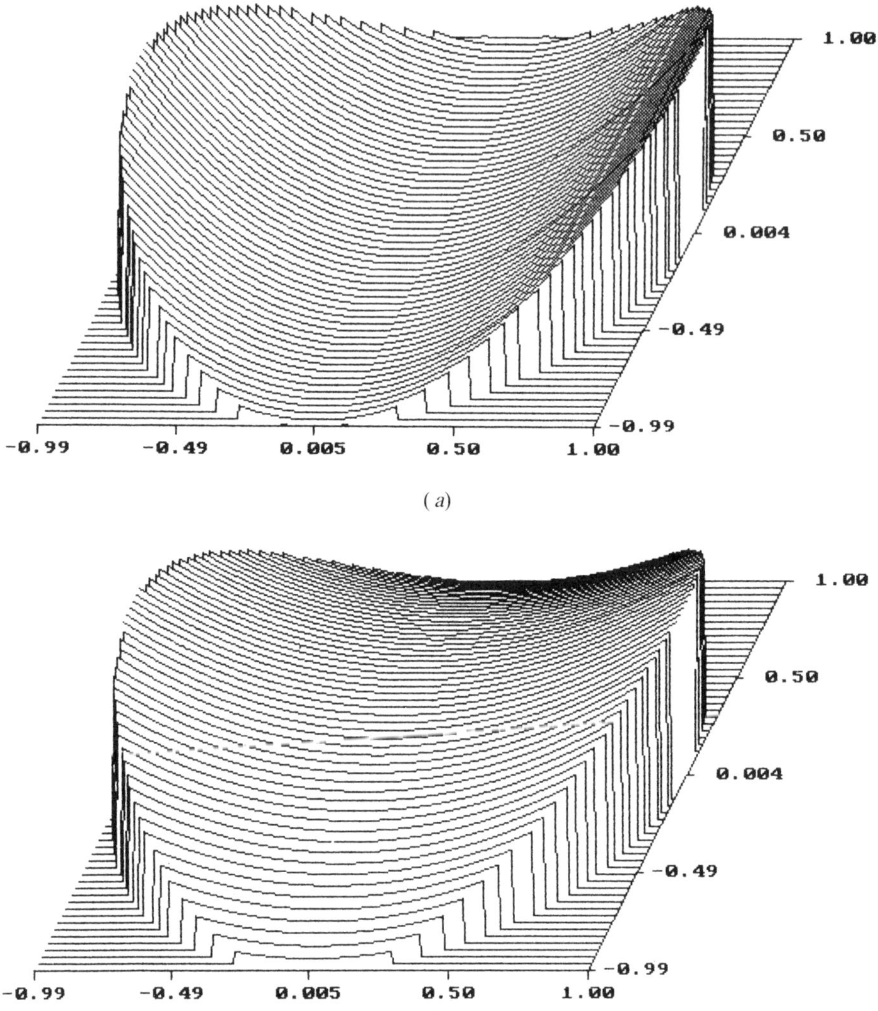

Figure 7.11. (a) Wavefront shape for astigmatism in star space. (b) Wavefront shape for astigmatism with defocus term added.

it is the field dependence that is of importance: The wavefront shape is spherical. Specifically:

$_2W_{20}\tau r^2$ **(field curvature).** This is a longitudinal defocus term, but one that depends on the field position. This means that it is not possible to have the entire image on a plane, but instead, the image is stigmatic along a curved surface. Thus if its required to have the image on a plane, we will observe blurring. This is shown in Figure 7.12.

Figure 7.12. Field curvature

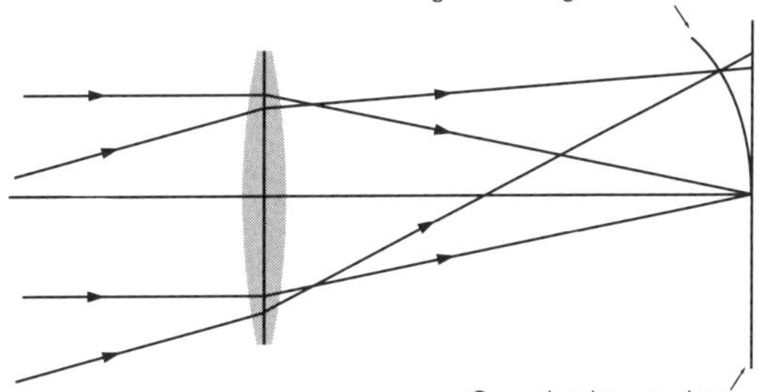

Figure 7.13. Distortion

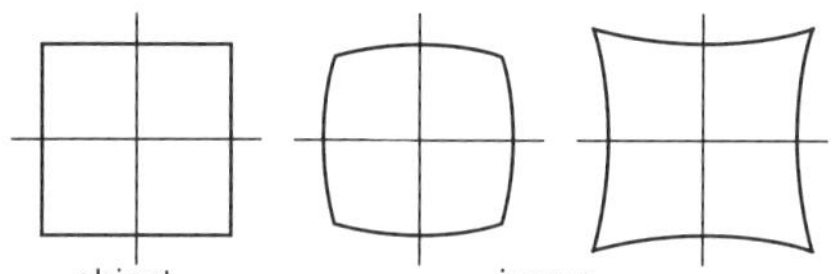

$_3W_{11}\tau^3 r \cos\theta$ **(distortion).** This is a transverse defocus term that depends on the field position. It is equivalent to a change in magnification along the image plane. The result is that the image of an extended object, although stigmatic, appears geometrically distorted. We distinguish between barrel and pincushion distortion, depending on the sign of $_3W_{11}$. If $_3W_{11} > 0$, then we have barrel distortion. The effect is shown in Figure 7.13. Where positional measurements must be taken from a photographic plate (e.g., in photogrammetry), this aberration must be eliminated or carefully considered.

The aberrations discussed above are the ones constrained by the relation $l + m \leq 4$. They are called *primary* or *third-order* aberrations, or *Seidel* aberrations. We prefer the term *primary*. Given the constraint $l + m \leq 4$, the term *fourth order* would apply more appropriately than *third order*; the latter refers to the ray aberration, which is the derivative of the wave aberration (the differentiation lowers the power of the terms by one). Thus one often hears, for example, that spherical aberration varies as the third power of the aperture: That refers to the ray aberration. Whether one uses the term *primary* or *third-order*, the term *higher-order* aberrations is used to refer unambiguously to aberrations for which $l + m \geq 4$.

Higher-order aberrations do not normally have individual names, but are classified broadly into three types: spherical aberration (independent of θ), comatic (odd powers of $\cos\theta$), and astigmatic (even powers of $\cos\theta$).

Despite the fact that the primary aberrations do not describe completely the image degradation, they are very useful as a diagnostic tool in lens design. As we will see in the next chapter, they can be computed by tracing only the PMR and the PPR, that is, with paraxial ray tracing only. Finally, they do provide a first approximation to the actual aberrations present, and represent the real aberrations adequately for medium aperture and field systems, just as paraxial optics represent the imaging properties adequately for infinitesimal aperture and field.

Apart from the power-series expansion of Eq. (7.36), the wavefront aberration can be represented with any set of orthogonal polynomials over the exit pupil. Particularly useful are the *Zernike polynomials*, which provide an alternative description to the Seidel one*. The different Zernike polynomials can also be thought of as aberration types, but they are more complicated functions of r and θ than the Seidel terms, at least for higher orders. Their primary usefulness lies in the fact that they can provide a more accurate representation of the wavefront aberration during interpolation with a limited number of terms and can also represent aberration types that are due to manufacturing defects. They are very frequently used in optical interferometric testing.

7.3.3. Ray Intersection Patterns and Spot Diagrams

Consider a bundle of rays emanating from a single object point and traced through the system to the image plane. In the presence of aberrations, the points of intersection of these rays with the image plane will form a pattern that is characteristic of the type of aberration present; this is the basic idea of a *spot diagram*. It is important to learn enough about the various patterns, so as to be able to recognize the type of aberration present, though this may be difficult for aberration combinations. Also, the spread of the rays is used as a measure of the ray aberration. In the previous section, we defined the ray aberration as the distance between the actual and the Gaussian ray intersections with the image plane. This definition, based on a single ray, does not give an adequate representation of the image quality, since it depends on the choice of ray. For example, for an axial object point, if the chosen ray forms a very small angle with the axis, then little or no aberration would be expected. Thus we must use a number of rays, if we are to represent the actual image blur produced by the aberration.

The choice of rays may depend on the application, but the most common and useful choice is to connect the object point with a set of regular grid points over the entrance pupil, corresponding to uniform illumination of the pupil by the object.

We can also look at the ray aberration by considering first the wavefront aberration, and then deriving from that the ray pattern in the image plane. This can be done in two ways: (1) analytically, as shown below, and (2) by tracing

* See, for example, C-J. Kim and R. R. Shannon, "Catalog of Zernike polynomials," in *Applied Optics and Optical Engineering*, Vol. X, R. Shannon and J. Wyant, Editors, Academic Press, 1987.

Figure 7.14. Ray intersection pattern for spherical aberration at the Gaussian image plane (no defocus). The outer ring is produced by rays from the rim of the pupil. The ray density decreases with increasing radius.

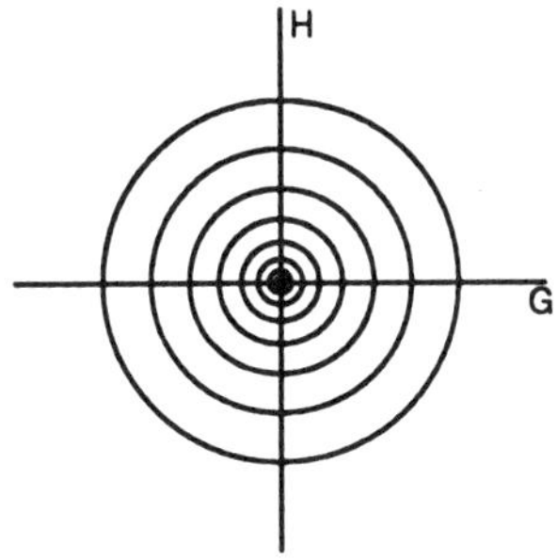

rays from a regular grid over the exit pupil, the direction of the rays being given by the fact that they are normal to the wavefront.*

We now caluclate analytically the ray intersection patterns. Recall Eq. (7.32),

$$\delta G = \frac{\partial W}{\partial x}, \qquad \delta H = \frac{\partial W}{\partial y}$$

where $x^2 + y^2 = r^2$, $x = r \sin \theta$, $y = r \cos \theta$, and we have dropped the primes for simplicity. The field dependence of the aberration is not relevant here; we can think of this as determining the ray aberration for a given field point. Thus we abbreviate the notation of the aberration coefficients by dropping the first subscript. We have:

Spherical aberration. For spherical aberration alone, the aberration polynomial becomes simply $W = W_{40}r^4 = W_{40}(x^2 + y^2)^2$. Therefore,

$$\delta G = \frac{\partial W}{\partial x} = 4W_{40}x(x^2 + y^2)$$

$$\delta H = \frac{\partial W}{\partial y} = 4W_{40}y(x^2 + y^2)$$

$$\delta G^2 + \delta H^2 = (4W_{40}r^3)^2 \tag{7.47}$$

This is the equation of a circle. The radius of the circle depends on r, the radial distance measured at the pupil. In other words, rays emanating from a ring of radius r at the pupil form a circular pattern of radius $4W_{40}r^3$ at the same plane. Of course, rays from the entire pupil must be considered in order to establish the full pattern; that is r must be seen as a variable spanning the range 0 to 1. Figure 7.14 gives an approximate idea of the ray intersection pattern. Each of the rings shown is produced by one ring of rays from the exist pupil; the rings over the exit pupil are uniformly spaced; that is, r varies from 0 to 1 in steps of 0.1 Figure 7.15 shows a spot diagram for spherical aberration. This gives a

* In the second case, note that a regular grid over the entrance pupil is not necessarily imaged to a regular grid at the exit pupil unless there is no distortion for pupil imagery (see Section 10.5). But this difference is often ignored, since the ray aberration does not give an exact representation of the image of a point in any case.

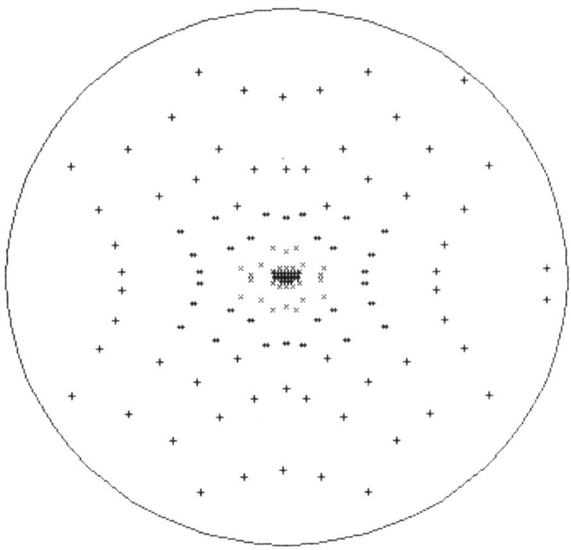

Figure 7.15. Spot diagram for spherical aberration (Gaussian image plane).

little better idea of the variation of irradiance across the image. The fourfold symmetry that is observed in the pattern is an artefact arising from the square-grid sampling of the pupil. No such symmetry exists in the real pattern.

Spherical aberration and defocus. In this case, the wave aberration polynomial is

$$W = W_{20}(x^2 + y^2) + W_{40}(x^2 + y^2)^2$$

After differentiation and some algebra, we obtain

$$\delta G^2 + \delta H^2 = 4r^2(2W_{40}r^2 + W_{20})^2 \tag{7.48}$$

As in the previous case, this the the equation of a circle; however, the radius of the circle now depends on W_{20}. So for different amounts of defocus, we obtain a different size of the blur spot. The maximum radius of the pattern, R_{max}, will depend on the ratio W_{20}/W_{40}, as well as r. The minimum value of R_{max} defines the geometrical best focus (or circle of least confusion). This occurs for $W_{20}/W_{40} = -3/2$. The value of R_{max} in this case is equal to W_{40}. This can be compared with the value at the paraxial focus ($W_{20} = 0$), which is $4W_{40}$ [See Eq. (7.47)].

Different values of W_{20} correspond to different reference spheres. The implication of the above discussion is that a judicious choice of the reference sphere can reduce the aberration, by providing a better fit to the wavefront. If we were to then judge the best focus according to the fit between wavefront and reference sphere, it would be reasonable to employ the criterion that the best focus should minimize the rms wavefront aberration. We state here, without proof, that this criterion is satisified for $W_{20} = -W_{40}$. This is often called the best diffraction focus. The corresponding R_{max} is $2W_{40}$, but the peak irradiance of the pattern is higher than in the best geometrical focus case.

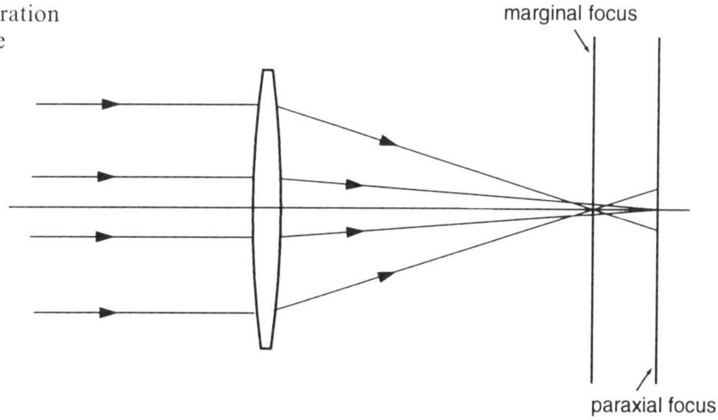

Figure 7.16. Spherical aberration as a longitudinal effect. The marginal rays focus closer to the lens than the paraxial rays: This is called *undercorrected* spherical aberration. The opposite condition is *overcorrected.*

The marginal rays ($r = 1$) come to a sharp focus for $W_{20} = -2W_{40}$. These various amounts of wavefront defocus translate into axial distances through the formulas shown in Appendix 4. For a single positive lens, Figure 7.16 shows the path of the paraxial (or small-angle) and the marginal rays. It can be seen that the marginal rays form a large blur spot at the paraxial focus, while the paraxial rays form a large blur spot at the marginal focus. From this simple construction, we can see that the circle of least confusion lies between the paraxial and the marginal forus and is closer to the latter.

Coma. In this case, the aberration polynomial is, excluding field dependence,

$$W = W_{31}r^3 \cos\theta = W_{31}y(x^2 + y^2)$$

It is left as an exercise to show that

$$\delta G^2 + (\delta H - 2W_{31}r^2)^2 = (W_{31}r^2)^2 \tag{7.49}$$

This is the equation of a circle of radius $W_{31}r^2$, centered at $(0, 2W_{31}r^2)$. As r changes from 0 to 1, we have the pattern indicated in Figure 7.17. The angle a indicated in the figure can be found from the diameter of the last circle, which is the one for $r = 1$, and is centered at $2W_{31}$. Therefore, $\sin a = W_{31}/2W_{31} = \frac{1}{2}$; so $a = 30°$, independent of r. The total length of the pattern is called tangential coma and is equal to $3W_{31}$. The diameter of the last circle is called *sagittal coma*, and is equal to $2W_{31}$. Figure 7.18 shows a spot diagram for coma.

Figure 7.17. Ray intersection pattern for coma.

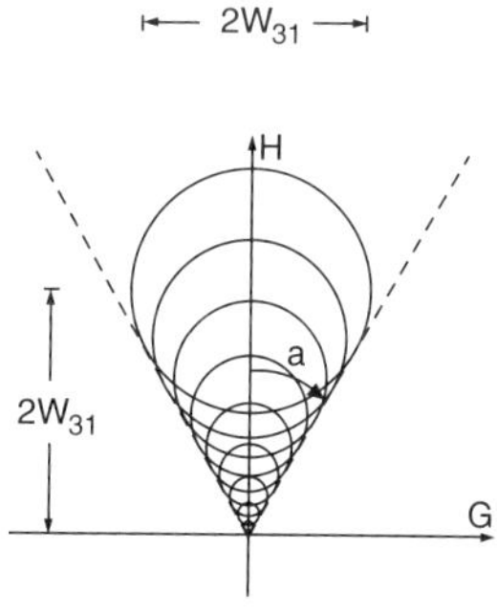

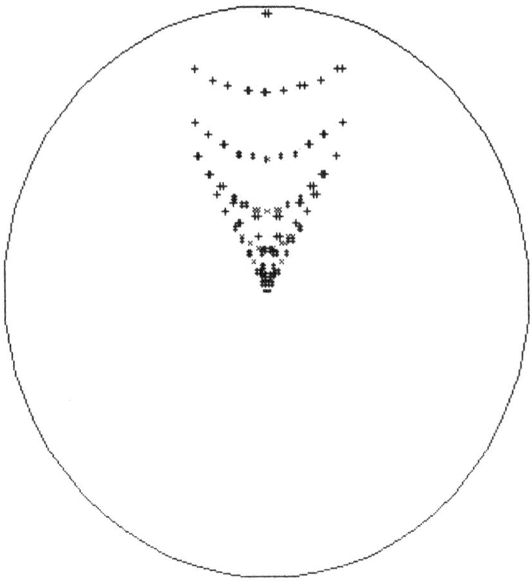

Figure 7.18. Spot diagram for coma.

Astigmatism. The effect of astigmatism is best appreciated when considered together with defocus. The wavefront aberration polynomial is then

$$W = W_{20}r^2 + W_{22}r^2 \cos^2 \theta = W_{20}(x^2 + y^2) + W_{22}y^2$$

Therefore

$$\delta G = 2W_{20}x = 2W_{20}r \sin \theta$$

$$\delta H = 2y(W_{20} + W_{22}) = 2r \cos \theta(W_{20} + W_{22})$$

$$\left(\frac{\delta F}{2W_{20}r}\right)^2 + \left(\frac{\delta H}{2r(W_{20} + W_{22})}\right)^2 = 1 \qquad (7.50)$$

This is the equation of an ellipse, with semiaxes $2W_{20}r$ and $2(W_{20} + W_{22})r$. If $W_{22} < 0$, the major axis is along G. At the paraxial image plane ($W_{20} = 0$) we have $\delta G = 0$; so the image is a line of length $4W_{22}$. If we choose the focal plane so that $W_{20} = -W_{22}$, then $\delta H = 0$ and the image is a line of the same length as before, but oriented along the G axis. For $W_{20} = -W_{22}/2$, Eq. (7.50) becomes $\delta G^2 = \delta H^2 = (W_{22}r)^2$; that is, we now have a circle of diameter $2W_{22}$. The diameter of the circle is half the length of the lines; hence it represents the minimum spot size (circle of least confusion). Figure 7.19 shows the shape of the pattern through focus, and Figure 7.20 is a spot diagram for astigmatism. The two lines are oriented along the tangential and the sagittal (radial) directions. The effect of astigmatism is readily appreciated by considering the image of a wheel with spokes (Figure 7.21). The circumference is imaged perfectly at the tangential focal plane, while the spokes are imaged perfectly at the sagittal focal plane.

Figure 7.19. An astigmatic wavefront coming to a focus.

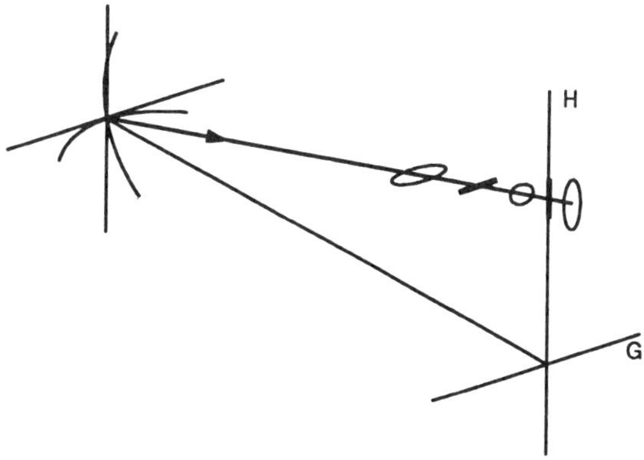

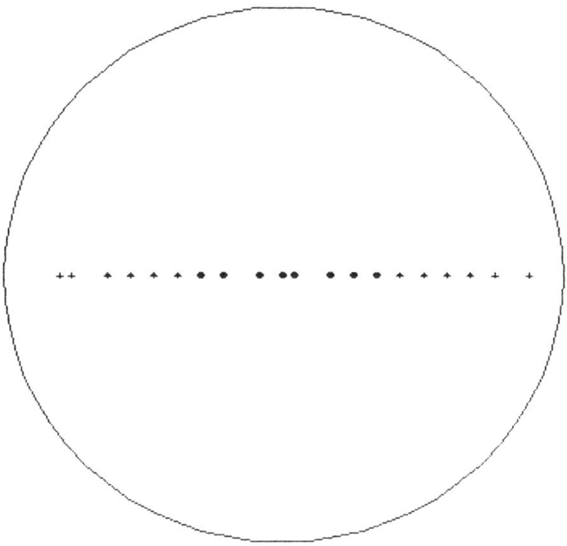

Figure 7.20. Spot diagram for astigmatism at the tangential focal plane.

Figure 7.21. Images of a wheel with spokes through a system suffering from astigmatism. Left: object; middle: sagittal image; right tangential image.

We emphasize that neither the spot diagrams nor the ray intersection patterns represent with accuracy the actual irradiance variation over the image plane; for that, a diffraction-based calculation is necessary.* Despite this drawback, spot diagrams are useful because their computation is fast; this is a help during system optimization, which can be a time-consuming procedure even with today's computers.

7.3.4. Longitudinal Aberration

So far we have encountered two ways to express aberration: as wavefront and as transverse ray aberration (TRA). In the case of spherical aberration or astigmatism, we are often interested in determining the distance between the paraxial and marginal foci, or the two astigmatic focal lines. Also, the value of spherical aberration and/or astigmatism is frequently expressed in dioptres; so we wish to know how to convert longitudinal aberration into wavefront aberration and vice vera.

Spherical aberration. The longitudinal spherical aberration (LSA) is the focus difference between the paraxial and marginal foci. From Eqs. (7.31) and (7.32), we have:

$$\delta H = \frac{\partial W}{\partial y} = nu\ \delta\eta \rightarrow \delta n = \frac{1}{nu}\frac{\partial W}{\partial y}$$

where we omit all primes for simplicity. Since the aberration is rotationally symmetric, δH alone suffices, and y is equivalent to r. Referring to Figure 7.22, $\delta\eta$ is the TRA as noted, while δl is the LSA. If the aberration (δu) is small compared to u, we can set

$$\delta l \cong -\frac{\delta\eta}{u}$$

so that

$$\delta l = -\frac{1}{nu^2}\frac{\partial W}{\partial y}$$

For spherical aberration, $W = W_{40}y^4$, and $\partial W/\partial y = 4W_{40}y^3$. Therefore, $\delta l = -(4/nu^2)W_{40}y^3$. The marginal focus is obtained for $y = 1$; so

$$\delta l = -\frac{4W_{40}}{nu^2} \qquad (7.51)$$

* For a comparison between spot diagrams and diffraction-based computations for all monochromatic aberrations, see N. D. Haig, *Interferrograms and Spread Functions of the Sixth-Order Aberrations*, SPIE Proc. Vol. 274, *Assessment of Imaging Systems: Visible and Infrared*, T. L. Williams, Ed. 1981, For beautiful pictures of the actual point images, see M. Cagnet, M. Françon and J. C. Thrierr, *Atlas of Optical Phenomena*, Springer Verlag, Berlin, 1962.

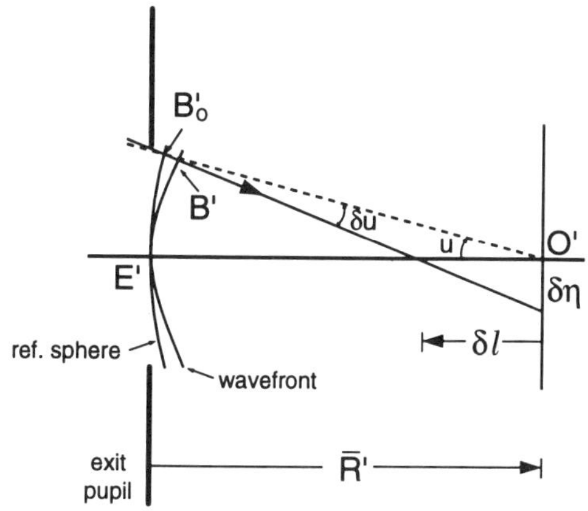

Figure 7.22. Longitudinal spherical aberration.

This result expresses the fact that a fixed amount of wavefront aberration (W_{40}) is more significant for a smaller pupil than it is for a larger pupil; for a small pupil, a given wavefront error results in a larger slope difference between wavefront and reference sphere than for a large pupil. From Figure 7.22, we can set $u^2 \cong (h_e/\bar{R})^2$; so Eq. (7.51) can be written as

$$W_{40} = -\tfrac{1}{4}nh_e^2\frac{\delta l}{\bar{R}^2} \cong \tfrac{1}{4}nh_e^2\left(\frac{1}{\bar{R}+\delta l} - \frac{1}{\bar{R}}\right)$$

The term in parentheses in the last expression is the LSA expressed as the difference in dioptres between the marginal focus and the paraxial focus. Setting that term equal to δD and $W_{40} \to \delta W$, we obtain

$$\delta W = \tfrac{1}{4}nh_e^2\,\delta D \tag{7.52}$$

Note that we have chosen to measure δD from the marginal focus, so that its sign is the same as that of δW; but the opposite convention can also be employed. Casual inspection of Eq. (7.52) might lead one to think that the LSA is inversely proportional to the square of the aperture—but this is wrong!. In fact, the LSA increases as the square of the aperture (r or u). Equations (7.51) and (7.52) apply to the total aberration (at the edge of the aperture); that is, W_{40} is fixed. If, however, we seek the variation of δl or δD with u (or r), then we must consider W_{40} as a variable; and it varies as the fourth power of r or u; so we can write $W_{40} = Cu^4$. In this case also, δl or δD is not the difference between marginal and paraxial focus, but the difference between the focus from a given pupil zone and the paraxial one. For this case, then we can write

$$\delta l = \frac{4}{n}Cu^2 \tag{7.53}$$

showing the quadratic dependence of δl on u. Thus, in summary, for spherical aberration, the wavefront term varies as the fourth power of the aperture, the

TRA varies as the third power of the aperture, and the LSA varies as the square of the aperture. Although the above simplified derivation breaks down for a telescopic system ($u \cong 0$), Eq. (7.52) is indepenent of u and in fact holds equally for focal and afocal systems.

Astigmatism. The derivation is similar to the one above. We have again

$$\delta l = \frac{1}{nu^2} \frac{\partial W}{\partial y}, \qquad \text{but with } \frac{\partial W}{\partial y} = 2W_{22}y$$

At full aperture ($y = 1$), we then obtain

$$\delta l = \frac{2}{nu^2} W_{22} \tag{7.54}$$

and using $u^2 = h_e^2/\bar{R}$, $\delta l/\bar{R}^2 \cong \delta D$ and $W_{22} \to \delta W$, we obtain

$$\delta W = \tfrac{1}{2} n h_e^2 \, \delta D \tag{7.55}$$

δD here expresses the difference between the T and S foci. Once again, Eqs. (7.54) and (7.55) show that a fixed wavefront error has more severe effect on a smaller pupil because it causes larger difference in slope between wavefront and reference sphere; but they do not imply that longitudinal astigmatism is inversely proportional to the square of the aperture. In fact, longitudinal astigmatism is independent of the aperture: this is because W_{22} varies as the square of u; so δl in Eq. (7.54) is constant.

Defocus and longitudinal color. Defocus (W_{20}) can also be expressed as a dioptre difference. Because of the quadratic dependence on r, the resultant equation is formally the same as Eq. (7.55); it is derived from first principles in Appendix 4. Also, longitudinal chromatic aberration is a wavelength-dependent defocus term; so the same equation applies there, too.

7.3.5. Aberration Tolerances

The amount of aberration tolerated in a system will of course depend on the requirements, but there is a set of aberration tolerances, called *Strehl tolerances*, that claim general applicability. The Strehl tolerances are equivalent to the quarter-wave criterion, according to which if the wavefront aberration is less than $\lambda/4$, it can be considered negligible. This is a somewhat arbitrary criterion, but it should be clear that since light is a wave, any effect less than a fraction of the wavelength should be negligible. Derivation of the Strehl tolerances requires diffraction theory, and is not attempted here.* These tolerances are:

* See K. Strehl, *Z. Instrumkde* **22**, 213 (1903); A. Maréchal, Thesis, University of Paris, 1948; W. B. King, *Applied Optics* **7**, 489 (1968).

Defocus	0.25λ
Primary spherical (with choice of focus)	1λ
Primary coma (with choice of lateral shift	0.6λ
Primary astigmatism (with choice of focus)	0.35λ

These tolerances apply to the point-imaging aberrations. Field curvature degrades the image of a point if the image must be on a plane. In that case, the defocus tolerance applies. It is to be noted that the above numbers represent wavefront aberration. The corresponding ray aberration can be found from Eq. (7.28), and it can be seen that its magnitude will depend of $\bar{R}$; therefore, no general set of ray aberration tolerances can apply.

Distortion can be expressed as a fraction $\delta\eta/\eta$, that is, the ray aberration divided by the corresponding image height. The tolerance for distortion will depend on the application; for visual images, distortion of a few percent is considered satisfactory.

Finally, since the importance of chromatic aberrations depends largely on the source spectral power distribution and the detector sensitivity, no general set of tolerances can apply. In the relatively common case where the detector is the human eye and the source is polychromatic, the following tolerances apply.* For longitudinal color, the tolerance is approximately one dioptre (expressed as the change in beam divergence for the F and C wavelengths in the eye space). For transverse chromatic aberration, the tolerance is about 2 minutes of arc, expressed as angular ray aberration in the eye space.

The reader should appreciate that the topic of aberration tolerancing does not lend itself to clear and unambiguous answers. For example, it is often said that if the aberration is within the Strehl tolerances it is not noticeable. But this clearly depends on the detector: The human eye can detect reliably even smaller amounts of aberration†.

What are some typical values of aberration? Again in wavefront terms, lenses that perform critical imaging functions may have near zero aberration on axis, and up to a few wavelengths at the edge of a 20° field. Lenses used for noncritical applications can have much larger aberrations of tens of wavelengths. When the aberration value is large, it is customary to use the ray aberration instead of the wavefront aberration. This is because the effects of diffraction (which can be conveniently folded into the wavefront aberration) are too small, compared with the effects of aberration.

---

* See: P. Mouroulis and G. C. Woo, "Chromatic aberration and accommodation in visual instruments," *Optik* **80**, 161–66 (1988), and P. Mouroulis, T. Kim, and G. Zhao, "Transverse color tolerances for visual optical systems." *Applied Optics* **32**, 7089–94 (1993).

† See N. D. Haig and G. J. Burton, "Effects of wavefront aberration on visual instrument performance, and a consequential test technique." *Applied Optics* **26**, 492–500 (1987)

7.3.6. Example: Computation of Wave and Ray Aberration

In this example, we wish to show how the wave and ray aberration may be computed from finite ray-tracing data. The finite ray-tracing equations used in this section are derived in Appendix 3, and must be studied before proceeding further. Aberration computation up to the primary (or third-order) level may be done manually or by computer and involves only paraxial ray tracing, as shown in the following chapters. The calculation of the total aberration through finite ray tracing is much more elaborate, and has not been performed manually since the advent of computers. However, in order to illustrate the steps of the process, we have chosen a simple example in which most calculations can be performed manually or with a small programmable calculator.

We consider a thick equiconvex lens, operating at symmetrical conjugates. We will trace only meridional rays, although the formulas of Appendix 3 are general. The lens specification is: $c_1 = 0.5\ \text{cm}^{-1}$, $c_2 = -0.5\ \text{cm}^{-1}$, thickness $d_1 = 0.8$ cm, $n = 1.52$ at a wavelength $\lambda = 0.56 \times 10^{-4}$ cm, stop position $d_2 = 1$ cm after the second surface, and object at $l_1 = -3.8462$ cm from the first surface. Figure 7.23 shows the geometry. The remaining Gaussian properties are: EFL $F = 2.064$ cm, paraxial image plane location $l' = 3.8462$ cm from the second surface, and radius of reference sphere $\bar{R} = 2.8462$ cm (distance between E' and O').

We trace a fan of 12 rays from the axial object point, from 1° to 12° inclination from the axis, in 1° increment. Figure 7.23 shows three representative rays, at 2°, 6°, and 12°. The first ray crosses the axis at 3.81 cm from the surface, very close to the paraxial prediction, but the other two rays deviate visibly.

The finite ray-tracing results are shown in Table 7.3. N is the z-direction cosine of the ray (z is the optical axis); Y_1, Y_2, and Y_3 are the ray intersection heights at the first surface, second surface, and reference sphere (treated as a dummy surface), respectively, TRA is the intersection of the ray with the paraxial image plane. All distances are in centimeters, except for the wavefront aberration W, which is customarily measured in units of wavelength.

In order to follow the ray-tracing computation (which is recommended), the following should be noted.

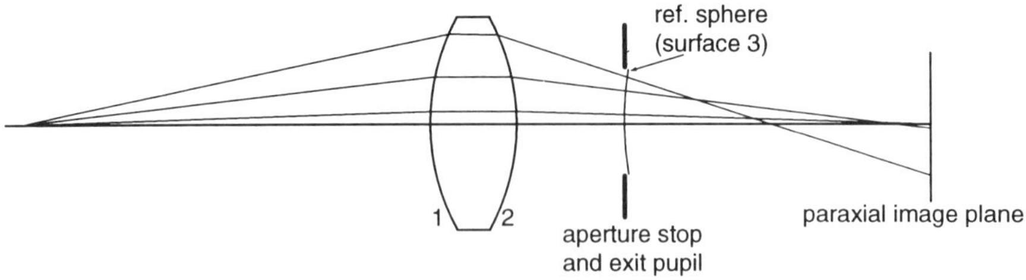

Figure 7.23. Finite ray tracing through a thick equiconvex lens.

TABLE 7.3.
Ray-tracing results

N	Y_1 (cm)	Y_2 (cm)	Y_3 (cm)	TRA (cm)	W (λ)
0.99985	0.06715	0.06714	0.04962	−0.0002	0.01
0.99939	0.13447	0.13437	0.09889	−0.0014	0.22
0.99863	0.20210	0.20179	0.14746	−0.0048	1.10
0.99756	0.27023	0.26949	0.19493	−0.0116	3.44
0.99619	0.33903	0.33759	0.24084	−0.0233	8.32
0.99452	0.40868	0.40625	0.28466	−0.0414	17.0
0.99255	0.47941	0.47565	0.32570	−0.0683	30.9
0.99027	0.55144	0.54601	0.36308	−0.1067	50.9
0.98769	0.62504	0.61765	0.39557	−0.1606	77.5
0.98481	0.70052	0.69096	0.42136	−0.2359	109
0.98163	0.77826	0.76652	0.43765	−0.3417	137
0.97815	0.85870	0.84511	0.43954	−0.4941	140

In the opening calculations, we compute the direction cosines of the rays, M and N, from the known angle. We then compute the intersection height Y_0 between the rays and the plane tangent to the first surface at the apex, from $Y_0 = -lM/N$.

The ray-tracing calculations comprise repetitive steps of refraction at the surface and transfer to the next surface. Our first operation is a transfer from the tangent plane to the first surface itself [Eqs. (A3.6)–(A3.10)]. Then the direction cosines of the surface normal are found, and from that, the cosine of the angle of incidence [Eqs. (A3.5), (A3.11), (A3.12)]. Finally, the direction cosines of the ray after refraction are given by Eq. (A3.3). Transfer to the next surface begins the next iteration.

In the closing calculations, we determine the intersection between the ray and the axis, and also the TRA, from the known paraxial image plane distance.

While the calculation of the intersection heights and the TRA involves a relatively straightforward application of the formulas of Appendix 3, the calculation of W deserves some additional explanation. In this simple case, it suffices to use Eq. (7.22). During transfer, we determine the geometrical path D traversed by the ray [Eq. (A3.10)], and from that we calculate the cumulative optical path (ΣnD) along the ray. This corresponds to the path $[B_0 \cdots B_0']$ of Eq. (7.22), which is then subtracted from the corresponding axial path. This gives W in centimeters. We then divide by λ in order to express W as a number of wavelengths; the result is shown in the last column of Table 7.3.

Since we are considering only the axial object point, only spherical aberration can exist, and it varies as the fourth (or higher) power of the aperture: This explains the rapid increase of W with increasing angle, seen in Table 7.3. However, we have not yet determined the coefficients

TABLE 7.4.
Polynomial terms

α	W_{40} (λ)	W_{60} (λ)	W_{80} (λ)
5°	8.5	0.8	0.14
8°	55	9	13
12°	274	−30	658

of the wavefront aberration polynomial. In this case, the polynomial is

$$W = W_{40}r^4 + W_{60}r^6 + W_{80}r^8 \tag{7.56}$$

where $r = Y_3/h'_e$ is the fractional pupil coordinate, W_{40} the usual primary spherical aberration term, and W_{60} and W_{80} are the next two orders of spherical aberration. A least-squares polynomial is then determined, using only the above three terms. The result depends on the maximum pupil height chosen, and it is instructive to consider the following three cases: $\alpha = 5°$, $8°$, and $12°$, corresponding to $h'_e = 0.2481$, 0.3961, and 0.5917. Note that α is the aperture angle in the object space, for which $u = \sin \alpha$. Thus we determine three different polynomials, by taking the W values from the last column of Table 7.3, and three different sets of $r = Y_3/h'_e$ values, for the three different values of h'_e. The resulting polynomial terms are shown in Table 7.4.

The following conclusions emerge: (1) At 12°, the aberration is so large that the concept of wavefront aberration is no longer useful (see also Figure 7.23). (2) At small angles, for example, up to 5°, primary aberration suffices to describe the image quality (higher orders are small). (3) As the aperture angle increases, the higher orders become more significant.

For the off-axis case, the ray-tracing procedure is the same, but the wavefront aberration computation becomes much more complicated, and must involve a careful analysis of the imagery between the pupils (see also Chapter 10). Also, the two-dimensional polynomial fitting is considerably more elaborate.* From this example, the reader might well wonder if anything is gained in practice by computing the wavefront aberration. The answer depends on the type of lens under design, the way the lens was specified, and the preference of the designer. For example, in several cases it suffices to compute the spot diagram using finite ray tracing, and omit the wavefront aberration calculation. But notice that although in this example we were able to calculate the aberration in detail, we learned nothing about how it can be reduced! In the following chapters, we show how we can obtain aberration correction, and how the wavefront aberration facilitates that process, by allowing us to compute the contribution of a single surface to the total aberration.

---

* See J. Macdonald and Y. Wang, "Sources of error in the rapid fitting of wave-aberration polynomials," *Optica Acta* **33,** 621–36 (1986).

PROBLEMS

1. Design achromatic doublets using three different pairs of glasses chosen from the glass chart of Figure 7.3.

2. Design an apochromatic doublet using Figures 3.3 and 3.5.

3. Design a telescope objective of $F = 50$ mm that produces a longitudinal chromatic error of about 1 D between the C and F wavelengths. The condition $|K_1| < 3|K_2|$ must be satisfied. What is the equivalent V value of the lens? It is possible to use a singlet instead of a doublet?

4. A camera lens comprises a thin element of $F = 100$ mm, placed 33.3 mm after the stop. The lens material has $V = 57$, and the image size is 50 mm. (a) Calculate the chromatic variation of focus and the chromatic variation of image height. Compare the latter with a film resolution of 50 cycles/mm. (b) For a relative aperture of $F/17$, determine the axial blur due to chromatic aberration when the d wavelength is in focus.

5. A Huygens eyepiece consists of two singlets of the same glass (517642). The focal length of the field lens is twice that of the eye lens. Determine the separation between the two lenses as well as the powers, for a total eyepiece power of one unit.

6. Derive the ray intersection pattern for a system suffering from spherical aberration ($W_{40}r^4$) and defocus ($W_{20}r^2$). Consider the following cases: (a) $W_{20} = -{}_2W_{40}$, (b) $W_{20} = -W_{40}$, (c) $W_{20} = -\frac{3}{2}W_{40}$, (d) $W_{20} = 0$. In all cases, plot the size of the intersection pattern as a function of r, where r varies from 0 to 1, and thus justify why case (c) is the best geometrical focus.

7. A two-lens system has $F_1 = F_2 = \frac{3}{2}F$, $F = 100$ mm, $F\# = 2.5$, FOV $= \pm 11.3^\circ$, $\bar{l}_1 = F_1/2$, object at infinity. It is found that the system suffers from one wavelength of primary coma (at full field). If it is assumed that other aberrations are small, determine the ray aberration and comment on the variation of ray aberration with field angle. (Note: It will be necessary to include W_{11} in your aberration polynomial.)

8. Repeat Problem 7, but assume that the system suffers from one wavelength of primary astigmatism. Calculate the TRA (transverse ray aberration) at the S focus and half-way between the two astigmatic lines.

9. Repeat Problem 7, but assume that the system suffers from one wavelength of spherical aberration. Calculate the TRA at the paraxial, the marginal, and the best geometrical focus.

10. Investigate the form of the image-plane ray intersection pattern associated with the general coma term

$$W(r,\phi) = W_{(2n+1),1} r^{2n+1} \cos \phi$$

and describe the effect on this pattern of the introduction of an additional aberration term of the form $W_{33}r^3 \cos^3 \phi$.

11. An optical system consists of a combination of spherical and cylindrical refracting surfaces whose centers of curvature lie on a common axis. The cylinder axes are each oriented in one of two mutually perpendicular directions (i.e., they are "crossed"). By identifying suitable symmetry conditions, show that a polynomial expansion of the wave aberration function contains the general term

$$A(\sigma^2)^i(\tau^2)^j(x^2)^k(y^2)^l(\sigma x)^m(\tau y)^n$$

where A is the coefficient, (i, j, k, l, m, n) are integers that may take values from zero to infinity, (σ,τ) are the object point coordinates, and (x,y) are the coordinates of the intersection of the ray with the pupil. Hence write down the forms of the primary aberration terms for such a system, and briefly describe the characteristics of those terms that affect the axial image. (A primary aberration has a field and aperture combined power dependence of ≤ 4.)

12. Repeat Problem 11 for an optical system comprising cylindrical lenses with their axes aligned. Show that the aberration polynomial contains terms of the form

$$(\sigma^2)^i(x^2)^j(\sigma x)^k(\tau - y)^{2l}$$

Computation of Primary Aberrations

8

"Further reduction of aberrations ... is best done using theoretical computation, though the theory has to be developed."

Ludwig Seidel, Zür Dioptrik, *Astronomische Nachrichten*, No. 1027, 289 (1856).

8.1. THE SEIDEL ABERRATION COEFFICIENTS

In this section we show that the total wavefront aberration can be calculated as a summation of the wavefront aberration contributions of the individual surfaces. Thus it is possible to see clearly how the aberration from one surface can be used to cancel the aberration of another. This fundamental physical insight is not possible through the concept of ray aberration.* Further, we will encounter the rather astonishing result that, at the primary aberration level, the computation of the wavefront aberration involves only paraxial quantities! We then apply this knowledge to the design of a simple camera lens, comprising a single element and a stop.

8.1.1. The Paraxial Refraction Invariants

The paraxial refraction formula is

$$n'u' - nu = -hc(n' - n)$$

This can be rewritten as

$$A \equiv n(hc + u) = n'(hc + u') \tag{8.1}$$

---

*Certain formulae are sometimes said to show the ray aberration contribution of a surface; in reality they show the converted wavefront aberration.

Therefore, the quantity $A = n(hc + u)$ is invariant upon refraction (it is not, however, invariant upon paraxial transfer). By using the paraxial refraction equation with PPR quantities, we obtain similarly

$$\bar{A} \equiv n(\bar{h}c + \bar{u}) = n'(\bar{h}'c + \bar{u}') \tag{8.2}$$

The physical significance of these two invariants is as follows. The PMR invariant, A, characterizes the effect of the surface on focusing (or convergence) of an axial ray bundle. Notice that for $A = 0$, we have $u' = u$, that is, no focusing. The amount of focusing is characterized by the bending of the PMR, $u' - u$. It can then be shown (in a few lines indeed) that

$$u' - u = A\,\Delta\left(\frac{1}{n}\right) \tag{8.3}$$

where $\Delta(1/n) \equiv 1/n' - 1/n$. Similarly, the PPR invariant, $\bar{A}$, characterizes the deviation of an extra-axial ray bundle, $\bar{u}' - \bar{u}$. Notice that if $\bar{A} = 0$, then there is no deviation, $\bar{u}' = u$. Also, the PPR deviation is given by

$$\bar{u}' - \bar{u} = \bar{A}\,\Delta\left(\frac{1}{n}\right) \tag{8.4}$$

As the aperture (u) or field angle ($\bar{u}$) increases, we can expect that higher-order aberrations will become more significant. Thus if A and/or $\bar{A}$ are greater than about 0.5, we can expect that the primary aberrations will not suffice to describe the wavefront shape.

8.1.2. Surface Contribution to the Wavefront Aberration

Recall Eq. (7.22): It shows that the wavefront aberration can be computed in principle by tracing a paraxial and a real ray up to the exit pupil sphere and subtracting the resulting optical path lengths. We shall show that it can also be computed on a surface-by-surface basis, by finding the increment to the total aberration due to a particular surface. Consider the refracting surface (somewhere in the middle of the system), and the incident and refracted wavefronts, (QV) and $(Q'V')$, respectively (Figure 8.1). BVP is the incident ray and $PB'V'$ is the refracted ray. The reference spheres are centered at O and O', respectively. Equating optical paths between wavefronts, we have

$$[Q\bar{P}Q'] = [VPV'] = [BPB'] - [BV] + [B'V'] = [BPB'] - W + W'$$

where W and W' are the aberrations of the incident and transmitted wavefronts, respectively. The aberration increment due to the surface is thus

$$\Delta W = W' - W = [Q\bar{P}Q'] - [BPB'] = [Q\bar{P}] + [\bar{P}Q'] - [BP] - [PB']$$

or

$$\Delta W = ([\bar{P}Q'] - [PB']) - ([\bar{P}Q] - [PB])$$

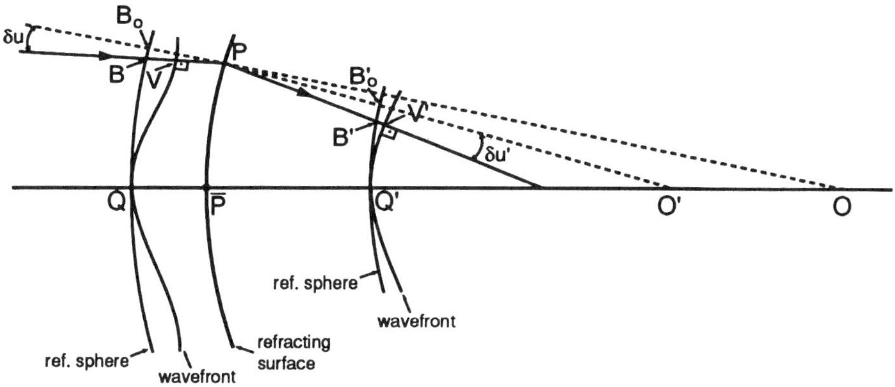

Figure 8.1. Passage of a wavefront through a refracting surface. The incident wavefront is assumed to have aberration (greatly exaggerated).

Now $(PB'_0) = (PB')\cos\delta u'$ and $(PB_0) = (PB)\cos\delta u$, where δu and $\delta u'$ are the angular aberrations; these will have the same aperture and field dependence as the transverse ray aberration. At the primary aberration level, W is at least of the fourth degree in r and τ, and the corresponding ray aberration is at least of the third degree in r and τ, $\delta u = O(3)$. Since $\cos\delta u = 1 - \delta u^2/2 + \cdots$, we have

$$(PB') = (PB'_0) + O(6) \qquad \text{and} \qquad (PB) = (PB_0) + O(6)$$

So

$$\Delta W = ([\bar{P}Q'] - [PB'_0]) - ([\bar{P}Q] - [PB_0]) + O(6)$$

and, at primary aberration level,

$$\Delta W = \Delta([\bar{P}Q] - [PB_0]) \tag{8.5}$$

This is an important result. Notice that it expresses the contribution to the total wavefront aberration arising from the particular surface on which the point P is located. (The wavefront $Q'V'$ can then be considered as the incident wavefront for the next surface, and so on.) It should be clear from the approximation involved in deriving Eq. (8.5) that this result cannot be extended to higher-order aberrations. The proof was given for axial conjugates, but the result is also valid off axis.

Notice the physical significance of this result. The approximation $[PB] \cong [PB_0]$, which is valid at the primary aberration level, allows us to compute the paths not along the finite ray BPB', but along the radii of the reference spheres, which are rays in the Gaussian approximation. This leads to the result that the aberration computation *at the primary level* can be performed using only paraxial (Gaussian) quantities.

8.1.3. The Seidel Aberration Formulae

Having found an expression for the surface-to-surface contribution to the primary aberration, we can express the paths in terms of the usual paraxial ray-tracing parameters, u, h, etc. The derivation was originally performed by L. Seidel in 1856, and the corresponding aberration coefficients (or sums) bear his name. The detailed derivation is not shown here,* but in order for the reader to appreciate what is involved in the derivation, we show a simplified argument for deriving the first Seidel sum at the end of this section. There are five sums, denoted as $S_{\mathrm{I}}, S_{\mathrm{II}}, \ldots, S_{\mathrm{V}}$. These are related to the primary wave aberration coefficients as follows:

$$\begin{array}{ll} \text{spherical aberration} & {}_0W_{40} = S_{\mathrm{I}}/8 \\ \text{coma} & {}_1W_{31} = S_{\mathrm{II}}/2 \\ \text{astigmatism} & {}_2W_{22} = S_{\mathrm{III}}/2 \\ \text{field curvature} & {}_2W_{20} = (S_{\mathrm{III}} + S_{\mathrm{IV}})/4 \\ \text{distortion} & {}_3W_{11} = S_{\mathrm{V}}/2 \end{array} \tag{8.6}$$

Thus, with the exception of field curvature, each of the primary aberrations is given directly by a corresponding Seidel sum. The formulae for the Seidel sums are:

$$\begin{aligned} S_{\mathrm{I}} &= -\sum_{i=1}^{k} A_i^2 h_i \,\Delta\left(\frac{u}{n}\right)_i \\ S_{\mathrm{II}} &= -\sum_{i=1}^{k} A_i \bar{A}_i h_i \,\Delta\left(\frac{u}{n}\right)_i \\ S_{\mathrm{III}} &= -\sum_{i=1}^{k} \bar{A}_i^2 h_i \,\Delta\left(\frac{u}{n}\right)_i \\ S_{\mathrm{IV}} &= -\sum_{i=1}^{k} H^2 c_i \,\Delta\left(\frac{1}{n}\right)_i \\ S_{\mathrm{V}} &= -\sum_{i=1}^{k} \left[\frac{\bar{A}_i^3}{A_i} h_i \,\Delta\left(\frac{u}{n}\right)_i + \frac{\bar{A}_i}{A_i} H^2 c_i \,\Delta\left(\frac{1}{n}\right)_i\right] \\ &= \sum_{i=1}^{k} \frac{\bar{A}_i}{A_i} [(S_{\mathrm{III}})_i + (S_{\mathrm{IV}})_i] \end{aligned} \tag{8.7}$$

where k is the total number of surfaces, and $\Delta(u/n)_i = u_{i+1}/n_{i+1} - u_i/n_i$, $\Delta(1/n)_i = 1/n_{i+1} - 1/n_i$. All the quantities in these expressions are the usual PMR and PPR quantities. Also, recall that $u_i' \equiv u_{i+1}$ and $n_i' \equiv n_{i+1}$.

---

* See W. T. Welford, *Aberrations of Optical Systems*, Adam Hilger, London, 1986.

Notice that if $A = 0$ at any surface, then S_V becomes indeterminate at that surface. In that case, we can use the following alternate form

$$S_V = -\sum_{i=1}^{k} \bar{A}_i \, \Delta(h\bar{u}^2 + \bar{h}^2 cu)_i \tag{8.8}$$

where the difference term is $(h_i\bar{u}_{i+1}^2 + \bar{h}_i^2 c_i u_{i+1}) - (h_i\bar{u}_i^2 + \bar{h}_i^2 c_i u_i)$.

When $S_{III} = 0$ (no astigmatism), then S_{IV} alone determines the value of the field curvature. However, when $S_{III} \neq 0$, we no longer have one but two image surfaces, the tangential (T) and the sagittal (S), corresponding to the two astigmatic focal lines. Those image surfaces have different curvatures; hence the term *field curvature* in the presence of astigmatism is ambiguous and thus avoided. The term *Petzval curvature* is used to denote S_{IV} alone and represents the field curvature in the absence of astigmatism. (The name derives from J. Petzval, who first discussed the condition for a flat image field.) Further discussion regarding the interaction between astigmatism and field curvature is to be found in Section 8.2.

It is important to notice that the first of the five sums, S_I, depends on PMR quantities only; hence primary spherical aberration is independent of the pupil positions. Three of the remaining four sums (corresponding to coma, astigmatism, and distortion) depend on the pupil location as they contain $\bar{A}$. But S_{IV}, unlike the other four sums, is only a function of the system constructional parameters and the optical invariant. It can then be regarded as intrinsic to the optical system, whereas the other four sums depend in general on the object/image and pupil positions. To understand this better, consider the case of a single surface and a thin lens. For a single surface, we have

$$S_{IV} = -H^2 c \, \Delta\left(\frac{1}{n}\right)$$

and for a thin lens of refractive index n,

$$S_{IV} = -H^2\left[c_1\left(\frac{1}{n} - 1\right) + c_2\left(1 - \frac{1}{n}\right)\right] = H^2(c_1 - c_2)\frac{n-1}{n}$$

or

$$S_{IV} = \frac{H^2 K}{n} \tag{8.9}$$

where K is the power of the lens. Noting that n can vary only within narrow limits, we can approximate $1/n \cong 0.65$; so $S_{IV} \cong 0.65 H^2 K$. To a first approximation, we can consider a general system as comprising thin lenses; so the total S_{IV} for the system will be

$$S_{IV} \cong 0.65 H^2 \Sigma K \tag{8.10}$$

Notice that the thin lenses do not have to be in contact and that ΣK is *not* the system equivalent power. Thus to a first approximation, the Petzval curvature depends only on the sum of the powers of the individual thin elements and the optical invariant. This shows why the reduction of field curvature requires a decision on the lens construction to be made at an early stage of the design. Most lens systems comprise predominantly positive elements; so in order to reduce S_{IV}, we need some negative elements in the system.

We shall give a somewhat abbreviated and simplified argument to justify the derivation of the expression for S_I. It is first necessary to prove the following theorem: At the primary aberration level, the aberration is independent of the choice of reference sphere (sphere radius).

Referring to Figure 8.2, let two reference spheres be centered at O' and pass through E_1 and E_2, respectively. The wave aberration referred to sphere 1 is $W_1 = [B_1 V_1]$, while the aberration referred to sphere 2 is $W_2 = [B_2 V_2]$. The difference,

$$W_1 - W_2 = [B_1 V_1] - [B_2 V_2] = [B_1 B_2] - [V_1 V_2]$$

Also,

$$(B_1 B_2) \cos \delta u = (B_1 P_2) = (E_1 E_2)$$

since the spheres are concentric. Further, $[E_1 E_2] = [V_1 V_2]$, since these are measured between the same wavefronts. Also, the angular aberration, δu, will be of at least of third degree in r, τ. Therefore, $\cos \delta u \cong 1 - \delta u^2/2 = 1 + O(6)$. So $[B_1 B_2] = [V_1 V_2] + O(6)$, and at the primary aberration level, $[B_1 B_2] = [V_1 V_2]$; hence $W_1 = W_2$.

Consider now a refracting surface, with its apex at $\bar{P}$, as in Figure 8.3. For simplicity, we take the incident wavefront to be unaberrated, but the same result holds if it is aberrated. The refracted ray crosses the axis at some point near the Gaussian image point O'. Given the previous theorem, we can apply Eq. (8.5) to any two reference spheres, including spheres of zero radius, at O and O', so that $Q \to O$, $Q' \to O'$, and $B_0 \to O$, $B_0' \to O'$. Then

$$\Delta W = [\bar{P}O'] - [PO'] - [\bar{P}O] + [PO] + O(6)$$

or

$$\Delta W = n'l' - nl - [n'(PO') - n(PO)] + O(6) \tag{8.11}$$

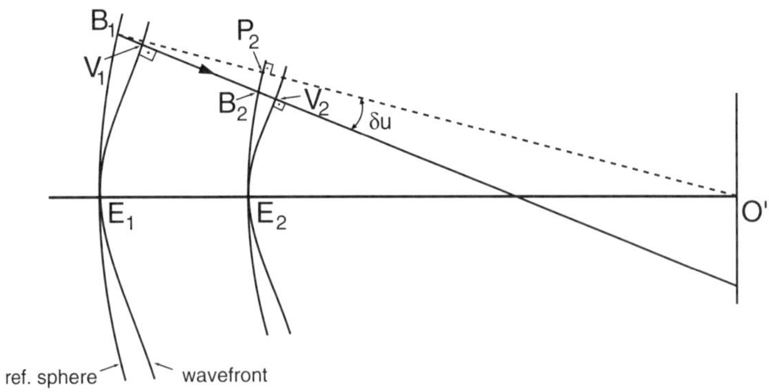

Figure 8.2. Two reference spheres centered at O'.

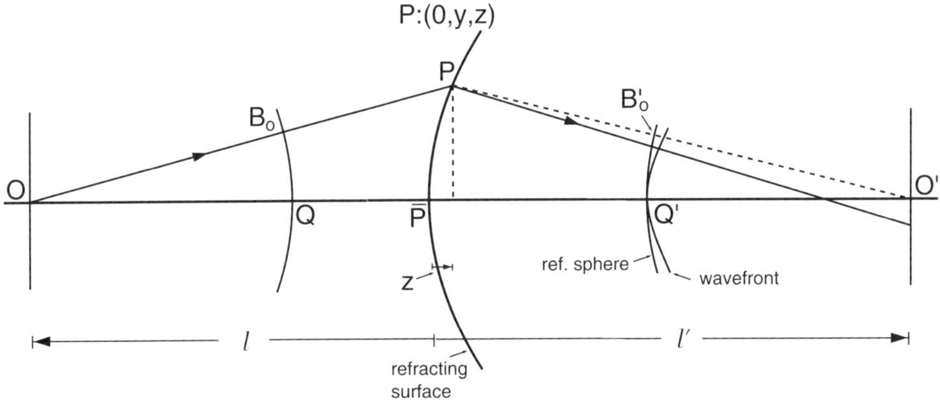

Figure 8.3. Primary wavefront aberration for axial conjugates.

The equation of the sphere (surface) can be expressed as a power series, as follows:

$$z = \tfrac{1}{2}c(x^2 + y^2) + \tfrac{1}{8}c^3(x^2 + y^2)^2 + O(6)$$

Along the y–z plane, this becomes

$$z = \tfrac{1}{2}cy^2 + \tfrac{1}{8}c^3y^4 + O(6) \tag{8.12}$$

$$z^2 = \tfrac{1}{4}c^2y^4 + O(6) \tag{8.13}$$

Now we need to determine (PO) and (PO') in Eq. (8.11). We have

$$(PO)^2 = (l - z)^2 + y^2 = l^2 - 2zl + z^2 + y^2$$

Substitution for z and z^2 from Eqs. (8.12) and (8.13) and some manipulation of terms yields

$$(PO)^2 = l^2\left[1 + \frac{y^2}{l}\left(\frac{1}{l} - c\right) + \frac{y^4c^2}{4l}\left(\frac{1}{l} - c\right) + O(6)\right]$$

We need the square root of this, which we take by applying the binomial expansion, $(1 + x)^{1/2} = 1 + x/2 - x^2/8 + \cdots$. We also multiply by n, to get the optical path, so

$$n(PO) = nl\left\{1 + \frac{1}{2}\left[\frac{y^2}{l}\left(\frac{1}{l} - c\right) + \frac{y^4c^2}{4l}\left(\frac{1}{l} - c\right)\right] - \frac{1}{8}\left[\frac{y^4}{l^2}\left(\frac{1}{l} - c\right)^2\right]\right\} + O(6) \tag{8.14}$$

A similar expression can be obtained for $n'(PO')$, but with l' in place of l. Notice that the expression

$$D = n\left(c - \frac{1}{l}\right) = n'\left(c - \frac{1}{l'}\right) \tag{8.15}$$

is another paraxial refraction invariant [this can be easily proven from $n'/l' - n/l = c(n' - n)$]. Upon substitution for $n(PO)$ and $n'(PO')$ into Eq. (8.11), and using the invariant D, we find that several terms cancel; the only terms remaining are the ones in the second set of square brackets of Eq. (8.14). We have

$$\Delta W = \frac{1}{8}\frac{y^4}{l'}\frac{D^2}{n'} - \frac{1}{8}\frac{y^4}{l}\frac{D^2}{n} + O(6)$$

or

$$\Delta W = \frac{1}{8} y^4 D^2 \, \Delta\left(\frac{1}{nl}\right) + O(6) \tag{8.16}$$

Using the refraction invariant $A = n(hc + u)$, and $u = -h/l$, the reader can verify that

$$D^2 \, \Delta\left(\frac{1}{nl}\right) = -A^2 h \, \Delta\left(\frac{u}{n}\right) \tag{8.17}$$

so that

$$\Delta W = -\tfrac{1}{8} A^2 h \, \Delta\left(\frac{u}{n}\right) y^4 \tag{8.18}$$

For axial conjugates, the total aberration is

$$W = {}_0W_{40} y^4 + O(6)$$

so that

$${}_0W_{40} = -\tfrac{1}{8} A^2 h \, \Delta\left(\frac{u}{n}\right) = \frac{S_{\mathrm{I}}}{8} \tag{8.19}$$

A stricter derivation of this and the other four Seidel sums can be found in Welford's book, referenced previously.

8.1.4. Special Aberration-Free Cases. The Aplanatic Meniscus

The expressions for the Seidel sums suggest immediately some cases where one or more aberrations can be made zero.

If $A = 0$ ($u = u'$), the surface does not focus the beam. This means that the object is at the center of curvature of the surface; so the PMR is normal to the surface. In that case, Eq. (8.7) gives $S_{\mathrm{I}} = S_{\mathrm{II}} = 0$; that is, there is no spherical aberration or coma.

If $\bar{A} = 0$ ($\bar{u} = \bar{u}'$), the PPR is not deviated by the surface, which means that the local pupil is at the center of curvature of the surface. In that case, we have $S_{\mathrm{II}} = S_{\mathrm{III}} = S_{\mathrm{V}} = 0$; that is, the coma, astigmatism, and distortion are all zero.

If $h = 0$ (local object on the surface), then $S_{\mathrm{I}} = S_{\mathrm{II}} = S_{\mathrm{III}} = 0$. A field lens is an example of this case; thus the field lens contributes only field aberrations (field curvature and distortion). We have encountered field lenses as part of the eyepiece of a visual telescope or microscope. In general, field lenses are used also for pupil matching in cascaded lens systems. Another important type of field lens is the so-called *field flattener*. This is a (normally negative) thin singlet lens, that may be placed in contact with (or close to) the final image, thus having little effect on the Gaussian properties of the system. This lens has the effect of flattening the field, by introducing power of the opposite sign from that of the rest of the system [see Eq. (8.10)]. In the simplest possible case, suppose that we are imaging a distant object on a screen, using a single positive lens; even if one can somehow correct the other aberrations, there will be positive field curvature that is proportional to the power of the lens. A negative field flattener can reduce the sum of the powers, and hence the field curvature, to near zero. However, it should be clear that field flatteners may not be practical in many cases. For example, mechanical or other constraints may not allow any lenses near the final image. Also, the size of the field flattener must be the same as the size of the image, which may be uncomfortably large.

If $\Delta(u/n) = 0$, then $S_{\mathrm{I}} = S_{\mathrm{II}} = S_{\mathrm{III}} = 0$. The condition $\Delta(u/n) = 0$ is called the *aplanatic condition*; a surface that satisfies it is called an *aplanatic surface*. There is an important limitation that restricts the usefulness of the aplanatic condition: Object and image are constrained to lie on the same side of the surface; so if the object is real, the image must be virtual. This is because

$$\Delta\left(\frac{u}{n}\right) = 0 \rightarrow \frac{u'}{n'} = \frac{u}{n} \rightarrow nl = n'l'$$

for $u = -h/l$. This shows that l and l' have the same sign. Further, using $n'/l' - n/l = c(n' - n)$ and solving for l and l', we obtain,

$$l = \frac{n' + n}{cn}, \qquad l' = \frac{n' + n}{cn'} \tag{8.20}$$

which shows that l and l' are on the same side of the surface as the center of curvature (same sign as c).

The *aplanatic meniscus* is a meniscus lens, as shown in Figure 8.4. The object is at the center of curvature of the first surface ($S_{\mathrm{I}} = S_{\mathrm{II}} = 0$), while the second surface is made to satisfy the aplanatic condition ($S_{\mathrm{I}} = S_{\mathrm{II}} = S_{\mathrm{III}} = 0$). Thus, overall, we have a lens with no spherical aberration or coma. As explained, such a lens cannot form a real image if used at its design conjugates. It is very useful, however, in reducing the numerical aperture of a beam, so that a real image can be formed by a subsequent lens operating at a smaller aperture; this lens is then easier to correct for aberrations. The ratio of the image-space beam aperture to that of the object space beam aperture is given by the inverse of the lens refractive index μ. This follows immediately from the aplanatic condition and the fact that the first surface does not have an effect on the beam aperture. If m aplanatic menisci are cascaded, the beam will have its numerical aperture reduced by $1/\mu^m$.

All the above cases involve primary aberrations only, with the exception of the aplanatic condition, which ensures freedom from higher orders of spherical aberration and coma, but not astigmatism.

Figure 8.4. The aplanatic meniscus.

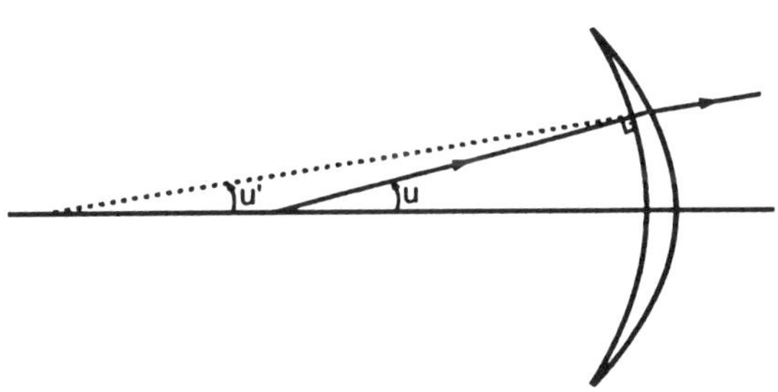

8.1.5. Chromatic Aberrations

We examined chromatic aberration from the ray viewpoint in the previous chapter. That analysis introduced the basic concepts, but did not lend itself to generalization. We overcome this difficulty by using the wavefront aberration, which allows us to determine the contribution of a particular surface to the total aberration, in the same way as was done for the monochromatic aberrations.

Longitudinal chromatic aberration is the chromatic variation of the focus term ${}_0W_{20}$, and is denoted as $\delta_0 W_{20}$. Similarly, transverse chromatic aberration is the chromatic variation of image height, $\delta_1 W_{11}$. There are two sums, denoted by C_{L} and C_{T}, that determine the chromatic aberrations in a way similar to the Seidel sums for the monochromatic aberrations. We have

$$\begin{aligned} \delta_0 W_{20} &= C_{\mathrm{L}}/2 \\ \delta_1 W_{11} &= C_{\mathrm{T}} \end{aligned} \tag{8.21}$$

where

$$\begin{aligned} C_{\mathrm{L}} &= \sum_{i=1}^{k} A_i h_i \, \Delta\left(\frac{\delta n}{n}\right)_i \\ C_{\mathrm{T}} &= \sum_{i=1}^{k} \bar{A}_i h_i \, \Delta\left(\frac{\delta n}{n}\right)_i \end{aligned} \tag{8.22}$$

where $\Delta(\delta n/n)_i \equiv \delta n_{i+1}/n_{i+1} - \delta n_i/n_i$. The index n implies the middle wavelength (e.g., n_d), and δn is the refractive index difference (dispersion) of the material at the two extreme wavelengths chosen (e.g., $n_F - n_C$).

The longitudinal and transverse chromatic aberrations are shown in Figure 8.5, in terms of wavefronts. All wavefronts shown are spherical: The only change is that a different spherical wavefront corresponds to each wavelength. The reference sphere can be any one of these wavefronts, for example, the one corresponding to n_d. The chromatic aberration is the optical path length between the two wavefronts, measured along the ray. The "ray" may correspond to either one of the two wavelengths: The path length would be $n(P_1 P_2)$ or $(n + \delta n)(P_1 P_2)$. Since δn is much smaller than n, the two are approximately equal.

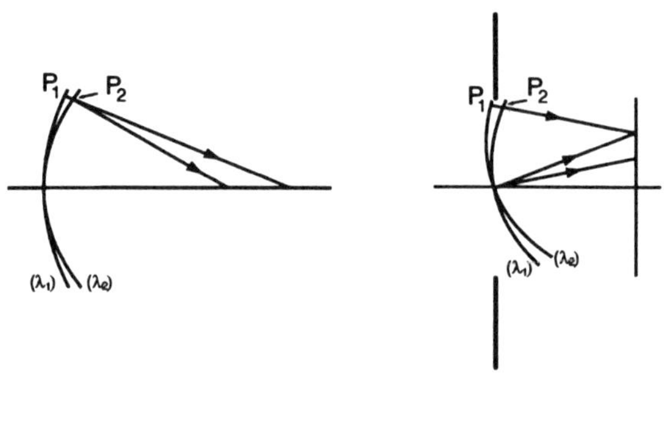

Figure 8.5. Longitudinal and transverse chromatic aberration.

We give here a simple argument to justify the form of C_L. Notice first that in Eq. (8.5) we can take Q and Q' to be on the surface, so that (referring to Figure 8.3), $(PQ) = (PQ') = 0$, and

$$\Delta W = -\Delta(n(PB_0)) = n(PB_0) - n'(PB'_0)$$

We can now use the defocus equations (A4.4) and (A4.5) (Appendix 4). Notice that those equations can be used to give the difference between any two spherical surfaces. If we choose one of the two surfaces to be the wavefront and the other one to be the refracting surface, we can rewrite (Eq. (A4.4) as

$$\Delta W = \tfrac{1}{2}nh^2\left(c - \frac{1}{l}\right)$$

giving the optical path difference between the surface and the wavefront, that is, PB_0. For simplicity, assume that there is no chromatic aberration before the surface. Let p_1 and p'_1 be the optical paths $[PB_0]$ and $[PB'_0]$, respectively, for wavelength 1, and p_2, p'_2 the corresponding paths for wavelength 2. We then have

$$p_1 = \tfrac{1}{2}nh^2\left(c - \frac{1}{l}\right), \qquad p'_1 = \tfrac{1}{2}n'h^2\left(c - \frac{1}{l'}\right)$$

$$p_2 = \tfrac{1}{2}(n + \delta n)h^2\left(c - \frac{1}{l}\right), \qquad p'_2 = \tfrac{1}{2}(n' + \delta n')h^2\left(c - \frac{1}{l'}\right)$$

The chromatic aberration will then be

$$\delta_0 W_{20} \equiv \delta(\Delta W) = (p_1 - p'_1) - (p_2 - p'_2)$$

or

$$\delta_0 W_{20} = \tfrac{1}{2}h^2\,\delta n'\left(c - \frac{1}{l'}\right) - \tfrac{1}{2}h^2\,\delta n\left(c - \frac{1}{l}\right)$$

where use was made of the refraction invariant $D = n(c - 1/l)$ in cancelling terms. Use of the same invariant gives

$$\delta_0 W_{20} = \tfrac{1}{2}h^2 D\,\frac{\delta n'}{n'} - \tfrac{1}{2}h^2 D\,\frac{\delta n}{n}$$

and since $hD = A$, we finally have

$$\delta_0 W_{20} = \tfrac{1}{2}Ah\,\Delta\left(\frac{\delta n}{n}\right)$$

which is the desired result from Eqs. (8.21) and (8.22). An argument along similar lines can be made to justify the transverse chromatic aberration coefficient, C_T.

8.1.6. Design Example: A Simple Camera Objective

We wish to design a simple camera objective, based on a single acrylic lens, with spherical surfaces. The following specifications are given:

$$F = 100 \text{ mm}$$
$$F\# = 17$$
$$\eta' = 50 \text{ mm (film semidiagonal)}$$
$$n = 1.5$$
$$V = 57$$
$$l = \infty, \qquad u_1 = 0$$
$$\text{film resolution} = 50 \text{ cycles/mm}$$

An aperture stop is to be included between the lens and the film. To minimize aberrations, we decide to employ the following tricks: (1) the PPR hits the first surface normally, and (2) the second surface is made aplanatic. To simplify the manual calculation, the lens is to be treated as thin.

With this information, we are asked to complete the Gaussian design, calculate the values of the primary aberrations, and obtain an estimate of the resolution at the different parts of the image, as determined by the various aberrations.

A sketch of the lens is shown in Figure 8.6. Before completing the Gaussian design, let us briefly look at the aberrations. For the first surface, if the PPR is undeviated, we have $\bar{A} = 0$; so $S_{\text{II}} = S_{\text{III}} = S_{\text{V}} = 0$. For the second surface, $\Delta(u/n) = 0$; so $S_{\text{I}} = S_{\text{II}} = S_{\text{III}} = 0$. Therefore, overall, we have $S_{\text{II}} = S_{\text{III}} = 0$, which is not a bad start. We have:

$$F\# = F/2h \rightarrow h = 2.94 \text{ mm}$$
$$u' - u = -hK \rightarrow u' = -0.0294$$
$$H = u'\eta' = -1.47 \text{ mm}$$
$$H = \bar{h}u_1 - h\bar{u}_1 \rightarrow \bar{u}_1 = 0.5$$

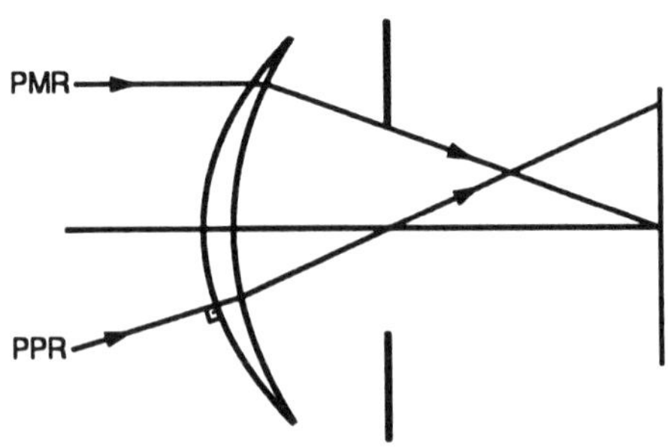

Figure 8.6. Camera objective lens.

In this example, we use the notation $u \equiv u_1$, $u' \equiv u'_2$, etc.; that is, if there is no subscript, the quantity refers to the entire lens, and if there is a subscript, it refers to a surface. The aplanatic condition for the second surface is

$$u'_2 - u'_1/n = 0 \rightarrow u'_2 \equiv u' = u'_1/n \rightarrow u'_1 = -0.0441$$

Also

$$nu'_1 - u = -hc_1(n-1) \rightarrow c_1 = 0.045\ \text{mm}^{-1}, \qquad r_1 = 22.2\ \text{mm}$$

$$\bar{h}c_1 + \bar{u}_1 = 0 \rightarrow \bar{h} = -11.11\ \text{mm}$$

$$\bar{u}' - \bar{u} = -\bar{h}K \rightarrow \bar{u}' = 0.611$$

$$H = -h'_e\bar{u}' \rightarrow h'_e = 2.405\ \text{mm}$$

$$\bar{R}' = -h'_e/u' \rightarrow \bar{R}' = 81.82\ \text{mm}$$

$$K = (n-1)(c_1 - c_2) \rightarrow c_2 = 0.025\ \text{mm}^{-1}, \qquad r_2 = 40\ \text{mm}$$

This completes the Gaussian design. We now proceed to calculate the aberrations. This is normally done by computer, but a manual calculation is instructive at this stage. We have:

$$S_{\text{I}} = (S_{\text{I}})_1 = -A_1^2 h\,\Delta(u/n)_1, \qquad \text{with } A_1 = hc_1 + u \text{ and } \Delta(u/n)_1 = u'_1/n$$

so

$$S_{\text{I}} = 0.00152\ \text{mm} = 1.52\ \mu\text{m}$$

$${}_0W_{40} = \tfrac{1}{8}S_{\text{I}} = 0.19\ \mu\text{m} \cong 0.3\lambda, \qquad \text{for } \lambda \cong 0.56\ \mu\text{m}$$

A look at the Strehl tolerances (Section 7.3.4) would convince us that 0.3λ of spherical aberration is practically undetectable. We could, in any case, estimate the resolution degradation caused by the aberration (at optimum focus W_{20}), using reasoning similar to that for the other aberrations below. Since S_{II} and S_{III} are zero, we estimate the effect of S_{IV} next. From Eq. (8.9),

$$S_{\text{IV}} = 0.0144\ \text{mm} \cong 27\lambda \qquad \text{and} \qquad {}_2W_{20} = \tfrac{1}{4}(S_{\text{III}} + S_{\text{IV}}) \cong 7\lambda$$

This is substantial. If this were the only aberration, we would have

$$W = {}_2W_{20}\tau^2 y^2 \text{ (meridional section)} \rightarrow \delta H = \frac{\partial W}{\partial y} = 2\,{}_2W_{20}\tau^2 y$$

For the worst-case ray $y = 1$ (edge of aperture), and at the edge of the field, $\tau = 1$, so that $\delta H \cong 14\lambda$. The non-normalized transverse ray aberration is $\delta\eta = \delta H/u' \cong 0.238$ mm (absolute value). This corresponds to a resolution of only about 4 cycles/mm, considerably less than the film resolution of 50 cycles/mm. Because of the τ^2 dependence, the resolution will improve quadratically as we move toward the middle of the image. For example, at $\eta' = 25$ mm ($\tau = \frac{1}{2}$), we would have 16 cycles/mm, and at image heights of less than around $\frac{1}{3}$ of the total, the resolution would be limited by the film. However, there are more aberrations to consider. We have

$$S_V = (S_V)_2 = -0.136 \text{ mm}$$

$${}_3W_{11} = \tfrac{1}{2}S_V = -0.068 \text{ mm}$$

The corresponding normalized ray aberration (linear in y) is $\delta H = \partial W/\partial y = {}_3W_{11}$. The actual ray aberration is $\delta\eta' = \delta H/n'u' = 2.31$ mm. Recall that distortion does not cause image blur, but represents a change in magnification over the image. Therefore, its effect is best appreciated if it is expressed as a fraction, through division by the Gaussian image height. So we have for the fractional distortion

$$\frac{\delta\eta'}{\eta'} = 0.046 \cong 5\%$$

the positive sign indicating pincushion distortion. This value is non-negligible, but it is tolerable for this type of lens. So we finally calculate the chromatic aberrations from Eqs. (8.22) and (8.21).

$$C_L = 0.00152 \text{ mm} \rightarrow \delta_0 W_{20} = 0.76\ \mu\text{m} \cong 1.3\lambda$$

This might be significant, but it is much less than the transverse chromatic aberration

$$C_T = \delta_1 W_{11} \cong 11\lambda$$

As before, we can obtain the denormalized TRA at full aperture and field, which is found to be about 0.2 mm, leading to a resolution of about 5 cycles/mm at the edge of the field. This is only marginally better than the number obtained for the field curvature.

Conclusion. This design is limited by field curvature and transverse chromatic aberration. With a single positive lens, these aberrations are inevitable and cannot be corrected further to any significant degree. At the edge of the field, the resolution is limited by both aberrations approximately equally. However, field curvature improves quadratically as we move toward the middle of the field, while transverse color improves only linearly. Thus at the middle of the field, the resolution due to field curvature would be 16 cycles/mm, while that due to color only 10 cycles/mm. At $\frac{1}{4}$ of the full field, the field curvature is no longer important, but transverse color still limits the camera performance (20 cycles/mm). An achromatic doublet should be capable of better performance, but the additional element will increase the cost. (Of course, we assume that the film is approximately equally sensitive to all wavelengths in the visible part of the spectrum.)

8.2. ASTIGMATISM AND FIELD CURVATURE

We discuss here the interaction between astigmatism and field curvature. This is a fundamental aspect of the aberration correction of a system, and a clear understanding is needed. Recall that the field curvature term is given by

$$_2W_{20}\tau^2 r^2 = {_2W_{20}}\tau^2(x^2 + y^2) = \tfrac{1}{4}(S_{III} + S_{IV})\tau^2(x^2 + y^2)$$

while the astigmatism term is

$$_2W_{22}\tau^2 r^2 \cos^2\theta = {_2W_{22}}\tau^2 y^2 = \tfrac{1}{2}S_{III}\tau^2 y^2$$

The sum of these two terms within the aberration polynomial can then be written as

$$_2W_{22}\tau^2 r^2 \cos^2\theta + {_2W_{20}}\tau^2 r^2 = \tfrac{1}{4}(3S_{III} + S_{IV})\tau^2 y^2 + \tfrac{1}{4}(S_{III} + S_{IV})\tau^2 x^2$$

These two terms are called the tangential and sagittal field curvature, respectively. They represent the focal shifts in the y and the x directions that produce the tangential and sagittal astigmatic lines. If either one of these terms is zero, the corresponding field curve is flat. From the preceding equation, we define tangential and sagittal astigmatism, W_T and W_S, as

$$W_T = \tfrac{1}{4}(3S_{III} + S_{IV}), \qquad W_S = \tfrac{1}{4}(S_{III} + S_{IV}) \tag{8.23}$$

The circle of least confusion lies in between the two astigmatic lines: The corresponding wavefront aberration is $\tfrac{1}{4}(2S_{III} + S_{IV})$.

As mentioned earlier, the designer has more degrees of freedom in controlling S_{III} than S_{IV}. This means that the designer has the option of introducing astigmatism (or not minizing it) in order to flatten the fields. Since S_{IV} is more or less fixed for a given system (and most often positive), we can plot W_T and W_S as a function of S_{III}, which results in Figure 8.7. It should be made clear

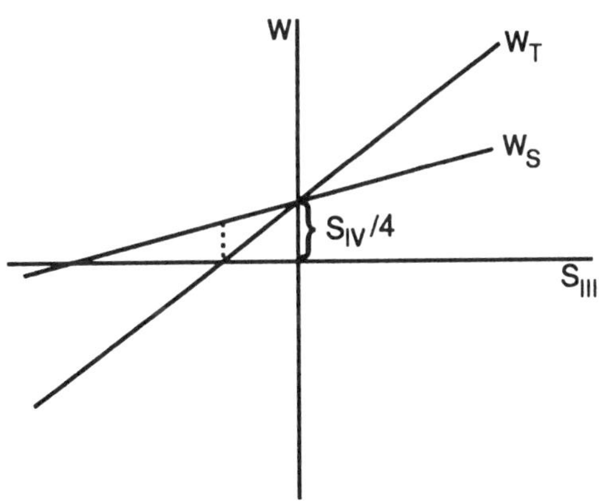

Figure 8.7. Sagittal and tangential astigmatism as a function of S_{III}.

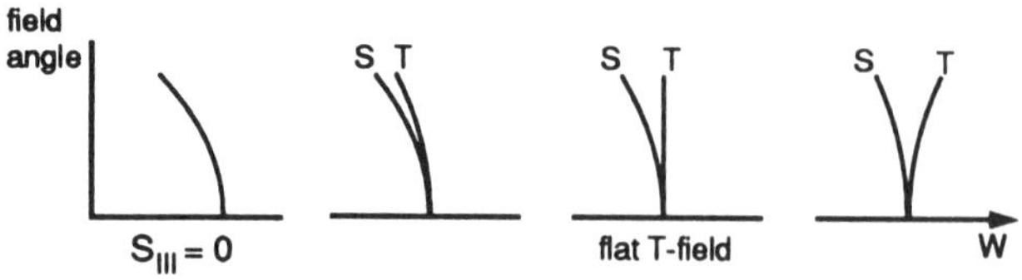

Figure 8.8. Introduction of astigmatism serves to flatten the field. As S_{III} increases, the T curve moves three times faster than the S curve (compare with Figure 8.7).

here that S_{III} is not variable between arbitrary limits; rather, different values of S_{III} represent different potential designs of the same system, the difference arising from the stop location. Because of the steeper slope of W_T, the condition $W_T = 0$ (flat T field) is often (but not always) desirable, as it leads to small values of W_S and total aberration. By contrast, the condition $W_S = 0$ leads to a large value of W_T.

The amount of aberration can be plotted against field angle (or normalized coordinate), as in Figure 8.8. The curves shown are called the *astigmatic field curves*. Within the primary aberration approximation, they are all spherical (higher-order aberration can in fact be beneficial in reducing the effect of lower orders, by making the fields curve backwards and become flatter overall). Figure 8.8a shows no astigmatism ($S_{III} = 0$); so the Petzval sum S_{IV} gives a true representation of the field curvature. As astigmatism is added (with the sign of S_{III} being opposite to that of S_{IV}), the S and T fields separate, while at the same time, the overall "field" becomes flatter. The designer must decide on the best compromise between astigmatism and field curvature depending on the detailed system requirements.

8.3. PRIMARY ABERRATIONS OF A REFLECTING PRISM (PLANE-PARALLEL PLATES)

A reflecting prism can degrade the wavefront if its surfaces are not perfectly plane, if there are any refractive index inhomogeneities, etc. Here we ignore these effects, by assuming an ideal prism. In that case, the effect of the prism is the same as that of a block of glass with plane and parallel faces. The thickness of the block d is calculated from the tunnel diagram for the prism. The aberrations can then be calculated easily from Eqs. (8.6), (8.7) and (8.21), (8.22). For simplicity, we take the prism to be in air; the general case is derived with similar ease. We have

$$c_1 = c_2 = 0, \qquad A_1 = u_1 \equiv u = nu'_1 = u'_2 \equiv u' = A_2$$

$$S_I = -u_1^2 h_1\left(\frac{u'_1}{n} - u_1\right) - u_2'^2 h_2\left(u'_2 - \frac{u'_1}{n}\right) = u_1^2\left(\frac{u'_1}{n} - u_1\right)(h_2 - h_1)$$

Also, $h_2 - h_1 = du'_1 = du_1/n$; so finally

$$S_{\rm I} = -\frac{n^2 - 1}{n^3} u^4 d \tag{8.24}$$

and similarly,

$$S_{\rm II} = \frac{\bar{u}}{u} S_{\rm I}$$

$$S_{\rm III} = \left(\frac{\bar{u}}{u}\right)^2 S_{\rm I} \tag{8.25}$$

$$S_{\rm V} = \left(\frac{\bar{u}}{u}\right)^3 S_{\rm I}$$

with $S_{\rm IV} = 0$. Notice that if $u = 0$, all four aberrations become zero.

For a numerical example, take $n = 1.5$, $u = 0.125$ ($F\# = 4$). Then $S_{\rm I}/d = 9 \times 10^{-5}$; so for $d = 1$ cm $= 10^4$ μm, $S_{\rm I} = 0.9$ μm $\rightarrow {}_0W_{40} = 0.11$ μm $\cong 0.2\lambda$, which is generally negligible. However, if the prism is thick (e.g., 5 cm or more), the spherical aberration would be significant. Also, for a more strongly focused beam (e.g., $F\# = 1$), then $u = 0.5$, and the spherical aberration is approximately 50 times larger, because of the u^4 dependence. Thus a glass block or prism cannot normally be used in a strongly focused beam, unless care is taken to balance the aberrations it produces with those of other elements in the system. Where image quality is paramount, as in microscopy, high-power oil-immersion objectives that operate at a numerical aperture of greater than 1 ($F\# < 0.5$) have some built in spherical aberration to counteract the spherical aberration of the protective cover slip ($d \cong 0.2$ mm) that is often placed over the sample. Objectives for use without cover slip are corrected for spherical aberration.

8.4. PRIMARY ABERRATIONS OF A SPHERICAL MIRROR

It is a simple matter to calculate the aberrations of a spherical mirror, based on the previous formulae. We have $n' = -n \rightarrow \Delta(u/n) = -(u' + u)/n = 2hc/n$, from the mirror refraction formulae. Therefore,

$$S_{\rm I} = -\frac{2A^2h^2c}{n} \tag{8.26}$$

with

$$S_{\rm II} = \frac{\bar{A}}{A} S_{\rm I}, \qquad S_{\rm III} = \left(\frac{\bar{A}}{A}\right)^2 S_{\rm I} \tag{8.27}$$

Also

$$S_{\rm IV} = \frac{2H^2c}{n} \tag{8.28}$$

with $S_V = (\bar{A}/A)(S_{III} + S_{IV})$ as usual. In the special case where the object is at infinity ($u = 0$), we have

$$S_I = -2nc^3h^4 \tag{8.29}$$

proof of which is left as an exercise. The other expressions do not simplify very much because of the presence of $\bar{A}$. There are two other special cases worth noting: (1) if the stop is at the mirror, and (2) if the stop is at the center of curvature. In the first case, we have $\bar{h} = 0$; so $\bar{A} = n\bar{u}$. The spherical aberration is not affected by the stop location. For the other aberrations, we have

$$S_{II} = -2n\bar{u}^2h^2c = 2AHhc$$

$$S_{III} = -2n\bar{u}^2h^2c = -\frac{2}{n}H^2c = -S_{IV} \tag{8.30}$$

and therefore

$$S_V = 0 \tag{8.31}$$

If the stop is at the center of curvature, then the PPR is reflected back on itself, and $\bar{h}/R = \bar{h}c = -\bar{u} \rightarrow \bar{A} = 0$; so in this case, there is no coma, astigmatism, or distortion.

PROBLEMS

1. A concave mirror has a radius of curvature of 40 cm, and an aperture (diameter) of 10 cm. Determine the magnitude of the primary aberrations of the mirror as a function of the field size, in the following two cases. (a) Stop at the mirror, (b) stop at the center of curvature ($u = 0$, $\bar{u} = 0.05$).

2. Is it possible to design a thin-lens achromatic doublet with zero Petzval sum? What limitations are imposed by the range of available glasses? What if the achromatic requirement were removed?

3. Design an aplanatic meniscus for the specification: $u_1 = \sin\alpha = -0.5$, $\lambda = 500$ nm, $l_1 = -10$ mm, $n = 1.5$, $H = -0.25$ mm, stop in contact with surface 2. Use the Seidel aberration computation program of Appendix 5 to verify your calculation.

4. Calculate the spherical aberration caused by a microscope cover slip that is 0.15 mm thick ($n = 1.5$) for an objective NA of 0.26 and 0.8. Comment on how necessary it is to correct for this in the design of the objective.

5. A right-angle prism ($n = 1.5$, side $= 1\ \text{cm}^2$) is used in a converging beam (single reflection). What is the minimum beam $F\#$ if the prism is to contribute no more than 1λ of spherical aberration? If the stop is taken to be at the first face and $\bar{u} = 0.01$, calculate the off-axis aberrations.

6. Design a singlet photographic objective with the following specifications: $F = 100$ mm, $F/10$, FOV: $\pm 5.7°$, object at infinity, $n = 1.5$, $V = 60$, $\lambda = 0.0005$ mm. Each surface of the singlet must have either $A = 0$ or $\bar{A} = 0$ or $\Delta(u/n) = 0$. How many designs can exist? Which one do you think is the best?

7. Investigate and comment on the extent to which the aberrations of a spherical mirror can be corrected by a plane glass plate.

8. Prove Eq. (8.17).

9. Show that the optical invariant H can be written in terms of the paraxial refraction invariants A and $\bar{A}$ as

$$H = A\bar{h} - \bar{A}h$$

10. Show that the *angular aberration* along the Y meridian can be written as

$$\delta u = \frac{1}{n}\frac{\partial W}{\partial Y}$$

following the notation of Chapter 7 and omitting primes. Hence show that the angular spherical aberration is given by $(4/n)(W_{40}/h_e)$, and the angular lateral color by $\partial W_{11}/h_e$. (Note: The concept of angular aberration is not particularly meaningful when applied to other Seidel aberration types.)

11. Can a Cassegrain mirror be corrected for both spherical and coma?

Aberrations of a Thin Lens in Air

9

"... I soon learnt that rectilinear images could not easily be obtained without dividing the positive lens into two positive lenses equal or unequal in power and placing the negative lens between them."

H. Dennis Taylor, in "Optical Designing as an Art," *Trans. Opt. Soc.* **24**, 143 (1923).

In this chapter, we will derive expressions for the primary aberrations of a thin lens in air, thus gaining some appreciation of the effect of the lens shape on aberrations. We will also apply the concepts to the design of an aberration-corrected doublet and other cases, so as to appreciate what is involved in correcting the aberrations of a system. Finally, we will examine in detail the effect of the location of the aperture stop on aberrations.

9.1. CENTRAL ABERRATIONS (STOP AT THE LENS)

We begin by examining the aberrations of a thin lens with the stop at the lens. We will call these "central" aberrations and denote them by the subscript c.

9.1.1. Shape-Independent Aberrations

We have already encountered the Petzval sum for a thin lens [Eq. (8.9)]. That was found to be

$$S_{\mathrm{IV}} = S_{\mathrm{IVc}} = \frac{H^2 K}{n} \tag{9.1}$$

Clearly, the value of S_{IV} does not depend on the specific curvatures of the lens surfaces, but only on its power; this is why we call it shape independent. As was discussed in Chapter 8, the Petzval sum for an entire system depends only on the sum of the powers of the individual elements. But since the powers are usually set by the Gaussian requirements of the system, the Petzval sum can be very difficult to correct. It is advisable to consider the value of the Petzval sum at the Gaussian design stage: If it is excessive, then alternative design approaches may have to be sought.

Let us consider central astigmatism next. We have

$$S_{\mathrm{IIIc}} = -\bar{A}_1^2 h\,\Delta\left(\frac{u}{n}\right)_1 - \bar{A}_2^2 h\,\Delta\left(\frac{u}{n}\right)_2$$

where $h_1 = h_2 = h$, since the lens is considered thin, and $\bar{A}_1 = \bar{A}_2 = \bar{u}$ for $\bar{h} = 0$. Also, $H = -h\bar{u}$; therefore, $\bar{A} = H/h$. Upon substituting for $\Delta(u/n)$ and using the thin-lens formula, we find

$$S_{\mathrm{IIIc}} = H^2 K \tag{9.2}$$

So central astigmatism is also hard to correct; but as we will see later, astigmatism can be controlled through the stop position.

To calculate central distortion, consider Eq. (8.8). We have $\bar{h} = 0$, and $\Delta(h\bar{u}^2)_1 = h(\bar{u}_1'^2 - \bar{u}_1^2) = h\bar{u}_1^2(n^2 - 1) = -\Delta(h\bar{u}^2)_2$ since $\bar{u}_1 = \bar{u}' \equiv \bar{u}_2'$. Therefore

$$S_{\mathrm{Vc}} = 0 \tag{9.3}$$

As we will show in the subsequent sections, spherical aberration and coma depend on the lens shape. The chromatic aberrations, however, do not. We have

$$C_{\mathrm{L}} = C_{\mathrm{L}_1} + C_{\mathrm{L}_2} = h\frac{\delta n}{n}(hc_1 + u_1 - hc_2 - u_2') = h\frac{\delta n}{n}[h(c_1 - c_2) + u_1 - u_2']$$

$$= h\frac{\delta n}{n}[hK/(n-1) + hK] = h^2 K\,\delta n/(n-1)$$

so that

$$C_{\mathrm{L}} = C_{\mathrm{Lc}} = \frac{h^2 K}{V} \tag{9.4}$$

Following the same steps, the reader can show that

$$C_{\mathrm{T}} = \frac{h\bar{h}K}{V} \tag{9.5}$$

so that

$$C_{\mathrm{Tc}} = 0 \tag{9.6}$$

Thus a field lens ($h = 0$) does not contribute to the primary chromatic aberrations. If $h \neq 0$ but the stop is at the lens (e.g., the objective of a Keplerian telescope), then the lens does not contribute to transverse color. So we have concluded the calculation of all the shape-independent aberrations.

Example: Alignment tolerances for a collimator. You are an optical consultant. One day you receive a telephone call from High Price Laser Optics, Inc. Sales of their HeNe laser beam expanders are losing out to a rival manufacturer, and they are trying to cut costs, in order to make them more competitive. Owing to a slump in demand for water pistols brought about by a drought, the Precision Toy Company of Dragsville is willing to manufacture the lens mount for the collimator component of the beam expander, a thin doublet corrected for spherical aberration and coma that is located at the aperture stop of the system. It has a focal length of 200 mm and a diameter of 80 mm. The current standard machining tolerance on the squareness of the mount to the optical axis is 5 minutes of arc, when made by HPLO, but Precision Toy can only manage 25 minutes of arc. Will this still give diffraction-limited performance? Your caller must have a reply within an hour, as Precision Toy is about to issue redundancy notices to their staff.

Solution. The numerical aperture for the lens is NA $= 40/200 = 0.2$. This beam is sufficiently focused to allow us to ignore any differences between Gaussian and uniform beams at first approximation. The problem specifies that there is no spherical aberration and no coma; therefore, only astigmatism remains. Notice that any tilt of the lens with respect to the optical axis is equivalent to the object point being off-axis.* The amount of central astigmatism is H^2K from Eq. (9.2). At the stop, $H = -h\bar{u}$, with $h = 40$ mm and $\bar{u} = 25$ mm. So $|H| = 0.29$ mm, and ${}_2W_{22} = S_{\text{III}}/2 = 0.2\ \mu\text{m} \cong 0.31\lambda$ for $\lambda = 0.633\ \mu\text{m}$. A look at the Strehl tolerances (Section 7.3.5) shows that 0.31λ of astigmatism is just within the tolerance (this is not to say that the effect is entirely negligible, but rather, that no more than the usual amount of marketing exaggeration would be needed in order to sell the Precision Toy collimator as diffraction limited). Also, if the beam is more Gaussian than uniform, its energy would be concentrated toward the middle of the aperture, thus reducing the effect of the aberration.

9.1.2. The Shape Factor and the Magnification Parameter

Before examining the dependence of aberrations on the surface curvatures, we need to introduce two auxiliary parameters. The shape factor of a thin lens, X, is defined as

$$X = \frac{c_1 + c_2}{c_1 - c_2} \tag{9.7}$$

The reason for introducing X is that c_1 and c_2 are not independent, as their difference is proportional to the lens power. Therefore, only one parameter is enough to describe the effect of the lens; the shape factor X is a symmetrical parameter and leads to simpler formulas.

Figure 9.1 shows different lens shapes and their corresponding X values.

* Of course, this is true if the lens is considered in isolation. In general, if the lens is part of a larger lens system, then misalignment destroys the rotational symmetry of the system.

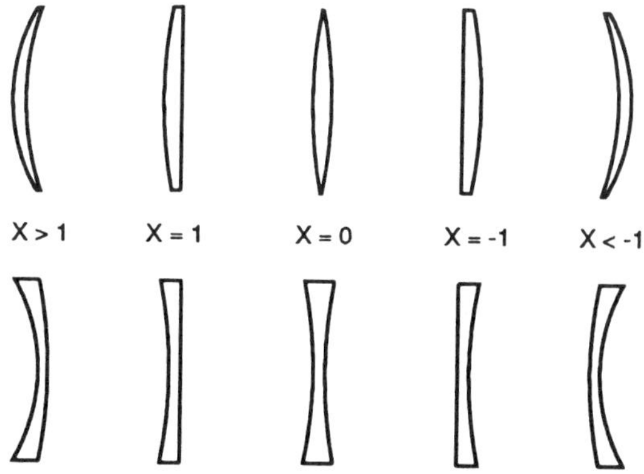

Figure 9.1. Lens shapes corresponding to different values of the shape factor X.

Notice that X can also be written as

$$X = \frac{n - 1}{K}(c_1 + c_2) \tag{9.8}$$

The shape factor is also often called *bending factor* for reasons that should be obvious from Figure 9.1.

The magnification parameter is defined as

$$Y = \frac{u + u'}{u - u'} \tag{9.9}$$

where $u \equiv u_1$, $u' \equiv u'_2$, and u is the PMR angle variable. Y is also called the *conjugate parameter*. The conjugate positions that correspond to various values of Y are shown in Figure 9.2. Notice that

$$Y = \frac{u + u'}{hK} \tag{9.10}$$

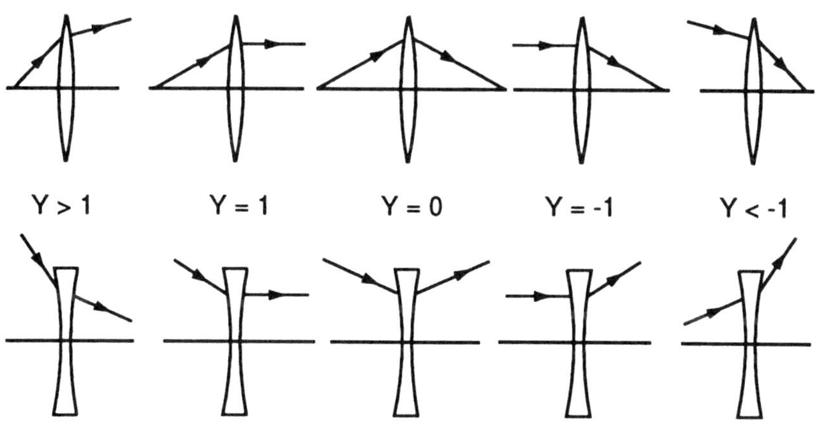

Figure 9.2. Conjugate locations for different values of the conjugate parameter Y.

so this parameter too is dimensionless, symmetrical and normalized with respect to K. The name magnification parameter can be justified through the relation

$$M = \frac{Y+1}{Y-1} \tag{9.11}$$

proof of which is left as an exercise. Also, the following relations are easily proven:

$$\begin{aligned} c_1 &= \frac{K(X+1)}{2(n-1)} \\ c_2 &= \frac{K(X-1)}{2(n-1)} \\ u \equiv u_1 &= \frac{hK}{2}(Y+1) \\ u' \equiv u_2' &= \frac{kK}{2}(Y-1) \end{aligned} \tag{9.12}$$

9.1.3. Shape-Dependent Aberrations

We are now ready to calculate the two shape-dependent central aberrations. We have

$$S_{\mathrm{Ic}} = S_{\mathrm{I}_1} + S_{\mathrm{I}_2} = A_1^2 h\,\Delta\left(\frac{u}{n}\right)_1 + A_2^2 h\,\Delta\left(\frac{u}{n}\right)_2 \tag{9.13}$$

with $A_1 = hc_1 + u$ and $A_2 = hc_2 + u'$. The variable u_1' that appears in $\Delta(u/n)$ can be found from $nu_1' - u = -hc_1(n-1)$. Thus Eq. (9.13) contains only h, n, c_1, c_2, u, u'. The values of the last four of these can be substituted into Eq. (9.13) from Eq. (9.12), in terms of X and Y. If we do that, it follows in (admittedly more than) "a few lines" that

$$S_{\mathrm{I}} = S_{\mathrm{Ic}} = \frac{h^4K^3}{4}\left[\frac{n+2}{n(n-1)^2}\left(X + \frac{2(n^2-1)}{n+2}Y\right)^2 + \left(\frac{n}{n-1}\right)^2 - \frac{n}{n+2}Y^2\right] \tag{9.14}$$

This equation is of the form $S_{\mathrm{I}} = a(X+b)^2 + c$; so a plot of S_{I} as a function of X will be a parabola. We choose to examine the behavior of S_{I} with X rather than with Y, because Y is normally fixed by the Gaussian system specification, whereas the lens shape can vary. The vertex of the parabola is at the point $X = -b$; that is

$$X_{\mathrm{vx}} = -\frac{2(n^2-1)}{n+2}Y \tag{9.15}$$

The value of X at the vertex represents the best shape; that is, it gives the minimum $|S_I|$, assuming of course that the parabola does not cross the X axis. The value of S_I at the vertex is

$$S_{Ivx} = \frac{h^4K^3}{4}\left[\left(\frac{n}{n-1}\right)^2 - \frac{n}{n+2}Y^2\right] \tag{9.16}$$

The parabola can cross the axis, thus giving zero aberration, but only for unusually high values of X (strong meniscus lens). Some other conclusions can be drawn by examining the properties of Eq. (9.14), for example, that the width of the curve is independent of Y, and that S_I decreases monotonically with n (so it is advantageous to use high-index glass, except that such glass may have other undesirable properties in practice). The properties of the spherical aberration parabola are shown in Figure 9.3. But there are also two more specific useful conclusions, which should be memorized.

1. For $Y = -1$ (object at infinity) and for $n \cong 1.686$, we have $2(n^2 - 1)/(n + 2) = 1$, and $X_{vx} = 1$; that is, the best shape is a planoconvex lens, with the curved surface toward the infinite conjugate. From this we can also derive the following approximate rule of thumb: A 1 in. (or 25 mm)

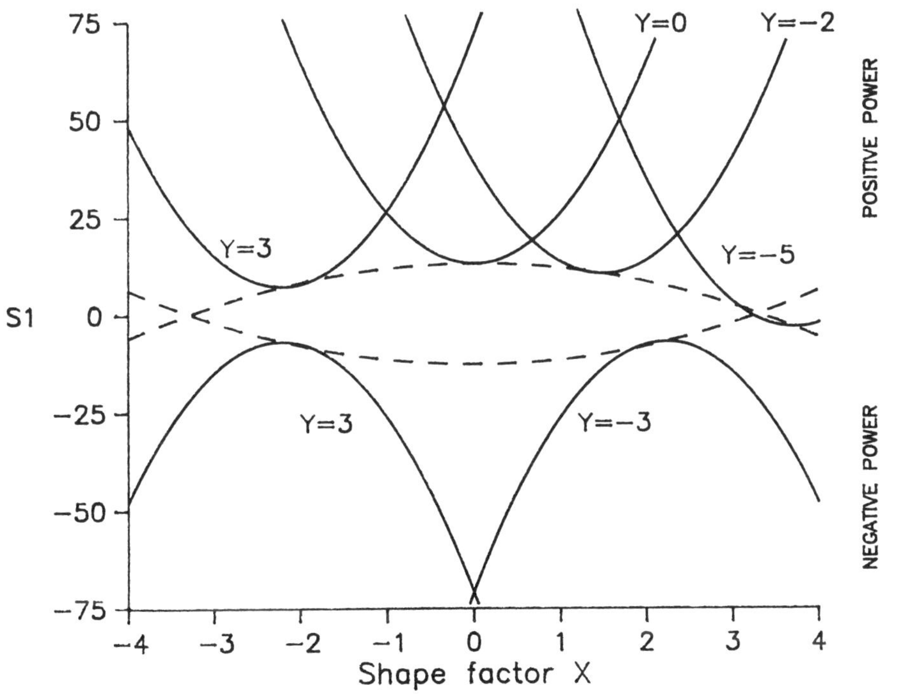

Figure 9.3. Properties of the spherical aberration parabola. The vertices of the parabolas for different Y values fall on two other broader parabolas (dashed). Notice that the parabola crosses the horizontal axis only for large (absolute) values of Y.

focal length lens at $F/5$ has approximately one wavelength of spherical aberration at best shape (spherical aberration will increase in proportion to the focal length, and in inverse proportion to the fourth power of the $F\#$). Notice that most singlet off-the-shelf lenses are usually made from borosilicate crown glass, with $n \cong 1.52$; so the best shape for $Y = -1$ is not exactly planoconvex; however, the difference is small.

2. For $Y = 0$, the best shape is $X = 0$. Therefore, an equiconvex lens is the best choice for symmetrical conjugates ($M \cong -1$, or $l \cong -2F$. $l' \cong 2F$).

Another qualitative rule of thumb contains both these cases. This rule is that the best shape tends to be that which distributes the ray bending equally between the two surfaces of the lens. To appreciate this, consider the planoconvex lens with the flat surface toward the infinite conjugate (i.e., the "wrong" way); in that case, the first surface does not bend the PMR at all (PMR perpendicular to the surface); all the work is done by the second surface. But if the lens is turned around, then both surfaces are contributing approximately equal amounts of ray bending. This is also easy to see in the case of an equiconvex lens with symmetrical conjugates.

We now consider the case of coma. An equation similar to Eq. (9.13) can be written for S_{II}, while S_{IIc} is obtained by setting $\bar{u} = 0$. The same substitutions of Eq. (9.12) are then made, to arrive at the following equation

$$S_{\mathrm{IIc}} = -\tfrac{1}{2}Hh^2K^2\left(\frac{n+1}{n(n-1)}X + \frac{2n+1}{n}Y\right) \tag{9.17}$$

This is a linear equation in X. The slope of the line is always positive if we choose the sign of H to be always negative, corresponding, for example, to $u > 0$ and $\eta < 0$ (notice that the sign of H has no physical significance; only H^2 does). Thus there is always a shape that gives zero coma. Setting $S_{\mathrm{IIc}} = 0$ in Eq. (9.17), we obtain

$$X_0 = -\frac{(2n+1)(n-1)}{n+1}Y \tag{9.18}$$

For $Y = 0$ we also have $X = 0$; so the equiconvex shape minimizes coma (as well as spherical aberration) at $M = -1$. In choosing the best shape lens, it is often found that the X-axis intercept of the coma line is close to the vertex of the spherical aberration parabola (see Figure 9.4); so it is possible to find a good compromise shape that will give small spherical aberration as well as coma. Coma is a more objectionable aberration; so it is typically advantageous to minimize S_{II} instead of S_{I}. Also, if we are near the vertex, the parabola has a small slope; so it varies slowly with X, whereas the slope of the coma line is larger and independent of X.

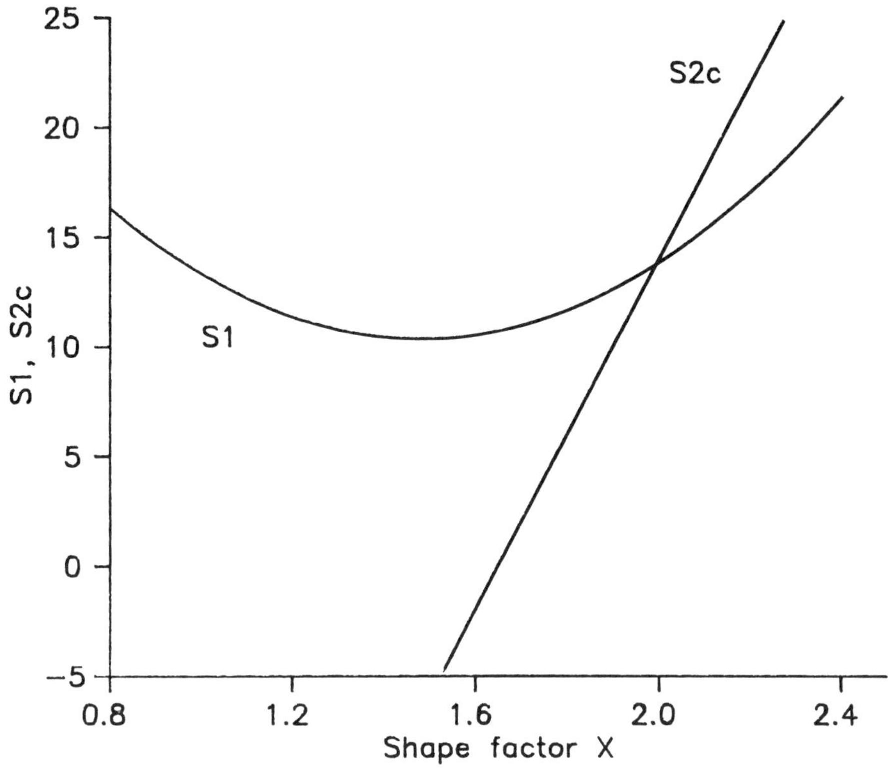

Figure 9.4. Example of spherical aberration and central coma for a single-element lens.

9.1.4. The Corrected Doublet

With two thin lenses in contact, it is possible to eliminate both spherical aberration and coma, by judicious choice of the lens shapes. Such a doublet is also called *aplanatic*. In addition, it is possible to have zero chromatic aberration, since the latter is controlled only by the dispersion and the power of the elements, as we have already seen. As an example, let us design a thin achromatic doublet of power K with the object at infinity. We require zero targets for spherical aberration and central coma.

Remember that nothing can be done about S_{IV} and S_{IIIc} since they are shape independent. However, distortion and transverse color are zero when the stop is at the lens. Also, longitudinal color can be made zero with the right combination of dispersion and power. From the achromatic doublet discussion, we recall that if we choose $V_2 = 2V_1$, then $K_1 = 2K$ and $K_2 = -K$.

We have a choice whether the first element will be the crown or the flint. The crown is more commonly chosen, so $V_1 > V_2$; the doublet is then called *crown leading*. Since the object is at infinity, we have $Y_1 = -1$. Also, $Y_2 = -2K/K_2 + 1$. At this point, we may want to re-examine our choice of dispersions, to make sure that Y_2 is not very large, as this might lead to substantial higher-order aberrations, which are excluded from our calculation. For $K_2 = -K$ we have $Y_2 = -3$. If we had chosen $V_2/V_1 = \frac{2}{3}$, then Y_2 would be -2, but the powers of the elements would be larger. Thus we compromise to either of the above or any in between choice.

Thus we know Y_1, Y_2, K_1, K_2, h. The required field of view ($\bar{u}$) determines H. Also, the V values will determine the choice of refractive indices from the glass chart (some limited flexibility is possible here). Thus we have determined all the parameters that appear in the first and second Seidel sums. Our equations for zero aberration will then be

$$\begin{aligned} \sum S_{\rm I} &= S_{{\rm I}_1} + S_{{\rm I}_2} = 0 \\ \sum S_{\rm IIc} &= S_{{\rm IIc}_1} + S_{{\rm IIc}_2} = 0 \end{aligned} \tag{9.19}$$

These two equations can be solved simultaneously for X_1 and X_2 [and from Eqs. (9.12) we can then obtain the surface curvatures]. There are always two real solutions. Substitution of some typical numbers gives the shapes shown in Figure 9.5a, where the lenses are shown with finite thickness for obvious reasons. Historically, the solution with the strong meniscus lens is called a *Gauss doublet*, and the other is the *Fraunhofer doublet*. The Gauss solution is relatively rare; derivatives of that design are used in exotic, large-aperture systems (for which, however, the primary aberration approach is inadequate). It has the disadvantage that most of the aberration arises from the first surface and is cancelled by the last one, which tends to lead to substantial higher-order aberration. In the Fraunhofer solution, the aberration arises from the second surface and is cancelled by the third. It is a good practice, whenever possible, to correct the aberrations of a surface as soon as possible after the surface.

It is also possible to obtain a flint-leading solution. The corresponding shapes, with all parameters the same as in the previous example, are shown in Figure 9.5b.

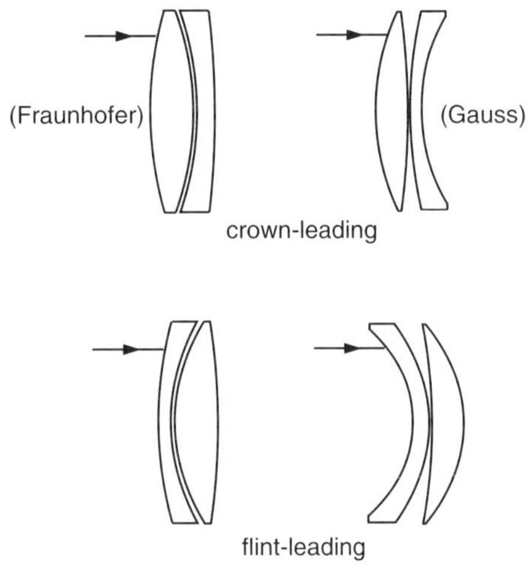

Figure 9.5. Contacted doublets corrected for spherical aberration and central coma.

Although in the example we took the object to be at infinity, a solution can be obtained for any object location. The general analytical solution of Eq. (9.19), although relatively easy to obtain, is rather cumbersome. Instead of reproducing it here, we give it in the form of a simple computer program in Appendix 5.

It is often advantageous to eliminate the air space between the two lenses, so that we have a cemented doublet. Such doublets are of the Fraunhofer type. Exact solution for zero aberration is not always possible because X_1 and X_2 cannot vary independently, but for appropriate combinations of refractive indices, solutions with practically zero spherical and coma can be found.* A specific example is given in Section 10.2.

9.1.5. A Practical Aberration Primer

This section shows some common rules of thumb that are very useful in putting together optical systems with off-the-shelf lenses. The optical designer is almost always concerned with a cost versus performance compromise, and off-the-shelf lenses are much less expensive than custom-made ones. Several optics companies offer a selection of lenses from which it is sometimes possible to construct an acceptable design. Typically, there is a collection of planoconvex, equiconvex, planoconcave, equiconcave, meniscus, and cemented doublet (achromatic) lenses. The doublets are normally optimized for use with one distant (infinite) conjugate. The rules for selecting and orienting these lenses have already been discussed or can be derived from the theory presented. They are summarized below:

1. When one conjugate is distant or at infinity, choose a plano lens with the curved surface toward the distant conjugate.
2. When the conjugates are approximately symmetrical about the lens, choose an equiconvex lens.

In both cases, we can gain by using more than one lenses. Starting from the second, notice that an equiconvex singlet lens can be thought of as comprising two planoconvex elements, if we split it in half (Figure 9.6). However, as the figure indicates, these planoconvex elements are then used the "wrong way around"; that is, they violate rule 1. If they are turned so as to have the flat surface toward the finite conjugate, we can expect better performance. The corollary is that two planoconvex lenses, with their curved sides facing each other, do a better job at symmetrical conjugates than a single equiconvex lens. Simple condenser lenses are often designed as two planoconvex lenses oriented as above.

Returning now to the first rule, it is clear that a doublet lens will outperform the singlet. Commercially available achromatic cemented doublets are of the

---

* See M. Iqbal Khan and J. Macdonald, "Cemented Doublets: A method for rapid design," *Optica Acta* **29**, 807–22 (1982).

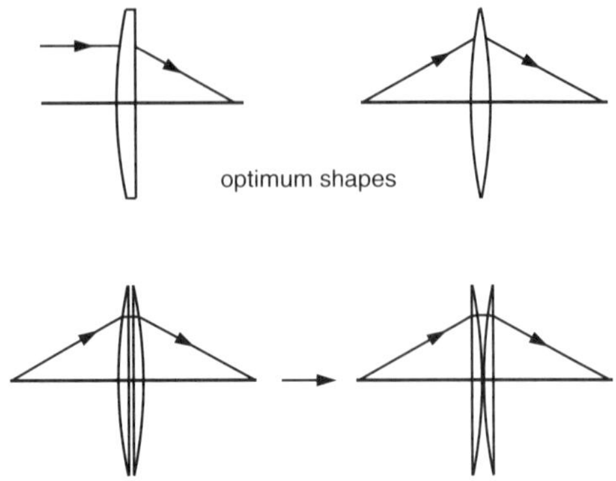

Figure 9.6. Optimum shapes. The two planoconvex lenses (with the curved surfaces in contact) provide better performance than a single biconvex element for symmetrical conjugates.

Fraunhofer type, which has the first surface more curved than the last one. Therefore, when orienting these lenses, the most curved surface must face the long conjugate. These doublets can also be used for imaging at symmetrical conjugates, if two of them are oriented with their curved faces facing each other.

It should be emphasized that these rules apply if the stop is at the lens. For a remote stop, the shape or orientation that optimizes the off-axis aberrations has to be established through detailed calculation. As an example, consider the singlet camera lens shown in Figure 5.5a: We obtain better performance with the plano surface toward the infinite conjugate if the F number is sufficiently high (see also the camera example and problems of the previous chapter).

Even though a doublet can give adequate correction for primary spherical aberration and coma, it cannot correct astigmatism and field curvature. For that, an additional element is required. The simplest possible case is to use the doublet with a field flattener; this will improve field curvature, but not astigmatism. It is shown in the next section that a properly designed separated triplet can provide adequate correction for astigmatism as well.

The occasional user of optics often wants to know what kind of resolution to expect from a doublet versus a more complex lens. Unfortunately, there can be no simple answer to this question. Lens manufacturers will sometimes, but certainly not always, give specifications on the minimum resolution, and even then not for all field points. Multielement lenses are subject to manufacturing aberrations, the most important being tilt and decentering of the elements; so a mass-produced lens may not approximate its design performance well. The performance tolerances vary by manufacturer, and it can be said that if one pays more, the extra money does not necessarily buy a better design, but tighter tolerances. In terms of performance versus price, lenses developed for the amateur photographic market are a rare bargain.

If we restrict ourselves to on-axis performance only, then it is possible to give some crude guidelines based on the required $F\#$. A properly designed lens of the type listed here will give nearly diffraction-limited performance.

$F\# > 10$: single element (near optimum shape)

$5 \le F\# \le 10$: doublet

$F\# < 5$: multielement lens

These figures are only approximate; so their use is limited to back-of-the-envelope calculations. Also, the performance of singlets and doublets deteriorates rapidly off-axis for a field angle of even a few degrees. When high performance at any appreciable field angle is required, a multielement lens is a must.

9.2. THIN-LENS ABERRATIONS WITH A REMOTE STOP

9.2.1. The Eccentricity Parameter

Consider the simple case of a single thin lens with the stop at the lens, as in Figure 9.7. We define a stop shift to be such that the axial ray pencil is unchanged in every respect: Therefore, the new stop position necessitates a change in the size of the stop, so as to leave the numerical aperture unchanged. Thus h and u (and, of course, η) are unchanged by the stop shift, while $\bar{h}$ and $\bar{u}$ are changed. The same conclusion holds in a multielement system: Only the pupil ray quantities are affected.

What is the effect of shifting the stop on the aberrations? Recalling the Seidel sum formulae, we see that S_{I}, S_{IV}, and C_{L} are independent of the stop position: But the other sums contain the quantity $\bar{A}$, which depends on the pupil ray parameters. Thus in order to find the effect of the stop shift on aberrations, we need to find first its effect on $\bar{A}$.

The eccentricity parameter E is defined as

$$E = \frac{\bar{h}}{h} \tag{9.20}$$

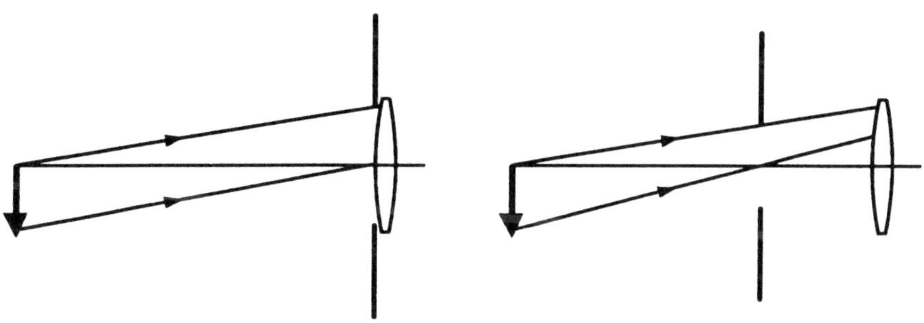

Figure 9.7. Nature of a stop shift: The PMR (axial pencil) is not affected by the stop shift, but the PPR is.

We can measure the stop shift by using the difference in E; that is, the quantity

$$\Delta E = \frac{\Delta \bar{h}}{h} \tag{9.21}$$

is a measure of the stop shift. Recall the two-ray form of the optical invariant: $H = n(\bar{h}u - h\bar{u})$. Now $A = n(hc + u) \rightarrow u = A/n - hc$, and similarly, $\bar{u} = \bar{A}/n - \bar{h}c$. Substitution into H gives

$$H = A\bar{h} - \bar{A}h \tag{9.22}$$

$H(=nu\eta)$ is not affected by the stop shift; so

$$\Delta H = 0 = \bar{h}\,\Delta A + A\,\Delta\bar{h} - h\,\Delta\bar{A} - \bar{A}\,\Delta h.$$

But $\Delta h = \Delta A = 0$ during stop shift; so

$$A\,\Delta\bar{h} - h\,\Delta\bar{A} = 0 \rightarrow A\frac{\Delta\bar{h}}{h} = \Delta\bar{A}$$

or

$$\Delta\bar{A} = A\,\Delta E \tag{9.23}$$

9.2.2. Stop-Shift Effects for a Single Surface

Suppose that we move the stop, so that $E \rightarrow E + \Delta E$, and $S_{\rm II} \rightarrow S_{\rm II}^*$, $S_{\rm III} \rightarrow S_{\rm III}^*$, etc., where the star denotes the aberration after the shift. We then have the following results, from Eq. (9.23)

$$S_{\rm II}^* = -A(\bar{A} + \Delta\bar{A})h\,\Delta(u/n) = S_{\rm II} - A\,\Delta\bar{A}\,h\,\Delta(u/n) = S_{\rm II} - \Delta E\,A^2h\,\Delta(u/n)$$

or

$$S_{\rm II}^* = S_{\rm II} + \Delta E\,S_{\rm I} \tag{9.24}$$

For $S_{\rm III}$ we have:

$$S_{\rm III}^* = -(\bar{A} + \Delta\bar{A})^2h\,\Delta(u/n) = S_{\rm III} - 2\,\Delta E\,A\bar{A}h\,\Delta(u/n) - \Delta E^2\,A^2h\,\Delta(u/n)$$

or

$$S_{\rm III}^* = S_{\rm III} + 2\,\Delta E\,S_{\rm II} + \Delta E^2\,S_{\rm I} \tag{9.25}$$

It is left as an exercise also to prove that

$$S_{\rm V}^* = S_{\rm V} + \Delta E(3S_{\rm III} + S_{\rm IV}) + 3\,\Delta E^2\,S_{\rm II} + \Delta E^3\,S_{\rm I} \tag{9.26}$$

$$C_{\rm T}^* = C_{\rm T} + \Delta E\,C_{\rm L} \tag{9.27}$$

This concludes the calculation for all the aberrations that are affected by stop shift. In the next section, we expand these results to a thin lens and a general system. This expansion is made very simple by the following theorem: The difference in the eccentricity parameter between two successive surfaces is independent of stop position. The corresponding formula is

$$E_{i+1} - E_i = -\frac{H\,d_i}{n_1'\,h_i h_{i+1}} \tag{9.28}$$

where the paraxial quantities on the right-hand side have their usual meaning. Notice that none of them depends on the stop position. Equation (9.28) is known as the Seidel difference formula.

To prove Eq. (9.28), write $H = n(\bar{h}u - h\bar{u}) = nu\eta$, so

$$H/nu = \bar{h} - h\bar{u}/u = \eta \rightarrow E - \bar{u}/u = \eta/h$$

or

$$E = \eta/h + \bar{u}/u = \eta'/h + \bar{u}'/u'$$

where the last equality can be proven via a similar route. Therefore, we can write

$$E_i = \frac{\eta'_i}{h_i} + \frac{\bar{u}'_i}{u'_i} \quad \text{and} \quad E_{i+1} = \frac{\eta_{i+1}}{h_{i+1}} + \frac{\bar{u}_{i+1}}{u_{i+1}}$$

But $\eta'_i = \eta_{i+1}$, $\bar{u}'_i = \bar{u}_{i+1}$, $u'_i = u_{i+1}$, so that

$$E_{i+1} - E_i = \eta_{i+1}\left(\frac{1}{h_{i+1}} - \frac{1}{h_i}\right) = \frac{\eta_{i+1}}{h_i h_{i+1}}(h_i - h_{i+1})$$

Further, using $h_i - h_{i+1} = -d_i u_{i+1}$ and $u_{i+1}\eta_{i+1} = H/n'_i$, Eq. (9.28) follows.

9.2.3. Stop-Shift Effects for a General System and a Thin Lens

The significance of the Seidel difference formula is as follows: Since the difference in E is independent of stop position, if the stop is moved such that $E_i \rightarrow E_i^* = E_i + \Delta E$, then $E_{i+1} \rightarrow E_{i+1}^* = E_{i+1} + \Delta E$; that is, the change in E at a surface due the shop shift is the same for all the surfaces in the system. (In other words, following the stop shift, $E_{i+1}^* - E_i^* = E_{i+1} - E_i \equiv \Delta E$.) This allows us to generalize immediately Eqs. (9.24)–(9.27), since ΔE will be the same for all surfaces. We have:

$$\sum S_{\text{II}}^* = \sum S_{\text{II}} + \Delta E \sum S_{\text{I}} \tag{9.29}$$

$$\sum S_{\text{III}}^* = \sum S_{\text{III}} + 2\,\Delta E \sum S_{\text{II}} + \Delta E^2 \sum S_{\text{I}} \tag{9.30}$$

$$\sum S_{\text{V}}^* = \sum S_{\text{V}} + \Delta E(3 \sum S_{\text{III}} + \sum S_{\text{IV}}) + 3\,\Delta E^2 \sum S_{\text{II}} + \Delta E^3 \sum S_{\text{I}} \tag{9.31}$$

$$\sum C_{\text{T}}^* = \sum C_{\text{T}} + \Delta E \sum C_{\text{L}} \tag{9.32}$$

where the summation extends over all the system surfaces. For a thin lens, the results derive directly from summing two single-surface contributions. But instead of comparing between two arbitrary stop positions, we choose to compare between the cases where the stop is remote or at the lens (central aberrations). In that case, ΔE becomes simply $E(=\bar{h}/h)$. So we have

$$S_{\text{II}} = S_{\text{IIc}} + ES_{\text{I}} \tag{9.33}$$

$$S_{\text{III}} = S_{\text{IIIc}} + 2ES_{\text{IIc}} + E^2 S_{\text{I}} \tag{9.34}$$

$$S_{\text{V}} = E(3S_{\text{IIIc}} + S_{\text{IV}}) + 3E^2 S_{\text{IIc}} + E^3 S_{\text{I}} \tag{9.35}$$

$$C_{\text{T}} = EC_{\text{L}} \tag{9.36}$$

Thus for the thin lens, the effect of the stop shift is to raise the power of the dependence of the aberrations on X and Y, the shape and magnification parameters. Recall that S_{IIc} is linear in X and Y. However, S_{II} contains S_I, which is quadratic in X and Y; so S_{II} also becomes quadratic. Similarly, all of the other monochromatic aberrations are then represented by parabolas.

9.2.4. Stop-Shift Theorems

A number of powerful theorems can be deduced from the stop shift formulae: These are summarized here. Notice that the word *aplanatic* implies zero spherical aberration and coma when it applies to a complete lens system, despite the fact that the aplanatic condition for a single surface $[\Delta(u/n) = 0]$ also gives zero astigmatism.

1. For a system with no spherical aberration, coma is independent of the stop position. This follows immediately from Eq. (9.29).
2. For an aplanatic system, astigmatism is independent of the stop position. This follows immediately from Eq. (9.30).
3. For an aplanatic system with a flat T field, all primary aberrations are independent of the stop position. For distortion, this follows from Eq. (9.31), noting that a flat T field means $3S_{III} + S_{IV} = 0$. For the other aberrations, this is the same as before.
4. For a system free of longitudinal color, the transverse color is independent of stop position [from Eq. (9.32)].
5. It is impossible to design an anastigmat comprising any number of thin lenses in contact. The term *anastigmat* implies a lens with spherical, coma, and astigmatism all zero. To show this, notice that E is the same for all lenses, since they are thin and in contact. For $S_I = 0$ and $S_{II} = 0$ we have $S_{II} = S_{IIc} + ES_I \rightarrow S_{IIc} = 0$. But $S_{III} = S_{IIIc} + 2ES_{IIc} + E^2S_I = S_{IIIc}$. Further, $S_{IIIc} = H^2 \sum K \neq 0$; so $S_{III} \neq 0$. This theorem explains why all but the cheapest camera objectives are designed as more than one separated elements.
6. An achromatic system of two separated components requires both components to be individually achromatic. We have: $\sum C_L = 0 = C_{L_1} + C_{L_2} \rightarrow C_{L_2} = -C_{L_1}$ and $\sum C_T = C_{T_1} + C_{T_2} = E_1C_{L_1} + E_2(-C_{L_1}) = C_{L_1}(E_1 - E_2)$, which can only be zero if $E_1 = E_2$ or if $C_{L_1} = 0$. The first condition is rejected, as it implies that the lenses are in contact. Therefore, $C_{L_1} = 0$, which then gives $C_{L_2} = 0$, and so also $C_{T_1} = C_{T_2} = 0$.

The importance of these theorems lies in the fact that they provide limits to the amount of correction that is possible, and also provide a guide for how to achieve that correction. The budding designer who is unaware of these restrictions may end up spending considerable amounts of time trying to improve a system that cannot be improved.

9.2.5. Example: The Petzval Projection Lens

We give here an example of how the thin-lens aberration formulae can be applied to the design of a system. This example will not be taken to a conclusion. It will be more instructive to follow what seems to be a reasonable approach, and consider what to do when things go wrong. The lesson to learn is that useful information must be extracted from attempts that do not work. In real life, the lens would probably be designed more efficiently by modifying an existing published design of similar specification, but that approach fails for unusual systems. Also, the "from scratch" method illustrates the application of aberration theory. Additional design examples are given in the next chapter.

The lens to be designed is a projection lens, used for example in a slide projector system. For such a lens, the object (slide) is nearby, and the image is distant. It is nevertheless convenient and customary to design it in reverse as a photographic objective (that is, with the object at infinity).

The thin-lens version of the system is shown in Figure 9.8. The Gaussian specification calls for $F = 100$ mm, $F\# = 2.5$, FOV $= \pm 11.3°$. There is some flexibility in choosing the powers and stop position, but to simplify the problem let us make two arbitrary choices:

$$K_1 = K_2 \qquad \text{and} \qquad E_2 = -E_1 \equiv E$$

The first of these gives $K = 2K_1 - dK_1^2$ for the system power. Notice that the condition $E_2 = -E_1$ does *not* imply that the stop is halfway between the lenses.

Now let us establish a design philosophy. As we already know, S_{IV} is intractable, while C_{L} and C_{T} can be made zero by achromatizing each element separately (this means two doublets). Notice that central astigmatism is fixed for both lenses $S_{\text{IIIc}_1} = S_{\text{IIIc}_2} = H^2K_1$. Therefore, we have four free aberration parameters, the values of S_{I} and S_{IIc} for each component. We should choose these four values so that they (1) result in the best overall correction/balance when summed over the two components, and (2) will be achievable in each of the components without the risk of inducing excessive higher-order aberration. We might try

$$S_{\text{I}_1} = S_{\text{I}_2} = 0$$

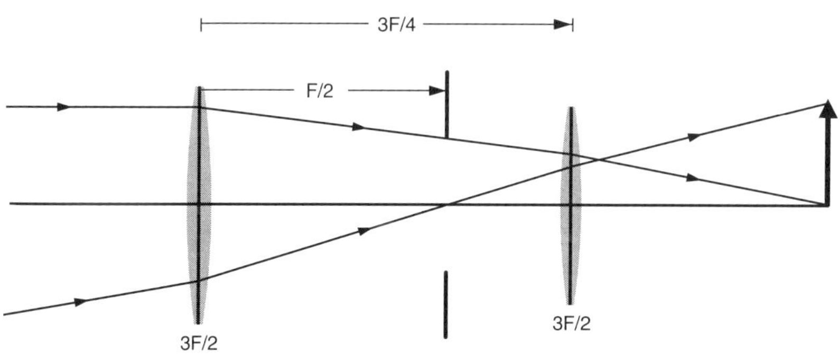

Figure 9.8. Gaussian predesign of the Petzval projection lens.

to begin with, but it would then be a big mistake to also choose $S_{\text{IIc}_1} = S_{\text{IIc}_2} = 0$, as this, although yielding zero totals for S_{I}, S_{II}, and S_{V}, would leave the massive positive central astigmatism unchecked (i.e., $S_{\text{III}} = S_{\text{IIIc}}$). Thus the choice

$$S_{\text{IIc}_1} = -S_{\text{IIc}_2} \neq 0$$

is better, with S_{IIc_1} chosen to mitigate the S_{III}, S_{IV} problem. We have

$$\sum S_{\text{IV}} \cong 0.65H^2(K_1 + K_2) = 1.3H^2K_1$$

from Eq. (8.10), and

$$\sum S_{\text{III}} = H^2K_1 + H^2K_2 + 2E_1S_{\text{IIc}_1} + 2E_2S_{\text{IIc}_2} + E_1^2S_{\text{I}_1} + E_2^2S_{\text{I}_2}$$

or

$$\sum S_{\text{III}} = 2H^2K_1 - 2E(S_{\text{IIc}_1} - S_{\text{IIc}_2}) = 2H^2K_1 - 4ES_{\text{IIc}_1}$$

Now we have to decide on the best balance of astigmatism and field curvature. Following the reasoning of Section 8.2, we opt for a flat T field, so that

$$W_{\text{T}} = \tfrac{1}{4}(3\sum S_{\text{III}} + \sum S_{\text{IV}}) = 0 \rightarrow 6H^2K_1 - 12ES_{\text{IIc}_1} + 1.3H^2K_1 = 0$$

from which

$$S_{\text{IIc}_1} = 0.61\,\frac{H^2K_1}{E}$$

$$W_{\text{S}} = \tfrac{1}{4}(\sum S_{\text{III}} + \sum S_{\text{IV}}) = \cdots = 0.22\,\frac{H^2K_1}{E}$$

Distortion (S_{V}) has so far been left unaccounted for. We must now accept what value exists since we have fixed the stop position and all the variable central aberrations. Fortunately, from Eq. (9.35) we find that

$$\sum S_{\text{V}} = -E(3S_{\text{IIIc}_1} - 3S_{\text{IIIc}_2}) + E(S_{\text{IV}_1} - S_{\text{IV}_2}) + 3E^2(S_{\text{IIc}_1} + S_{\text{IIc}_2}) = 0$$

using the above values. Thus our design will have:

$$S_{\text{I}} = S_{\text{II}} = W_{\text{T}} = S_{\text{V}} = C_{\text{L}} = C_{\text{T}} = 0 \qquad \text{and} \qquad W_{\text{S}} = 0.22\,\frac{H^2K_1}{E}$$

This looks good for now, but we must complete the Gaussian design. The reader can verify that the requirement $E_1 = -E_2$ leads to

$$d = \frac{2\bar{l}_1}{1 + 2\bar{l}_1K}$$

and since $K = 2K_1 - dK_1^2$, we obtain, by eliminating d,

$$K_1^2 + K_1\left(\frac{1}{\bar{l}_1} - K\right) - \frac{K}{2\bar{l}_1} = 0$$

There are infinite combinations of K_1 and d that satisfy our requirements; so we must choose $\bar{l}_1$. We take $\bar{l}_1 = F_1/2$ (somewhat arbitrarily), thus obtaining

$$K_1 = \tfrac{2}{3}K \qquad \text{or} \qquad F_1 = \tfrac{3}{2}F$$

$$d = \frac{F_1}{2} = \frac{3F}{4}, \qquad \bar{l}'_1 = \frac{F_1}{3} = \tfrac{2}{3}d$$

The required FOV gives $\bar{u} = 0.2$. Also, $h_1 = F/2F\# = 20$ mm, $u_1 = 0$. Upon completing the paraxial ray tracing, we find $H = -4$ mm, $E = -0.75$. This is all that is required for the primary aberration calculation. We have:

$$S_{\text{IIc}_1} = 0.61\,\frac{H^2K_1}{E} \cong 147.6\lambda = -S_{\text{IIc}_2}$$

$$W_S = 0.22\,\frac{H^2K_1}{E} \cong 53.2\lambda$$

Thus we have established all our aberration targets. Our two doublets can now be designed as follows: First, they must be achromatic; so this gives us the powers and glass types for each of the four singlets. The required shapes to yield the aberration targets for S_I and S_IIc can be calculated using a simple computer program to do a thin-lens bend (computing the aberrations as a function of shape) and choosing a suitable pair of shapes by inspection of the output tables. The reader can attempt to complete this example with the aid of the thin-lens bending program of Appendix 6.

A cemented doublet can be designed by a more elaborate procedure.* The cemented doublet would normally be preferable since, once cemented, it is easier to mount. Using these techniques, the system of Figure 9.9 can be obtained as a possible solution. A cemented doublet for the second component was not possible.

The extreme curves of the second component are very disturbing. Although the primary aberration targets are met, higher-order aberration is very likely, not to mention presenting difficulty in manufacture and assembly. It is sensible to halt this design direction at this point and look for ways to produce more

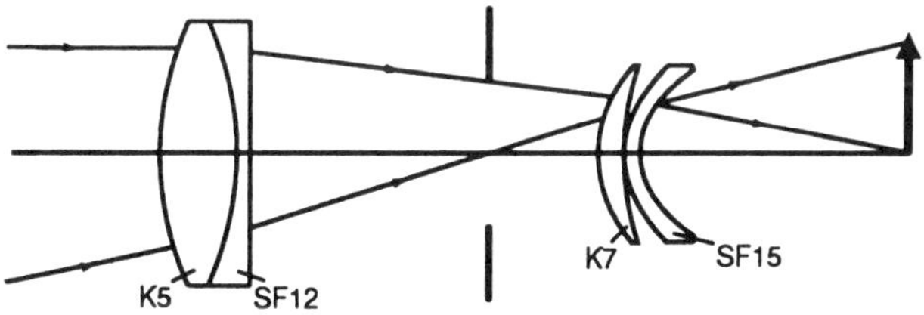

Figure 9.9. A completed (unsatisfactory) design of the Petzval projection lens.

* See M. Iqbal Khan and J. Macdonald, "Cemented doublets: A method for rapid design," *Optics Acta* **29**, 807–22 (1982).

moderate shapes. Inspection of the bending results reveals that it was the insistence on large negative coma at the second component that caused the problem. We may then have to abandon the zero-coma target, or allow some spherical aberration in order to reduce the residual coma. We may also find it necessary to abandon the $W_T = 0$ target as too ambitious.

In practice, what happens next depends mostly on the sophistication of the software at one's disposal. Access to professional optical design software offers the option of allowing the program to optimize the design automatically. Of course, in that case the normal procedure would be to modify a design from the many published ones, without going through the detailed calculation above. But for the student of optical design, there is value in understanding the underlying aberration theory in sufficient detail, and in developing an intuition based on knowledge, as opposed to mere imitation; the latter will ultimately prove very limiting.

9.3. THE TWO- AND THREE-COMPONENT SOLUTIONS WITH FIXED $\sum K$

We return here to the topic of Gaussian predesign. In Chapters 3, 4, and 5, we discussed the one- and the two-lens solution. As might be expected, the one-lens solution is often too simplistic. The two-lens solution is applicable to a large variety of problems, including the slide projector lens mentioned previously. But as we have seen, once the Gaussian design is complete, the Petzval curvature term is essentially fixed, as it depends only on the sum of the powers of the individual elements. We are therefore interested in developing a Gaussian thin lens solution that controls the sum of the powers of the elements.

9.3.1. The Two-Component Solution

The reader will recall the discussion of Section 4.10. In this case, we specify the following parameters: l_1, l'_2, M, $\sum K = K_1 + K_2$, and from these, we seek to determine K_1, K_2, and d. (Notice that simultaneous control of the pupil positions as in Section 5.2.4 is not possible.) The equations from Section 4.10 can be combined to give the following results:

$$
\begin{aligned}
d &= \left[2 - \left(\frac{h_1}{h_2} + \frac{h_2}{h_1}\right)\right] \Big/ \left[\sum K - \left(\frac{1}{l'_2} - \frac{1}{l_1}\right)\right] \\
K_1 &= \frac{1 - h_2/h_1}{d} - \frac{1}{l_1} \qquad\qquad (9.37) \\
K_2 &= \frac{1}{l'_2} - \frac{h_1/h_2 - 1}{d}
\end{aligned}
$$

where $h_1/h_2 = Ml_1/l'_2$.

9.3.2. The Three-Component Solution*

In the general case, three separated components provide an infinite number of possibilities; there is no unique solution unless a sufficient number of parameters are specified. A useful approach starts with the realization that the two-component solution may not provide sufficient control of S_{IV} if both lenses must be positive. A three-lens solution allows the introduction of negative power. We restrict the discussion to the case with the middle lens negative.

The device used to obtain a method of solution is to consider the apparently different problem of wishing to change the length d (i.e., the lens separation) of the two-lens solution, without otherwise altering the solution. This is equivalent to inserting a "black box" system between the two lenses to produce a simple "spacer" effect, as indicated in Figure 9.10. This can be achieved by inserting a unity-magnification three-lens telescope between the two lenses, while changing their separation at the same time. The details are given in the Hopkins article referenced previously. The result is as indicated in Figure 9.11. Notice that the lenses are in contact, with the exception, of course, of the middle one.

The original two-lens solution has K_1^*, K_3^*, d^* (these symbols are used instead of the more logical K_1, K_2, and d, in anticipation of the three-lens system for which we are reserving nonstarred quantities). The three-lens solution has total separation $d = d_{12} + d_{23}$, and $\sum K \equiv K_1 + K_2 + K_3$, where $K_1 = K_1^* + \delta K_1$, $K_3 = K_3^* + \delta K_3$. The three telescope lenses have powers δK_1, K_2, and δK_3. The difference in power sum between the three- and two-lens solutions is $\delta K = \sum K - (K_1^* + K_3^*)$. To obtain a solution, we specify the desired separation

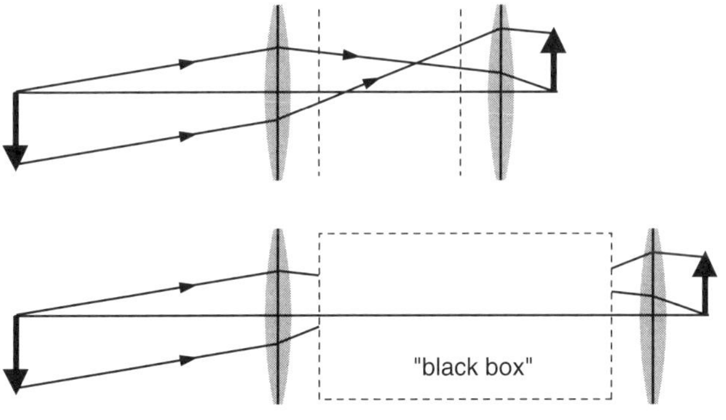

Figure 9.10. In order to arrive at a three-lens solution, we first determine a "black box" optical system that can elongate the two-lens solution while leaving it otherwise unchanged.

* See H. H. Hopkins. "The Gaussian optics of multilens systems," Proc. 1961 ICO Conference on Optical Instrumentation, pp. 133–59, K. J. Habell, Ed., Chapman and Hall Ltd., London, 1962.

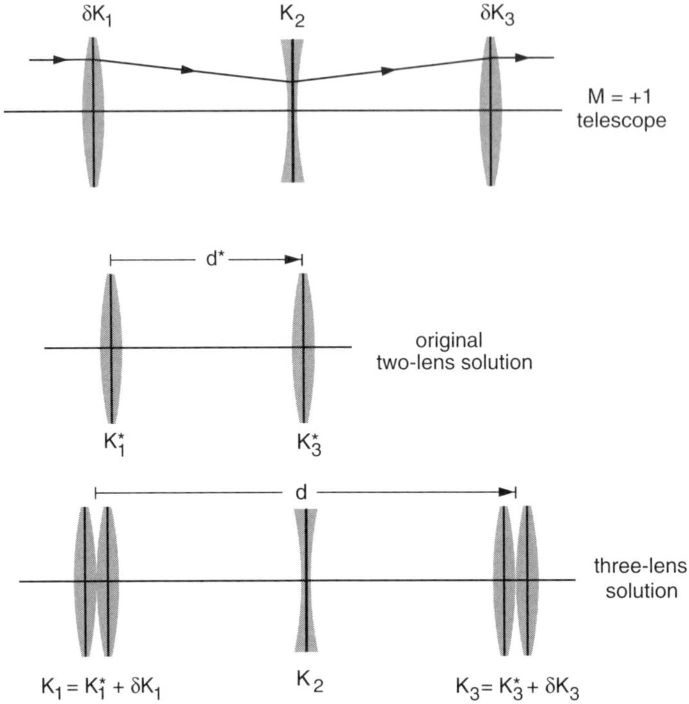

Figure 9.11. A unit magnification telescope inserted between the two lenses of the two-lens solution (middle) achieves the effect implied by Figure 9.10.

d and power increment δK. We then have:

$$K_2 = -\frac{d^* \, \delta K}{d - d^*}$$

$$d_{12} = \frac{d}{2}\left\{1 \pm \left[1 - \left(\frac{2}{d}\right)^2 \frac{d - d^*}{K_2}\right]^{1/2}\right\}$$

$$d_{23} = d - d_{12}$$

$$K_1 = K_1^* - \frac{d_{23}K_2}{d^*}$$

$$K_3 = K_3^* - \frac{d_{12}K_2}{d^*}$$

Of course, the solution does not exist unless the square root in the equation for d_{12} is real. A four-lens solution can also be obtained if the problem requirements cannot be satisfied with the three-lens solution.

9.3.3. The Cooke Triplet

As was pointed out, it is impossible to design an anastigmat comprising thin lenses in contact. If, however, the lenses are separated, then the design is possible, at least in theory. A typical triplet anastigmat is the so-called Cooke triplet, which comprises a negative element between two positive ones (see opening quotation of this chapter). The stop is internal, and for convenience it can be taken to coincide with the negative lens (Figure 9.12). The triplet provides just enough free parameters to allow control of all the five primary monochromatic aberrations, the two chromatic ones, and the power. The "free" parameters are the three element powers, the two separations, and the three bendings. If one adds the refractive indices and dispersions, then some additional flexibility is possible, especially if an additional parameter must be controlled (such as the back focal length).

It is possible in theory to solve the system of eight equations and unknowns, but since the equations are nonlinear, in practice the solution is obtained in an iterative and constrained fashion. There are several different ways of attacking the problem, and for a complete discussion, the reader is referred to Smith (1990), Begunov et al. (1988), or Kingslake (1962). Here we simply point out how one can think about this problem. A starting point can be the equations of the previous section, to provide a desired sum of powers. An alternative starting point is to write down as many of the eight equations as possible, starting with the simpler ones. Thus we have the following:

By summing the three paraxial refraction equations for each element, we obtain

$$u'_3 - u_1 = -(h_1K_1 + h_2K_2 + h_3K_3) = -h_1K$$

or

$$K = K_1 + \frac{h_2}{h_1}K_2 + \frac{h_3}{h_1}K_3$$

which expresses the total system power. The sum of the powers is intended to control S_{IV}; so we may write

$$\frac{K_1}{n_1} + \frac{K_2}{n_2} + \frac{K_3}{n_3} = -\frac{S_{IV}}{H^2}$$

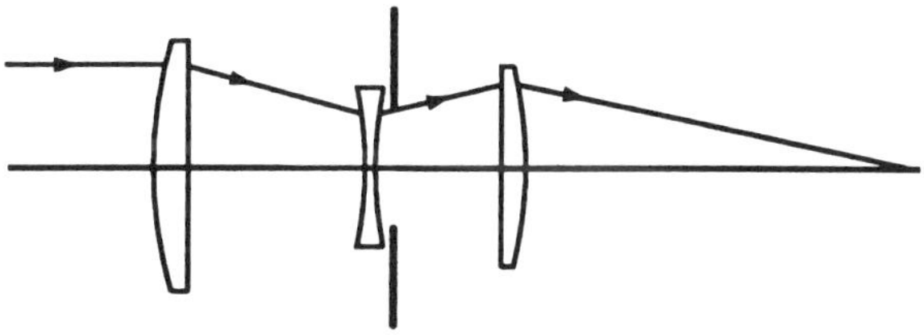

Figure 9.12. A version of the Cooke triplet anastigmat (several versions of this lens exist with somewhat different bendings).

The condition for zero longitudinal color is

$$\frac{K_1}{V_1} + \left(\frac{h_2}{h_1}\right)^2 \frac{K_2}{V_2} + \left(\frac{h_3}{h_1}\right)^2 \frac{K_3}{V_3} = 0$$

and the condition for zero transverse color gives (for $\bar{h}_2 = 0$)

$$\frac{K_1}{V_1} + \frac{h_3}{h_1}\frac{\bar{h}_3}{\bar{h}_1}\frac{K_3}{V_3} = 0$$

In order to control the remaining four Seidel sums, we can use graphs or tables of the S_{I}, S_{II}, etc. parabolas as a function of shape. But for the negative element, the aberrations are central. Therefore,

$$\sum S_{\text{V}} = S_{\text{V}_1} + S_{\text{V}_3} = 0$$

is the simplest of the four conditions. Also, S_{III_2} is fixed once we have decided on the power. But for the other two aberrations, all the individual element contributions vary as a function of shape.

Before proceeding further, a set of values for the powers can be tried by making some first (perhaps crude) assumptions. For example, we can assume that $\bar{h}$ will be considerably smaller than h, so that E^2 and E^3 can be neglected at the beginning. Also, we can set all refractive indices approximately equal. With those and similar assumptions, we can obtain a set of initial values for the powers and separations. When that is done, we can plot or tabulate the aberrations as a function of shape and then try to determine an acceptable solution. But we will not pursue this example further because it has been dealt with in great detail in the literature cited at the beginning of this section.

PROBLEMS

Note: For some of the problems there is no unique solution. Those problems are meant more as suggestions that can be used to put the concepts into practice. Any design problem can be attacked in as many ways as there are designers; in the end, a satisfactory solution must be obtained, one that satisfies the specifications or the customer. In our case, a solution may be deemed satisfactory if the student learned a lot on the way.

1. Using the computer program of Appendix 6, investigate the properties of the primary aberration parabolas for a single thin lens and a stop. Specifically with respect to the sign of the aberration (positive or negative), investigate the following four cases: (a) $E > 0$, $K > 0$; (b) $E > 0$, $K < 0$; (c) $E < 0$, $K > 0$; (d) $E < 0$, $K < 0$. (Note: This will be useful when attempting to cancel the aberrations of one element with those of the next.)

2. Use the program of Appendix 5 (or your own solution) to obtain aplanatic and achromatic contacted doublets for the following two pairs of glasses:

(487704, 785258) and (501564, 626390). Draw the resulting lenses to scale using a compass (and some reasonable thickness). Conclusions? (Take an object at infinity and assume a focal length of 2.5 cm.)

3. Select one of the doublets of Problem 2. Use the Seidel ray-tracing program of Appendix 5 (or your own computation) to verify that the spherical aberration and coma of the doublet are close to zero (set lens thickness and separation equal to zero). For a field angle of 4°, determine the off-axis aberrations. (For the Seidel aberration program, you will need the dispersion of the glasses in the form of $\delta n = n_F - n_C$. These are: 0.006924 for glass 487704 and 0.030468 for glass 785258.)

4. Continue Problem 3 by assuming a reasonable thickness for the different elements, and retrace the doublet. How have the aberrations changed? Has the power of the system been affected? (Note: See Chapter 10 for a systematic method of thickening the lens and a completed doublet design.)

5. Using the computer program of Appendix 6, perform the following operations: (a) Determine the spherical aberration and the off-axis aberrations of an equiconvex singlet with the object at infinity (stop at the lens). Assume reasonable values for the focal length and refractive index. (b) Bend the lens to show how the aberrations are minimized at the proper shape. (c) Split the lens into two thin contacted elements. Obtain correction for spherical aberration and coma. Determine astigmatism and field curvature. (d) Add a field flattener near the image plane and determine the improvement in field curvature.

6. Complete the third-order design of the Petzval projection lens started in Section 9.2.5 by changing the design approach as indicated there. (It is possible to arrive at a solution that does not involve extreme bendings.)

7. (a) A thin equiconvex lens operates at symmetrical conjugates ($M = -1$) and is made of crown glass ($n = 1.52$). The focal length is $F = 2.5$ cm. The stop is at the lens, and $h_e = 0.3$ cm. For an object height (η) of 0.2 cm, determine the monochromatic aberrations. (b) Design a thin "air-in-glass" lens with the same throw T as before (i.e., fill the air spaces with glass while turning the lens into an air gap, and calculate the curvatures). What effect does this have on the optical invariant H? On the image illuminance? Determine the monochromatic aberrations again and compare with the previous case. (c) Extend this example by choosing a real glass and calculating the chromatic aberrations in both cases.

8. (a) Starting from the Gaussian parameters of the two-lens Petzval solution of Section 9.2.5, determine a three-lens solution that reduces $\sum K$ by 50%. (b) Proceed to design a Cooke triplet derivative with minimum target values for all primary aberrations. A cemented doublet may be used as the first or the last element.

Optical Design

10

> "Geometrical Optics is either very simple, or else it is very complicated."
>
> **Richard P. Feynman,** in *The Feynman Lectures on Physics,* Addison Wesley, 1963.

In this chapter we expand the discussion of aberrations beyond the primary level, we discuss the process of optical design, and we show examples of completed or nearly completed designs. It should be clearly understood at the outset that this discussion is introductory, and limited by two constraints: (1) we exclude diffraction theory, which in many cases is a necessary part of the system evaluation, and (2) we exclude systems that may have exotic aperture and field specifications or may include unusual optical elements. In practice, such problems are most efficiently tackled with the help of professional design software. Upon achieving an understanding of the material contained in this book, the reader should find that many advanced features of such software become approachable, or even transparent. This is our attempted solution to Feynman's dilemma: Instead of covering those details that make geometrical optics "very complicated," we aim for an understanding of the basics that will serve as a springboard for attacking the more advanced problems.

10.1. THE OPTICAL DESIGN PROCESS

Determination of the primary aberration is normally only the beginning of a design; the effect of higher-order aberrations must be determined as well. This is done through finite ray tracing, using algorithms similar to that of Appendix 3.

The most basic decision that has to be made when faced with the problem of designing a lens is whether to use one or more lenses (components). As we have already seen, this decision is governed by the required Gaussian properties in the first instance, but also by the desired image quality.

The following steps are normally involved in designing a lens system. Some of these steps are discussed in the following sections in more detail.

1. Specification;
   size and position of conjugates, pupils
   mechanical properties, size
   illumination of image
   image quality (e.g., resolution, MTF allowable distortion)
   environmental properties (temperature, vibration, etc.)
2. Gaussian design
   All of the above requirements must be converted to usual Gaussian parameters, H, K, h, u, M, etc. How many components? (one-, two-, or three-lens solution, etc.). At this stage, a component is a single thin lens.
3. Thin lens, primary aberration design
   Approximate correction of primary aberrations. Requires specifying number of elements for each component, as well as the shape of each element. Insert real optical glasses (index, dispersion).
4. Thickening
   Insert lens thickness. Determine image quality; this can be done through a primary aberration calculation routine at this point, but eventually we must turn to ...
5. Finite ray tracing
   Trace rays and determine total aberration. This may involve ray aberration, wavefront aberration, or other means of image-quality assessment that suit the problem and the temperament of the designer.
6. Optimization
   The so-called merit function must be set up. It can be, for example, the sum of the squares of certain aberrations, or the rms blur spot at the image plane. The computer normally takes over at this point, by iteratively computing the merit function and modifying slightly the system until a satisfactory solution is obtained.
7. Design evaluation
   This step is often combined with optimization, but it may also be performed separately. The optimized design may have to be evaluated not only in terms of alternative image quality measures, but also for its sensitivity to manufacturing tolerances. In particular, the lens elements may be tilted in their mounts, or decentered (in which case the apex of a surface does not coincide with the optical axis). The aberrations caused by these errors are called manufacturing aberrations and are not fully described by the Seidel theory for axisymmetric systems.

It is often important to calculate the sensitivity of a system to tilt and decentering errors in order to ensure that the resulting mechanical tolerances can be met. In some cases, the sensitivity to manufacturing aberrations can become part of the merit function during optimization.

A very important tool of the designer is the library of past designs. Some very useful designs can be found in the *Military Handbook* 141, in Kingslake (1978), Smith (1992), Levi (1968), and Cox (1964). The design prescriptions from these references can be used as starting points for a given problem, without having to read through the whole book or even chapter. Also the patent literature contains many designs that can be used as starting points.

The reader should be warned that this process is not always followed: There are cases where the final solution is not a mere extension of a thin-lens predesign. An obvious example is provided by high-power microscope objectives: The first element of such objectives is usually "hyperhemispherical," that is, thicker than a hemisphere, with the front surface flat and the rear surface aplanatic. The action of such a lens is very different from any predictions made by the thin-lens approximation. Another example is lenses with very large field of view, which employ almost hemispherical meniscus elements.

10.2. MAKING THE SYSTEM REAL: THICKENING, TOTAL ABERRATION

The first nine chapters of this book are dedicated to discussing just the first three steps of the optical design process, as outlined in the previous section. After one has obtained a satisfactory thin-lens design, the next obvious step is to insert thicknesses.

In moving from a thin tens to a thick lens, there are several considerations. Unless the curvatures are modified, the addition of thickness will change the lens power. This may be acceptable, if the change is small and if the focal length (or magnification) does not have to be controlled precisely. The lens aberrations will also change, since the ray height is in general no longer identical at the two surfaces. Normally, the lens (singlet) is thickened by maintaining the power K and the shape factor X constant; in this way, both the magnification and the aberration correction are controlled. The reader can solve this problem using the equations of the previous chapter. Alternatively, one may use a simple computer program.* The desired clear aperture of the lens must also be considered. The larger the desired aperture, the thicker the lens must be. Finally, there are two practical considerations: The edge thickness (positive lens) and

* See G. W. Hopkins "Basic algorithms for optical engineering," in *Applied Optics and Optical Engineering*, Vol. IX, R. Shannon and J. Wyant, Editors, Academic Press, 1983.

the axial thickness (negative lens) must not be too small, in order to avoid manufacturing difficulties.

For a system comprising more than one separated singlet the total system Gaussian properties can be made to match those of the thin-lens version by taking into account the principal plane separation and location for the individual components. The separation between successive principal planes must be made equal to the distance between the corresponding thin elements.

If the thin-lens predesign contains two or more elements in contact, the performance of the corresponding real system may depart considerably from that of the thin-lens version, in terms of both the aberrations and even the Gaussian properties. Both the lens thickness and the additional separation at the center or the edge (that is inevitable for noncemented lenses) will serve to change the ray height considerably between the first and the last surface. We illustrate all the above concepts by designing two thick aplanatic doublets of both the Fraunhofer and Gauss types (with the stop in contact with the first surface).

Our starting point is the thin aplanatic doublet equations of Chapter 9. We choose two common glasses: BK7 and SF5, and a crown-leading solution. Thus we have $n_1 = 1.5168$, $V_1 = 64.17$, $n_2 = 1.6727$, $V_2 = 32.21$. Our target design has zero chromatic aberration, zero spherical aberration, and zero central coma. The desired focal length is 10 cm, and the object is at infinity. The equations of Chapter 9 (or the program of Appendix 5) give the following numbers (all lengths in centimeters):

Fraunhofer solution (thin):
crown lens (1): $c_1 = 0.16495$, $c_2 = -0.22356$
flint lens (2): $c_1 = -0.22390$, $c_2 = -7.4081\text{E-}2$

Gauss solution (thin):
crown lens (1): $c_1 = 0.52989$, $c_2 = 0.14138$
flint lens (2): $c_1 = 0.53112$, $c_2 = 0.68093$

There are two immediate conclusions: (1) The Fraunhofer solution has the two adjacent curvatures practically identical, so it can be made cemented, and (2) the Gauss solution has an extreme meniscus shape, which can be expected to be troublesome ($X = -8.1$ for the flint).

We now proceed to thickening. We decide on a clear aperture of 1.5 cm (diameter), giving an $F\#$ of about 6.6. We choose the axial thickness (d) for the various elements semiarbitrarily, and (using the Hopkins program referenced previously) we obtain the following values:

Fraunhofer solution (thick, cemented):
crown: $c_1 = 0.1666$, $c_2 = -0.2258$, for $d = 0.3$
flint: $c_1 = -0.2258$, $c_2 = -7.475\text{E-}2$, for $d = 0.2$

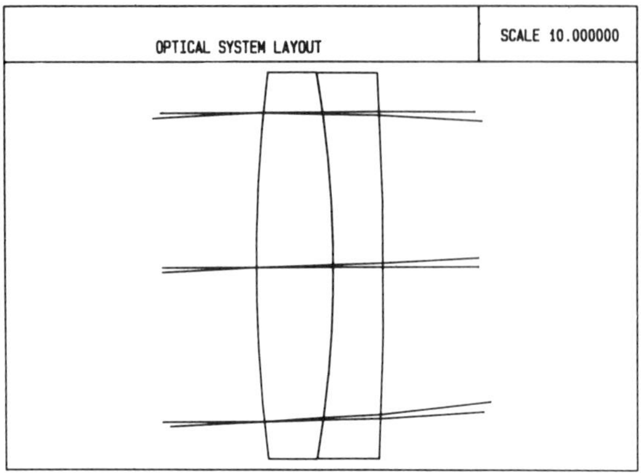

(a)

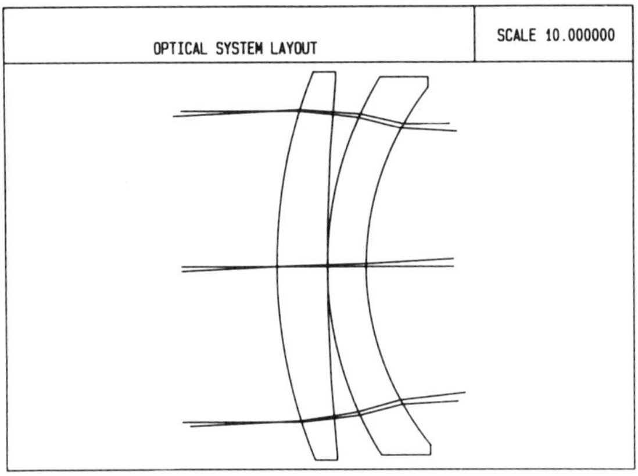

(b)

Figure 10.1. Initial aplanatic doublet solution. (a) Fraunhofer; (b) Gauss.

Gauss solution (thick):
crown: $c_1 = 0.5231$, $c_2 = 0.1396$, for $d = 0.2$
flint: $c_1 = 0.6453$, $c_2 = 0.8273$, for $d = 0.15$

The two lenses are shown in Figure 10.1. Notice that the flint element of the Gauss solution would be too thin for practical purposes, yet there is already considerable difference between the curvatures of the thin and thick versions of this lens. If it is made any thicker, the discrepancy between the thin-lens theory and the thick-lens result will become very large.

Although we have maintained the power and the shape factors of the individual elements, in putting them together we discover that the Gaussian properties have been affected. Again, the discrepancy is significant only for the Gauss doublet: Its focal length is 8.6 cm instead of the desired 10. By contrast, the thick Fraunhofer doublet has an EFL of 9.9 cm.

Next we calculate the aberration, to see if the thickness had any effect (this can be done with the help of the program of Appendix 5). We must specify a field size for the off-axis aberrations: We set that to be $\pm 2.9^\circ$ ($\bar{u} = 0.05$). Under these conditions, the Fraunhofer doublet is found to present no spherical aberration and no coma (less than 0.1λ). Also, the axial chromatic aberration is less than 0.2λ, and the transverse practically zero. Thus this doublet has met all the design targets. Its performance is limited only by astigmatism (about 1.2λ) and field curvature, which are impossible to correct with this design.

However, the Gauss doublet has not met the targets. In order to make a fair comparison between the two doublets, we first "scale" the lens so as to have the same focal length and F number as the Fraunhofer doublet. Scaling can be thought of as looking at the lens design through a magnifier: the radii, separations, aperture sizes. etc. are adjusted to fit the desired focal length and F number.

The scaled lens prescription is;

crown: $c_1 = 0.4499$, $c_2 = 0.1201$, $d = 0.23$

flint: $c_1 = 0.5550$, $c_2 = 0.7112$, $d = 0.17$

After scaling, we calculate the aberrations, whereupon we find that the lens has about 2λ of primary spherical aberration and 2λ of primary coma. In addition, there is approximately 4λ of longitudinal chromatic aberration. Thus the thickened version of the Gauss doublet has failed to meet the aberration targets.

In Chapter 9, we warned that lenses with large bendings, such as the Gauss doublet, can be expected to present substantial higher-order aberration. This is indeed the case here. The effect of the higher-order aberration can be estimated in one of two ways: Higher-order aberration coefficients can be determined in a way similar to that of the Seidel terms (noting of course that paraxial ray tracing is no longer enough), or the blur spot calculated from a spot diagram can be compared with the blur predicted by primary aberration theory. (In practice, the choice depends on the software used; both options should be available). In this case, we find that significant higher-order spherical aberration and coma exist, similar in size to the primary and of the same sign, thus adding to the primary effect. Therefore, the Gauss lens is unusable in this form.

Having thus demonstrated the agreement as well as the departure from the idealized theory of the previous chapters, it is appropriate to give a brief glimpse into the real world of optical design. We can consider the Gauss doublet as merely a starting point and submit it to optimization by a professional optical design program (optimizaion is discussed in some more detail in Section 10.4).

For simplicity, we allow only the four curvatures to vary,* but maintain the separations and glass types. We then have to set an optimization target. Because we have restricted the choice of variables to essentially the two lens bendings and the ratio of the powers, we cannot expect to control all the required parameters (two orders of spherical, two orders of coma, axial color, and focal length). For example, we may attempt to set as our target to minimize the sum of first- and second-order spherical aberration, the sum of first- and second-order coma, and the longitudinal color, while keeping the focal length constant. In that case, the optimization routine returns a solution with practically zero spherical aberration and about 2λ of total coma, but with no improvement on the chromatic correction. A different choice of optimization targets. based on rms blur at the image plane, gave a solution for which the chromatic aberration was down to 1.4λ, but with almost 4λ of total coma. The basic difference between the two solutions is that in the second case. the axial (polychromatic) correction was given more weight.

For the sake of continuing the example, we decide to forgo the chromatic correction and concentrate on the monochromatic only. In that case, the optimization routine returned the following solution

crown: $c_1 = 0.3589$, $c_2 = 0.1230$, $d = 0.23$

flint: $c_1 = 0.5969$, $c_2 = 0.6736$, $d = 0.17$

This new version has about -0.6λ of primary spherical, but it also has a similar amount (with opposite sign) of secondary spherical, giving essentially diffraction-limited axial performance. Similarly, primary and secondary coma are non-negligible but tend to cancel each other, though not as fully. There is about -1.2λ of primary coma and approximately half that (but positive) of secondary coma. Thus, although the Gauss doublet still fails to meet the primary aberration specification, it can be made to have good image quality over this aperture and field size, if the higher-order aberrations are made to cancel the primary ones partially. This is referred to as *aberration balancing*.

In general, higher orders of the chromatic aberrations also exist. Recall that the primary chromatic aberrations represent the chromatic variation of the Gaussian properties (image location and magnification). The next order of chromatic aberrations is the variation of the primary monochromatic aberrations with wavelength. Each of these terms represents a higher-order chromatic effect, but only one term has been given a name: *spherochromatism* is the chromatic variation of spherical aberration. Spherochromatism can be important in wide-aperture systems: In such a case, the designer must be

* One of the curvatures, normally the last, must be reserved to control the focal length. Thus only the other three curvatures are used for aberration correction.

aware that the chromatic correction based on the primary errors may not be sufficient.

10.3. DESIGN EXAMPLE: OPERATING SPECTACLES

The design was begun in Chapters 4 and 5, from which we take the following Gaussian specification.* It is a two-lens system, for which the image is located at the same place as the object, 25 cm in front of the first lens, and with a magnification of 2. From Chapters 4 and 5, we have (all lengths in centimeters):

$$K_1 = 0.187, \qquad K_2 = -0.298, \qquad d = 3, \qquad M = 2$$

$$\eta = -2.5, \qquad l_1 = -25, \qquad h'_e = 0.2, \qquad l'_2 = 1$$

from which

$$H = -0.0345, \qquad l'_1 = 6.8, \qquad l_2 = 3.8, \qquad l'_2 = -28$$

$$u_1 = 0.0138, \qquad u'_2 = 0.0069$$

and

$$\bar{l}_1 = 12.8 \text{ cm}, \qquad \bar{l}_2 = 0.7704, \qquad \bar{l}'_1 = 3.7704$$

$$h_e = 0.522, \qquad h_1 = 0.345, \qquad h_2 = 0.193$$

$$\bar{h}_1 = -0.848, \qquad \bar{h}_2 = -0.173, \qquad \bar{u}_1 = 0.066, \qquad \bar{u}'_2 = 0.1725$$

We then proceed with the aberration correction. Since we are using two separated singlets, we know that the lens cannot be made achromatic. Nevertheless, we must check whether judicious choice of glasses can make the chromatic aberration tolerable. We have:

$$\sum C_{\mathrm{T}} = h_1\bar{h}_1\frac{K_1}{V_1} + h_2\bar{h}_2\frac{K_2}{V_2} \overset{?}{=} 0$$

from which

$$V_2 \cong 0.18V_1$$

This range of V values is not possible: If we use two very common glasses such as BK7 and SF5, we have $V_2 \cong 0.5V_1$ and $\sum C_{\mathrm{T}} = -5.5\text{E-}4$. A better choice would be FK5 and SF57, with $n_1 = 1.48749$, $V_1 = 70.41$, $n_2 = 1.84666$, $V_2 = 23.83$. This pair gives $V_2 = 0.34V_1$, and $\sum C_{\mathrm{T}} = -3.6\text{E-}4$. Upon dividing this value by the exit (eye) pupil radius, we obtain the transverse color in angular terms to be about 6 minutes of arc. This is certainly not negligible, but it is tolerable for the edge of the field (see Section 7.3.4). Longitudinal color can be calculated similarly, and it is found to be less than the transverse. Significantly

* The reader is strongly advised to follow this example by using the software tools that are mentioned. There is little value in simply reading this section without following the calculation.

better correction is not possible with singlets. If it was required, we would have to achromatize at least the front lens, and make the negative element out of some low-dispersion crown glass: The design approach would then change considerably from the one to be shown here.

We are now ready to consider the monochromatic correction. The computer program of Appendix 6 can be used to obtain a table of the primary aberrations as a function of the lens shape. Using the tables, we look for a combination of shapes that gives a minimum overall aberration. A good match is obtained for $X_1 \cong 1$ and $X_2 \cong 2.55$, which, luckily, are also very reasonable shapes. For these values, we obtain

$$\sum S_{\mathrm{I}} \cong -7\text{E-}5, \qquad \sum S_{\mathrm{II}} = 1.1\text{E-}4, \qquad \sum S_{\mathrm{III}} = 1\text{E-}5, \qquad \sum S_{\mathrm{V}} = 0.0016$$

that is, about one wavelength of coma, 2.3% distortion, and little spherical or astigmatism. These are very promising numbers; so we decide to proceed further. The shape factors result in the following curvatures:

$$c_1 = 0.3836, \qquad c_2 = 0, \qquad c_3 = -0.6247, \qquad c_4 = -0.2728$$

where the subscripts 3 and 4 refer to the second lens. As an extra precaution against unwitting errors, we trace the system using the paraxial ray-tracing program of Appendix 5, using zero thickness for the lenses, and we obtain similar aberration values.

The next step is to thicken the lenses. We do so by using the program referenced in the previous section. We set a clear aperture approximately equal to $2(|h| + |\bar{h}|)$ and sufficient axial and edge thickness to make the lenses easily manufacturable. The resulting lens is shown in Figure 10.2. The separation between the lenses was adjusted to 2.8 (from 3) in order to preserve the required Gaussian properties.

We may now trace the system again using the paraxial ray-tracing program:

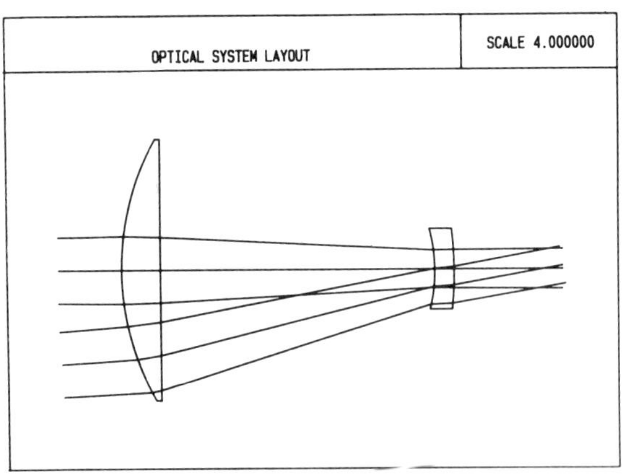

Figure 10.2. Initial third-order solution for the operating spectacles problem.

changes can be expected as a result of thickening. In this case, it turns out that the aberration correction depends rather critically on the lens separation, and that it is possible to get the same aberration correction as for the thin system, at the expense of departing slightly from our exact Gaussian specification. However, we cannot rely on the primary aberration alone. A finite ray-tracing program must be used next, and the sophistication of the available software will determine the subsequent steps.

To complete the example we used a commercial software package. The program showed that the limiting aberration was indeed transverse color, and confirmed that the lens had very good monochromatic correction, although secondary astigmatism contributed to rapid deterioration of image quality at the very edge of the field. Monochromatic aberration is more troublesome than chromatic for visual systems. Also, if one has access to an optimization routine, one cannot resist the temptation to optimize. Since we do not expect any large improvement in chromatic correction, we set up a simple merit function to be the sum of primary and secondary spherical aberration, coma, and astigmatism, and we specified the first three curvatures as free variables, while reserving the last curvature in order to maintain the desired Gaussian properties (i.e., the angle u'). In only four iterations, the program returned a system that looks very similar to the previous one (Figure 10.3), but has essentially diffraction-limited performance to the edge of the field. As a bonus, transverse color went down to 4.5 min, and the (already acceptable) distortion was almost halved.

Thus this example demonstrates simultaneously how much one can do with simple means, and how much more one can do with advanced optical design software. The primary aberration design produced quite acceptable image quality, but with only small changes, a better design was possible. The reader should clearly understand that the speed with which the optimization routine converged to an excellent solution was due to the fact that the starting point

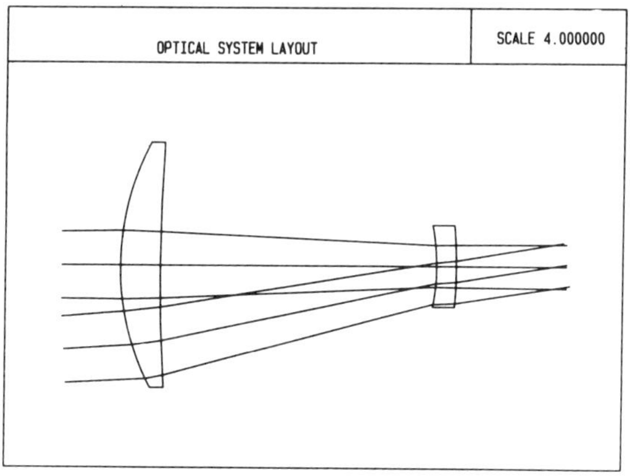

Figure 10.3. Further improvement of the operating spectacle design through finite ray tracing and optimization.

was already very close to that solution. If one had started from an arbitrary or badly chosen initial solution, the quality of the final solution would then be determined by the sophistication of the available software, which is to say that the designer would have little or no control.

10.4. OPTIMIZATION

Most fundamental research in optical design has revolved recently around the topic of optimization. The idea of optimization is very simple, but the mathematics involved can become very complex. The first step is to set up an "inverse" figure of merit for the optical system, called a merit function or error function. We call it "inverse" because usually a desirable system has a large figure of merit, whereas the optimum value for the lens merit function is a minimum.

The elements of the merit function are called *operands*; these might be specific aberration coefficients, spot size, ray aberration, chromatic terms, and even mechanical aspects such as weight and size, etc. The significance of each operand is denoted by a weighting factor, and the merit function M is the weighted sum of the squares of the operands, that is,

$$M = w_1 O_1^2 + w_2 O_2^2 + w_3 O_3^2 + \cdots$$

The values of the operands, and hence the merit function, are controlled by the system parameters, such as curvatures. separations, etc. Some of these parameters are specified as variables, while some may remain fixed. Optimization is an iterative process by which a change in the variables produces a change in the merit function, and the process is repeated until a minimum value of M is found, indicating that no further improvement is possible. Of course, this last statement can almost never be made with certainty. As can be seen from above, the merit function is a multidimensional function, and finding the global minimum of such a function involves a search of the entire parameter space, which in turn would require very long computing time for any but the simplest of systems.

Optimization methods can be broadly classified into two types: mathematical/descent methods and direct search methods. In the first case, the merit function is allowed only to decrease; that is, if a change in the variables causes an increase in the merit function, it is seen as moving in the wrong direction and rejected. When the program has reached a local minimum, changes in the variables will cause an increase in the merit function; so no further improvement will be possible. Thus descent methods are practically certain to converge to the local minimum that is closest to the original solution. Although this may not sound very promising, such methods are in fact the most broadly used, because of the speed of convergence. The operating spectacles design discussed in the previous section is a good example: From a good original solution, we obtain an even better one very quickly.

Direct search methods perform a search over as much of the parameter space as possible. This is called a global search and can result in several solutions, which, if not adequate by themselves, may be used as starting points for a descent method. Another possibility is this: Upon arriving at a (presumably local) minimum, the computer effects an intentionally large change in the parameters, large enough to "kick" the merit function out of the current local minimum and onto the slope of another, hopefully deeper minimum. This situation is shown schematically in Figure 10.4. If no such other minimum can be found after several tries along the direction of all variables, then the current minimum must be accepted as global.

Thus, when using descent methods, upon finding a minimum, the designer has to decide whether the solution is satisfactory or not. If not, a different starting design must be generated; hence the truism that the quality of the final design is proportional to the quality of the initial solution submitted for optimization to the computer. On the other hand, direct search methods appear to promise completely new and previously unthinkable designs, concocted entirely by the computer; it would seem that the starting design is immaterial and that as few constraints as possible should be set, so as to not restrict the range of potential solutions. However, things are not that simple. First. the computation time can become enormous; and if the entire parameter space is not covered, then discovery of a successful solution becomes a matter of luck. In any case, a comparison of the lenses designed after 30 or so years of computer optimization with the lenses designed using log tables reveals that the fundamental design principles remain the same, although the computer has provided some interesting variations, and has improved the performance of systems significantly.

A good automatic optimization routine is a necessary tool of the designer. The distinguishing characteristic of the good optical designer is still the ability to generate an initial solution that others cannot; but that solution must be further refined with the help of the computer. Furthermore, tolerance sensitivity analysis and diffraction-based image analysis are now commonplace, and they can only be done efficiently with the help of good software.

In using optimization routines, the beginner is often confounded by the fact that the optimization may change the Gaussian properties considerably, unless

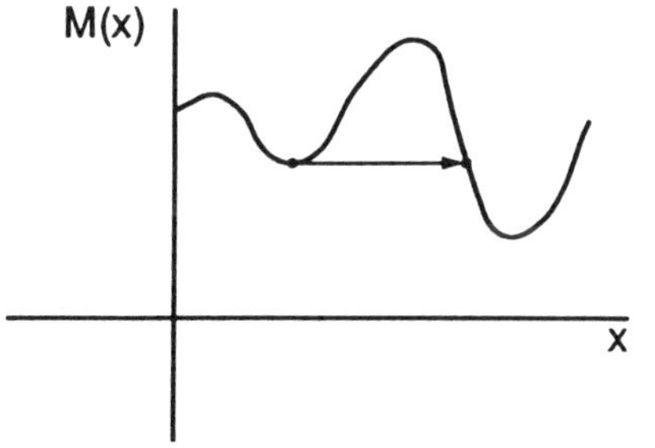

Figure 10.4. Illustration of a merit function with local minima. If the optimization has resulted in the solution represented by the shallower minimum, a large change in one or more of the variables may be used to send the merit function on the slope of the deeper (global?) minimum.

due care is taken. One invests a considerable amount of time in getting the Gaussian properties right (after thin-lens design, thickening, etc.); so it would seem reasonable to expect that the computer should "know better" than to change the focal length and F number of the system. But, in fact, this is not a reasonable expectation. There are several Gaussian parameters, such as F, $F\#$, throw T, back focal length, system length, eye relief, etc., all of which are controlled by the curvatures, thicknesses, and glass types, which the user is normally willing to accept as variables. But the computer cannot know in advance which of the Gaussian properties are critical to maintain; if all are considered critical, then automatic improvement of image quality may prove impossible. The user must specify which Gaussian properties are to be preserved, and there are two ways of doing that: (1) through a "solve," that is, by reserving one or more parameters to control specified Gaussian design values, and (2) by inserting these Gaussian parameters in the merit function as operands.

If the gaussian properties are not properly constrained, the user may well end up with a zero-power system comprising only flat surfaces (and thus exhibiting very little aberration!). Another frequently encountered tendency of optimization routines is to fill the airspaces with glass, that is, to come up with very thick lenses. This is due to the fact that a system with air-in-glass lenses has less aberration that the same system with glass-in-air lenses.

10.5. PUPILS AND PUPIL IMAGERY

In our discussion so far, we have assumed the entrance and exit pupils to be circular, and to have a well-defined location. More often than not, these assumptions do not hold. In this section we extend the discussion of the pupils to encompass those considerations.

10.5.1. Pupil Aberration

We have mentioned that perfect correction of aberrations may be possible for imagery between a single pair of conjugates, but not for two or more pairs. It follows that the exit pupil will in general be an aberrated image of the entrance pupil. If we consider the entrance pupil as an object, Seidel aberration coefficients can be defined in exactly the same way as for the original conjugates, by swapping the roles of the PMR and the PPR. So, for example, the spherical aberration of the pupil is given by

$$\bar{S}_{\mathrm{I}} = \sum \bar{A}^2 \bar{h}\, \Delta\left(-\frac{\bar{u}}{n}\right) \tag{10.1}$$

Other aberrations can be defined following the same relations as before. We have

$$\bar{S}_{\mathrm{II}} = \frac{A}{\bar{A}} \bar{S}_{\mathrm{I}}$$

$$\bar{S}_{\mathrm{III}} = \frac{A}{\bar{A}} \bar{S}_{\mathrm{II}}$$

$$\bar{S}_{\mathrm{IV}} = S_{\mathrm{IV}}$$

$$\bar{S}_{\mathrm{V}} = \frac{A}{\bar{A}} (\bar{S}_{\mathrm{III}} + \bar{S}_{\mathrm{IV}})$$

It can be seen that the original conjugates play the role of pupils for pupil imagery. Formulae similar to the stop-shift ones can then be developed, only these will now describe the effect of conjugate shift on the aberrations (see, e.g., Welford, 1986). These formulae are typically of lesser importance than the stop-shift formulae, because the object and image locations are not normally considered as variables.

The spherical aberration of the pupil, as defined previously, is the most important of the pupil aberrations. If it is large, the pupil location is ill-defined, even for axial imagery. This will be important if pupil matching is needed. But for a lens with no pupil matching at all, such as a photographic objective, pupil aberrations are of no consequence; that is, they do not have to be corrected. Their theoretical importance remains, as it arises from the fact that the pupil aberrations control the change of image aberrations during conjugate shift. For example, it can be shown that if the spherical aberration of the pupil is zero, then the image distortion is invariant with conjugate shift. This and other cases are discussed by Welford (1986).

There is an interesting theorem that relates the aberration correction of the pupil and object conjugates, as follows: If the system satisfies the sine condition (see Chapter 5), then there is no distortion of the pupils. This theorem can be proven in an elegant and simple way through the use of the canonical coordinates, introduced in Chapter 7. The reader should recall from Chapter 5 that the sine condition ensures the lack of aberration for imagery near the axis, and is expressed as $u' = \sin \alpha'$, or equivalently, $(\sin \alpha)/u = (\sin \alpha')/u'$, where u,u' are the PMR angle variables and α,α' the corresponding finite ray angles. Lack of pupil distortion is expressed as $(x, y) = (x', y')$, where (x, y) and (x', y') are the reduced (canonical) coordinates of a ray at the entrance and exit pupil, respectively. Thus a square grid at the entrance pupil is imaged as a square grid at the exit pupil. However, this holds only if we use pupil spheres, rather than pupil planes, as the pupil surfaces.*

* See H. H. Hopkins, "Geometrical and physical theory of image formation: a synthesis," SPIE Proc. **1780** *Lens and Optical Systems Design*, 2–13 (1992).

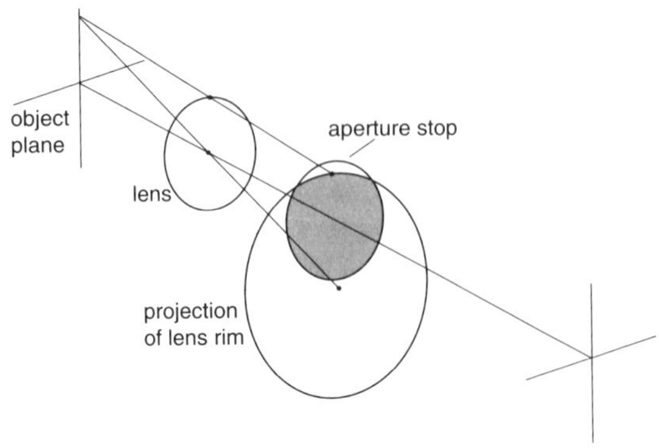

Figure 10.5. Off-axis pupil imagery and vignetting. The rim of the lens is projected at the plane of the aperture stop. The shaded part is the real pupil shape off-axis, for the case illustrated.

10.5.2. Off-Axis Pupil Shape and Vignetting

Viewed from an off-axis position, a circular pupil becomes elliptical. For small field angles this effect can be neglected, at least to a first approximation but for systems with large field of view the ellipticity of the pupil must be taken into account.

In the presence of vignetting at a lens rim, the pupil shape is even more complicated. Also, since the vignetting takes place far from the pupil, the location of the pupil is no longer well defined. Figure 10.5 gives an idea of the effect. If the field of view is sufficiently large and the vignetting severe, the pupil may even be entirely above or below the optical axis.

For systems without rotational symmetry, the entrance and exit pupils may not even exist; for example, the image of an aperture stop through a cylindrical lens can hardly qualify as an entrance pupil in the usual sense. However, for aberration calculations the pupils do not have to be precisely defined, and can be seen merely as a set of intermediate auxiliary planes (or spheres).*

It is important to know the precise shape and size of the pupil for several reasons. Clearly, we must know where rays are vignetted, so that we can decide on the appropriate clear aperture of the various elements. Also, radiometric or photometric calculations are directly affected by the size and shape of the pupils. But information on the pupil shape also enters into the wavefront aberration computation and subsequent diffraction-based image assessment calculations. For these reasons, an optical design program must perform what is known as *pupil exploration.* In broad terms, pupil exploration is a process of tracing rays at increasingly larger aperture angles, until they are vignetted; note is then made of where and how the ray was vignetted. The process is repeated through several azimuths over 180° (assuming meridional plane symmetry), until a complete picture of the pupil rim is obtained.

---

* See H. H. Hopkins, "Calculation of the aberrations and image assessment for a general optical system," *Optica Acta* **28**, 667–714 (1981).

10.5.3. Off-Axis Image Formation and Canonical Coordinates

The canonical coordinate normalization scheme introduced in Section 7.2.3 can be extended off-axis. In fact, part of its attractiveness is that the on-axis and off-axis normalizations are conceptually identical and formally very similar. Further, systems with no symmetry can be treated the same way as axisymmetric systems.* Apart from introducing the reader to these more advanced concepts, our additional motivation for discussing this topic is to extend the normalization to the relation between wave and ray aberration for extra-axial points. Use of the axial normalization at large field angles can lead to substantial errors.

The first step is to establish a base reference ray, called *principal* or *chief* ray: this is similar to but not identical with the paraxial pupil ray of previous chapters: The key difference is that it is a finite (real) ray. To establish this ray, we must first complete the pupil exploration for the particular field point of concern. Pupil exploration will provide us with the values of the angles α_{T_1} and α_{T_2} that the upper and lower marginal rays, respectively, form with the axis (Figure 10.6). Our reference or principal ray is then the "central" ray of the pencil, defined as the ray whose η-direction cosine, $\bar{M}$, is the arithmetic mean of the direction cosines of the upper and lower marginal rays.

$$\bar{M} = \tfrac{1}{2}(M_{T_1} + M_{T_2})$$

or (10.2)

$$\sin\bar{\alpha} = \tfrac{1}{2}(\sin\alpha_{T_1} + \sin\alpha_{T_2})$$

where the subscript T stands for tangential. It should be clear from this that the points $\bar{E}$ and $\bar{E}'$ in Figure 10.6 do not coincide with (and may even be far from) the intersections of the PPR with the axis; further, the principal ray will not normally pass through the center of the aperture stop.

Because of meridional plane symmetry, simpler considerations apply for the sagittal (ξ, z) plane, perpendicular to the (η, z) of Figure 10.6. The ξ-direction

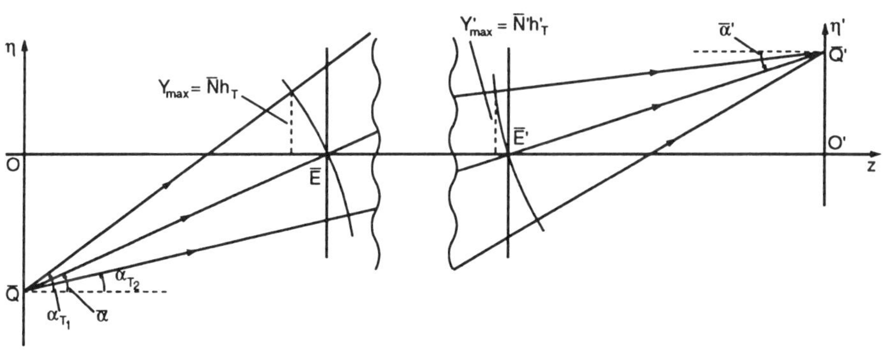

Figure 10.6. Off-axis canonical coordinate normalization.

---

* See H. H. Hopkins, "Image formation by a general optical system," *Applied Optics* **24**, 2491–519 (1985).

cosine of the principal ray is zero, and the left and right marginal rays form angles α_S and $-\alpha_S$ with the z axis. However, the angle α_S must still be established through pupil exploration.

We then proceed to extend the concept of the paraxial marginal ray for an off-axis point. The reader will recall that the PMR is not a real ray, but predictions made from it can be quite accurate in reasonably well-corrected systems, provided the convention $u = \sin \alpha$ is followed. In the off-axis case, the principal ray previously established becomes the new effective "axis." For the sagittal plane, we define a paraxial angle variable through

$$u_S = \sin \alpha_S \tag{10.3}$$

and we use this value to perform paraxial ray tracing, and thus determine a maximum exit pupil height $X'_{max} = h'_S = -\bar{R}'u'_S$ for the sagittal section. For the tangential section, the paraxial angle variable is

$$u_T = \frac{\sin \alpha_{T_1} - \sin \bar{\alpha}}{\cos \bar{\alpha}} \tag{10.4}$$

In this equation, $\cos \bar{\alpha} \equiv \bar{N}$ is a normalizing factor that arises from the obliquity of the entire ray pencil. The difference of the two sines in the numerator represents the (tangential) numerical aperture of the off-axis ray pencil.* Paraxial ray tracing now establishes the value of u'_T, from which the exit pupil height $Y'_{max} = \bar{N}'h_T = -\bar{N}'\bar{R}'u'_T$ is found (Figure 10.6).

To summarize, the off-axis normalization scheme is as follows (for both object and image spaces, with unprimed or primed quantities, respectively).

The X coordinate of the exit pupil reference sphere is divided by X'_{max} (corresponding to the sagittal finite ray marginal value), and the Y coordinate is divided by Y'_{max} (the tangential marginal value) to give:

$$x'_S = \frac{X'}{X'_{max}}, \qquad y'_T = \frac{Y'}{Y'_{max}} \tag{10.5}$$

The maximum value X'_{max} is given by the paraxial exit pupil height h'_S for the S ray, while Y'_{max} is given by $\bar{N}'h'_T$. The image coordinates are normalized by multiplication by the image-space numerical apertures in the respective sagittal and tangential sections. These are

$$\begin{aligned} (\mathrm{NA})_S &= n'u'_S = -\frac{n'h'_S}{\bar{R}'} \\ (\mathrm{NA})_T &= n'u'_T = -\frac{n'\bar{N}'h'_T}{\bar{R}'} \end{aligned} \tag{10.6}$$

*The reader should be clear that Eq. (10.4) is a definition. We could just as easily define u_T using $\sin(\alpha_{T_1} - \bar{\alpha})$, instead of $(\sin \alpha_{T_1} - \sin \bar{\alpha})$, corresponding to $\bar{\alpha}$ being the arithmetic mean of the angles α_{T_1} and α_{T_2}. While such a choice might appear more logical, it has several disadvantages, which become evident with a more detailed development.

The normalized transverse ray aberration is given by expressions analogous to Eqs. (7.30) and (7.31),

$$\delta G'_{\mathrm{S}} = n' u'_{\mathrm{S}}\, \delta\xi' = -\frac{n' h'_{\mathrm{S}}}{\bar{R}'}\, \delta\xi'$$
$$\delta H'_{\mathrm{T}} = n' u'_{\mathrm{T}}\, \delta\eta' = -\frac{n' \bar{N}' h'_{\mathrm{T}}}{\bar{R}'}\, \delta\eta' \tag{10.7}$$

Finally, the relations between wave and ray aberrations are given in normalized form as:

$$\delta G'_{\mathrm{S}} = \frac{\partial W}{\partial x'_{\mathrm{S}}}, \qquad \delta H'_{\mathrm{T}} = \frac{\partial W}{\partial y'_{\mathrm{T}}} \tag{10.8}$$

which are similar to Eqs. (7.32). To give a specific example, consider the transverse focal shift equations of Appendix 4. Equation (A4.9) holds, but Eqs. (A4.10–A4.12) apply only for focal shifts near the axis. In the general case, they become

$$\delta W_{\mathrm{T}} = -n u_{\mathrm{T}}\, \delta\eta\, y_{\mathrm{T}}$$

or

$$\delta W_{\mathrm{T}} = \delta W_{11}\, y_{\mathrm{T}}$$

with

$$\delta W_{11} = -n u_{\mathrm{T}}\, \delta\eta = n\bar{N}\, \frac{h_{\mathrm{T}}}{\bar{R}}\, \delta\eta$$

where primes have been omitted for convenience, as in Appendix 4.

10.6. ASPHERICS

Aspheric surfaces allow additional degrees of freedom in aberration correction. As has been mentioned already, conicoids of revolution can provide aberration-free imaging of an axial point: In aberration theory terms, this is equivalent to eliminating spherical aberration over a wide aperture. But in general, the aspheric shape can be considered as an extra degree of freedom over the spherical shape, and thus allows correction of one more aberration (per aspheric surface).

There are two issues to be considered: (1) ray tracing, and (2) aberrations. The ray-tracing algorithm of Appendix 3 can be used to trace rays not just for a system of spherical surfaces, but for any conicoid of revolution. However, the general aspheric surface is not quadratic, so a different ray-tracing algorithm is needed. If ρ is the radial distance from the z axis (the axis of rotational symmetry), then the equation of a conicoid of revolution with the apex at $z = 0$ can be written as

$$z_{\mathrm{con}}(\rho^2) = \frac{c\rho^2}{1 + \sqrt{1 - (1 + Q)c^2\rho^2}}$$

where c is the vertex curvature and Q the eccentricity parameter. The equation for the general aspheric surface is then written as

$$z = z_{\text{con}}(\rho^2) + A_4\left(\frac{\rho}{\rho_{\max}}\right)^4 + A_6\left(\frac{\rho}{\rho_{\max}}\right)^6 + \cdots$$

where $\rho_{\max}$ is a convenient normalizing constant, which may or may not coincide with the actual maximum radius of clear aperture. Thus we consider the aspheric surface to comprise a conicoid of revolution plus some higher-order terms, which are often referred to as "figuring" terms (borrowing from the optical fabrication terminology, which refers to the process of generating an aspheric surface as figuring). The conicoid is called "parent" conicoid.

Ray tracing always involves the same two steps, refraction and transfer. Snell's law has the same form independent of the surface shape, although the surface and the surface normal equations are different from the conicoid case of Appendix 3. Transfer for aspherics is an iterative process. Upon leaving the previous surface, the ray is transferred first to the parent conicoid, for which an analytical solution exists. Then the actual point of intersection with the aspheric surface is found by a series of successive approximations. As a result, ray tracing through an aspheric surface is considerably slower than through a spherical surface, although with fast computers the difference may not be significant. The full procedure is not described here; an algorithm is given by Smith (1990).

The second issue is the aberrations of aspherics. In order to discuss this, we separate the aspheric equation into spherical and aspheric parts, the latter being of the fourth or higher degree in ρ. The spherical part gives rise to the aberrations examined already, while the aspheric part will contribute additional amounts of various aberrations (but no new aberration types, since the surface is still one of rotational symmetry). For primary aberrations only, we state here without proof that when the stop is at the surface the aspheric term contributes only spherical aberration. But with a remote stop, the surface will contribute also coma, astigmatism, and distortion (see Smith, 1990, or Welford, 1986).

Aspherics are harder to manufacture than spherical surfaces; for this reason their use is not widespread. The most common method of mass producing aspheric surfaces is replication through injection molding. In this way the aspheric surface has to be generated only once, in order to create the mold. However, replicated surfaces are not of the same optical quality as polished surfaces; so until recently, their use was limited to illumination systems, which are not critical from an imaging viewpoint but require a large aperture. But manufacturing techniques are improving, and replicated aspherics are finding their way into camera objectives or other lenses that perform more critical imaging tasks. Since they permit correction of additional aberrations, aspherics can be used to reduce the number of surfaces in a system. Typically, no more than two surfaces within a system are aspheric. Most often, a single aspheric surface is used. Large astronomical mirrors are another example of aspheric shapes. In that case, the deviation from the sphere may be very small, on the order of wavelengths over an aperture of meters. This is accomplished with careful polishing of the surface.

10.7. A BRIEF GUIDE TO OPTICAL DESIGN SOFTWARE

The proliferation of optical design software justifies a brief discussion on what to expect from optical design programs. No recommendations on specific software products are appropriate here.

Optical design research is moving in the direction of making the design process accessible to an ever wider constituency. The hope is, perhaps, that any engineer will some day be able to sit in front of the computer, input the lens specifications, and obtain a solution, leaving the details to the machine. The development of expert systems that rely on vast libraries of past designs may achieve that goal, but we are not there yet. It is possible in some cases to give the computer a set of glass plates as input and have it come up with an acceptable design. However, the required computing power is beyond the means of most people.

There are some who think that the computer is making it unnecessary to learn geometrical optics and optical design practices. But there are serious difficulties with that argument, not the least of which is the fact that only sophisticated users can take advantage of sophisticated software. Even if one had a computer as user-friendly and powerful as the one aboard the *U.S.S. Enterprise*, that could be voice-commanded with the familiar "Computer!...," the engineer would still be faced with two unavoidable functions: (1) to provide the computer with a full system specification, including the Gaussian, radiometric, image quality, mechanical and environmental, etc. properties, as well as allowable cost, and (2) to provide the computer with a meaningful merit function so as to choose from the possible solutions.

It can be argued that the problem of optical design has not been simplified conceptually from the days of log tables. The computer has facilitated better designs by removing the tedium of repetitive calculations, by allowing the designer to pose "what if" questions, and by speedily calculating several alternative measures of image quality. In short, the computer has improved the quality of the design and the life of the designer. Also, the wide availability of optical design software means that one no longer needs to solve all equations by hand or write one's own code, and this has made optical design accessible to more people.

Let us assume that we can follow the design procedure of Section 10.1. How much help is available from the computer along the way?

1. Specification: Clearly, little help can be expected here, and none is available. Reduction of a given problem to a set of meaningful specifications is the principal function of the engineer.

2. Gaussian design: the authors know of no program that can help with this. This is one of the reasons why a good knowledge of the fundamentals of geometrical optics is essential.
3. Thin lens, primary aberrations, thickening: the simple programs given in this book are of some help. Also, the thickening program given by G. W. Hopkins (1983) is useful. Commercial software can be of help, but a systematic approach is not always offered. In fairness, it should be said that this step is often bypassed in favor of modifying a previously published similar design. This step is only needed if one cannot rely on past designs, such as if there are any unusual requirements.
4. Finite ray tracing: This is the level where the commercially available software becomes useful; so there is plenty of help here. Finite ray tracing is the primary function of every optical design program.
5. Optimization: A lot of help is available from virtually every commercial optical design program. This is perhaps the most critical function that differentiates one program from another.
6. Design evaluation: Again, the availability of several different image-quality measures (geometric and diffraction-based) and the accuracy with which these are computed are important in separating the best software from the rest.

In summary, commercial software typically becomes useful after one has obtained an initial solution down to the level of individual surface curvatures, surface separations, and glass types. A substantial amount of knowledge and labor is needed to get to that point.

The price of software spans a range of more than 100:1. Therefore, a wide range of capabilities should be expected. At the low end, programs can be expected to perform most of the following functions:

Finite and paraxial ray tracing

Tilted, decentered, cylindrical, and toric surfaces

General aspherics

Geometrical spot diagrams

Glass catalog

Third-order aberrations and chromatic aberrations

Elementary optimization

The accuracy of the ray-tracing computation is usually the same as that of more expensive programs. The merit function used for optimization is usually the rms spot size, computed geometrically. The rays used for this calculation may have to be input manually.

At the next price level, the analytical capabilities of the programs do not improve by much. The extra money usually buys better graphics, more diagnostic options, and perhaps some diffraction-based calculations (point spread function and MTF), though the accuracy of the calculation may not be the same as that of more powerful software.

One step above this level are programs that can handle almost all optical design tasks. Typically, in addition to the above functions, the following will be available:

Zoom lenses

Gradient-index lenses, holographic lenses

Gaussian beam propagation

Diffraction-based system analysis

Wavefront aberration analysis

Extensive lens library

Tolerance analysis

Advanced optimization routine

An in-built programming language may also be available, allowing the user to define new macros. The optimization routine normally allows user-defined merit functions as well as providing a default. It can include not only geometrically based criteria such as spot size, but also specific aberration coefficients, diffraction-based criteria, and even lens characteristics such as weight and size. In addition, the optimization routine itself is more advanced, allowing faster convergence as well as a more reliable exploration of the solution space.

At the high end of the spectrum, programs can offer such exotic features as GRIN rod array analysis, thermal and mechanical stability analysis and tolerancing, and partially coherent illumination, the importance of which can only be assessed by each user individually.

Clearly, as computing power improves, so do software capabilities; the listed guidelines are only applicable at the time of writing. Finally, a word of caution is perhaps warranted: In the authors' opinion, beginners may well find that they are not stretching the abilities of even simple software. Unless specific features are needed, one can do a lot of useful work with inexpensive programs. More advanced programs offer a lot of features, and it takes a while to learn how to utilize them fully. The selection of the program will inevitably be based on personal preferences, but it may be comforting to know that all advanced programs offer comparably high levels of performance.

Optical design is a lively research area. The student of geometrical optics and optical design should become aware of the annual conferences dedicated to the topic. Especially interesting and entertaining are the occasional design competitions, with topics such as designing a "nonlens," an optimized triplet, etc. These competitions are a good way of gauging the progress of optical design and finding out what design methods and software people use.

PROBLEMS

These problems assume access to finite ray-tracing software. Problems at this level of complexity cannot normally be attacked with primary aberration calculations only, although in some simple cases finite ray-tracing may be needed only as a confirmation of a primary aberration solution. Additional design problems can be obtained from several of the sources cited in the references. The reader is encouraged to seek out those sources and study (through ray tracing and aberration calculation) some of the designs contained therein.

1. Design a lens for focusing the output of a laser diode onto an optical fiber. The following specifications are given: $\lambda = 670$ nm, fiber NA $= 0.1$, core (mode) diameter 7 μm. The laser emits divergent light from a rectangular area of dimensions 0.5 μm × 5 μm; the divergence half-angles are 20° and 5° (smaller divergence in the direction of the larger side). It is required to couple the maximum amount of light possible into the fiber. Perform the design according to the following steps: (a) First design a collimating lens. Notice that the large aperture will necessitate the use of one or two aplanatic menisci, followed by a doublet (or some variation of this basic idea). (b) Then design an anamorphic beam expander, to be placed in the collimated part of the beam. This can be either a cylindrical telescope or a set of prisms. (c) Finally, design the focusing lens, paying attention to matching, as much as possible, the fiber NA and core size. Because of the limited fiber NA, a doublet may prove sufficient. Evaluate your design by using primary aberration theory, and also through finite ray tracing (using either your own or a commercial code). Investigate the effect of astigmatism inherent to the laser diode, by assuming that the wavefront emanating from the diode is not spherical, but astigmatic, with the S and T foci being separated by 5 μm.

2. Design a 20× beam expander for a nominally collimated laser beam. Diffraction-limited performance is expected. You are given $w_0 = 1$ mm, $\lambda = 633$ nm, beam waist at the output laser mirror. The formulas of Appendix 2 will be useful.

3. Design a Galilean telescope for visual use. Specifications are: object at infinity, 3× magnification, exit (eye) pupil diameter 3 mm, maximum telescope length 10 cm, field of view $\pm 10°$ (eye space). First, obtain the best possible solution using only two singlets. Then determine if significantly better correction can be obtained by adding a third, and then a fourth, element.

4. Use the Fraunhofer cemented doublet designed earlier in this chapter, and add a field flattener (planoconcave) lens, with the flat surface being 2 mm in front of the image. Is an improvement in off-axis image quality possible? Quantify the improvement by determining the field angle at which the doublet/flattener system has the same resolution as the doublet alone.

Appendix 1

Matrix Methods in Paraxial Optics

The matrix formulation of paraxial optics (Brouwer, 1964) has enjoyed varying degrees of popularity over the years. It is more often found in texts dealing with beam propagation through nonuniform media (Kogelnik, 1979; Haus, 1984) and Gaussian beams, with applications to optical resonator design. The paraxial ray equations describing gradient index media are examined in Chapter 6, while Appendix 2 gives Gaussian beam ray-tracing formulas that do not involve matrices.

To simplify the notation, we will use the reduced paraxial variables which are defined as

$$\hat{d} = \frac{d}{n}, \qquad \hat{l} = \frac{l}{n}, \qquad \hat{u} = nu$$

The reduced paraxial angle variable $\hat{u}$ is sometimes called the *optical direction cosine*. The paraxial refraction and transfer equations take the form:

$$\hat{u}'_j - \hat{u}_j = -h_j K_j$$

$$h_{j+1} = h_j + \hat{d}_{j+1}\hat{u}'_j$$

It is obvious that refraction affects only the angle u, while transfer (or translation as it is usually called in this formulation) affects only the height h. We can rewrite these equations as follows, for refraction and translation, respectively:

$$h'_j = 1 \cdot h_j + 0 \cdot \hat{u}_j$$
$$\hat{u}'_j = -K_j h_j + 1 \cdot \hat{u}_j$$

$$h_{j+1} = 1 \cdot h_j + \hat{d}_{j+1} \hat{u}'_j$$
$$\hat{u}_{j+1} = 0 \cdot h'_j + 1 \cdot \hat{u}'_j$$

These can now be rewritten in matrix form as follows:

$$\begin{pmatrix} h'_j \\ \hat{u}'_j \end{pmatrix} = \begin{pmatrix} 1 & 0 \\ -K_j & 1 \end{pmatrix} \begin{pmatrix} h_j \\ \hat{u}_j \end{pmatrix}$$
$$\begin{pmatrix} h_{j+1} \\ \hat{u}_{j+1} \end{pmatrix} = \begin{pmatrix} 1 & \hat{d}_{j+1} \\ 0 & 1 \end{pmatrix} \begin{pmatrix} h_j \\ \hat{u}'_j \end{pmatrix}$$

The two 2×2 matrices are called the *refraction* and *translation matrix*, respectively, and will be denoted by $\mathscr{R}$ and $\mathscr{T}$, respectively. Therefore, we can write:

$$\begin{pmatrix} h_j \\ \hat{u}'_j \end{pmatrix} = \mathscr{R}_j \begin{pmatrix} h_j \\ \hat{u}_j \end{pmatrix}$$
$$\begin{pmatrix} h_{j+1} \\ \hat{u}_{j+1} \end{pmatrix} = \mathscr{T}_{j+1} \begin{pmatrix} h_j \\ \hat{u}'_j \end{pmatrix} = \mathscr{T}_{j+1} \mathscr{R}_j \begin{pmatrix} h_j \\ \hat{u}_j \end{pmatrix}$$

Since the commutative law does not hold for matrix multiplication, the order of multiplication is essential. This result can now be obviously extended to many surfaces:

$$\begin{pmatrix} h_k \\ \hat{u}'_k \end{pmatrix} = \mathscr{R}_k \mathscr{T}_k \mathscr{R}_{k-1} \cdots \mathscr{T}_2 \mathscr{R}_1 \begin{pmatrix} h_1 \\ \hat{u}_1 \end{pmatrix} \tag{A1.1}$$

The product $\mathscr{R}_k \mathscr{T}_k \cdots \mathscr{R}_1 \equiv \mathscr{M}$, also a 2×2 matrix, will be called the *system matrix* (sometimes the general term *ray matrix* is used). It contains the powers of individual surfaces and the reduced separations, which include the refractive indices. It thus contains all the information about the constructional parameters of the system. Note that all the individual refraction and translation matrices have unit determinants; from a well-known matrix property, the system matrix $\mathscr{M}$ will also have unit determinant. This property provides a useful check on arithmetic.

The matrix $\mathscr{M}$ can be written as

$$\mathscr{M} = \begin{pmatrix} A & B \\ C & D \end{pmatrix} \tag{A1.2}$$

The constants A,B,C,D are called the *Gaussian constants* of a system, and characterize it completely in the same manner as the power and the location of the principal points of Chapter 4, and, as we might expect, they are directly related to those parameters. Since $\det(\mathcal{M}) = 1$, we have

$$AD - BC = 1$$

that is, only three of the four parameters are independent, as before. Specific relations between A,B,C,D and δ,δ',K are derived at the end of this Appendix; for now, we want to concentrate on the problem of finding the image of an arbitrary object using the matrix formalism.

A1.1. THE CONJUGATE MATRIX

We can include the object and image by simply using two more translation matrices, $\mathcal{T}_1$ and $\mathcal{T}_{k+1}$, from the object to the first surface and from the last surface to the image, respectively. In order to keep a consistent notation, we will denote the object and image heights by h_0 and h_{k+1}, instead of η and η' as before (Figure A1.1). Thus we have:

$$\begin{pmatrix} h_{k+1} \\ \hat{u}'_{k+1} \end{pmatrix} = \mathcal{T}_{k+1}\mathcal{M}\mathcal{T}_1\begin{pmatrix} h_0 \\ \hat{u}_0 \end{pmatrix} \equiv \mathcal{N}\begin{pmatrix} h_0 \\ \hat{u}_0 \end{pmatrix} \tag{A1.3}$$

where the matrix $\mathcal{N}$ is the conjugate matrix. It should be clearly understood that the translations $\mathcal{T}_1$ and $\mathcal{T}_{k+1}$ need not correspond to object and image planes; they can represent any two arbitrary planes in the object and image spaces, in which case, the name *conjugate matrix* becomes inappropriate. Some simplification in the form of $\mathcal{N}$ is possible if we refer to conjugate planes. Let us write $\mathcal{N}$ as:

$$\mathcal{N} = \begin{pmatrix} n_{11} & n_{12} \\ n_{21} & n_{22} \end{pmatrix}$$

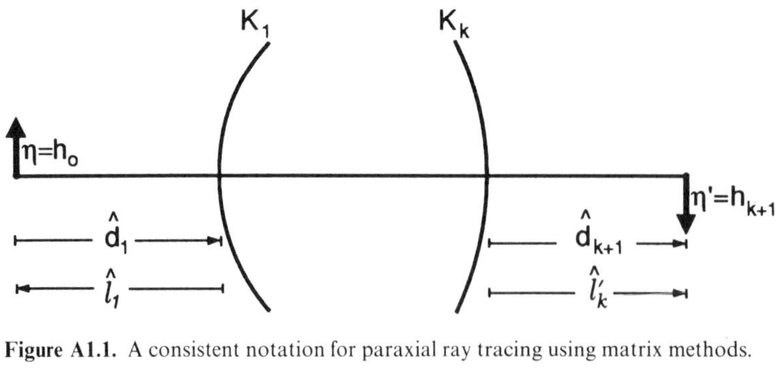

Figure A1.1. A consistent notation for paraxial ray tracing using matrix methods.

From Eq. (A1.3) we get $h_{k+1} = n_{11}h_0 + n_{12}\hat{u}_0$. If $n_{12} = 0$, then h_{k+1} depends only on h_0; therefore, all rays emanating from the point Q at height h_0 will meet at point Q', height h_{k+1}, independent of their angle with the axis $\hat{u}_0$. This means that Q and Q' are conjugates; therefore, the conjugate matrix is written as

$$\mathcal{N} = \begin{pmatrix} n_{11} & 0 \\ n_{21} & n_{22} \end{pmatrix}$$

Some special cases of interest are as follows:

1. If $n_{22} = 0 \rightarrow \hat{u}'_{k+1} = n_{21}h_0$; that is, the direction of the emergent ray is determined by the object height only. Therefore, all rays from point Q, height h_0, will emerge parallel; so Q must lie on the object-space focal plane.
2. If $n_{11} = 0 \rightarrow h_{k+1} = n_{12}\hat{u}_0$; that is, a parallel bundle of rays (constant $\hat{u}_0$) will come to focus at Q', height h_{k+1}. Therefore, Q' must lie on the image-space focal plane.
3. If $n_{21} = 0 \rightarrow \hat{u}'_{k+1} = n_{22}\hat{u}_0$; so the direction of the emergent ray depends only on the direction of the incident ray. A parallel bundle of rays will therefore emerge parallel; the system is afocal. This also suggests that n_{21} is proportional to the system power, which is proven below.

Let us write N explicitly and carry out the multiplication:

$$\mathcal{N} = \mathcal{T}_{k+1}\mathcal{M}\mathcal{T}_1 = \begin{pmatrix} 1 & \hat{l}'_k \\ 0 & 1 \end{pmatrix}\begin{pmatrix} A & B \\ C & D \end{pmatrix}\begin{pmatrix} 1 & -\hat{l}_1 \\ 0 & 1 \end{pmatrix}$$

where the minus sign in front of $\hat{l}_1$ is consistent with the sign convention established previously. In other words, the object distance is measured from the surface, but here we consider the object as another surface and measure from the object ($\hat{d}_1 \equiv -\hat{l}_1$). We then have:

$$\mathcal{N} = \begin{pmatrix} A + C\hat{l}'_k & [\hat{l}'_k(D - C\hat{l}_1) - A\hat{l}_1 + B] \\ C & D - C\hat{l}_1 \end{pmatrix}$$

From the conjugate condition $n_{12} = 0$ we obtain the image location:

$$\hat{l}'_k = \frac{A\hat{l}_1 - B}{-C\hat{l}_1 + D} \tag{A1.4}$$

Clearly, Eq. (A1.4) can also be solved for $\hat{l}_1$ in terms of $\hat{l}'_k$, giving a similar result. The conjugate matrix can then be written as

$$\mathcal{N} = \begin{pmatrix} A + C\hat{l}'_k & 0 \\ C & D - C\hat{l}_1 \end{pmatrix}$$

Since $n_{12} = 0 \rightarrow h_{k+1} = n_{11}h_0$; therefore, the magnification is

$$M \equiv \frac{\eta'}{\eta} \equiv \frac{h_{k+1}}{h_0} = n_{11} = A + C\hat{l}'_k \tag{A1.5}$$

Thus the problem of determining the size and position of the image in terms of object and system parameters has been completely solved.

A1.2. RELATION BETWEEN *A,B,C,D* AND δ,δ′,*K*

We will now extend the theory a little more to show its equivalence with the Gaussian optics approach developed in Chapter 4. Since $n_{11} = M$ and $\det(\mathcal{N}) = 1$, we have $n_{22} = 1/n_{11} = 1/M$; so $\mathcal{N}$ takes the form:

$$\mathcal{N} = \begin{pmatrix} M & 0 \\ C & 1/M \end{pmatrix} \tag{A1.6}$$

For $M = 1$, we obtain the positions of the principal planes from Eqs. (A1.4) and (A1.5) as:

$$\begin{aligned} \hat{l}_1 &= \frac{\delta}{n} \equiv \hat{\delta} = \frac{D-1}{C} \\ \hat{l}'_k &= \frac{\delta'}{n'} \equiv \hat{\delta}' = \frac{1-A}{C} \end{aligned} \tag{A1.7}$$

The condition $n_{11} = 0$ gives the image space focal plane (i.e., $\hat{l}'_k = F_b/n'$), where F_b is the back focal length of the system. We then have:

$$n_{11} = 0 = A + C\hat{l}_k = A + \frac{CF_b}{n'} \rightarrow \frac{F_b}{n'} = -\frac{A}{C} \tag{A1.8}$$

Also, the secondary focal length f' is given by $f' = F_b - \delta'$, and the EFL is $F = f'/n' = F_b/n' - \delta'/n'$; so from Eqs. (A1.7) and (A1.8) we get:

$$F = -\frac{A}{C} - \frac{1-A}{C} = -\frac{1}{C}$$

or

$$C = -K \tag{A1.9}$$

where K is the power of the system. Therefore, the conjugate matrix takes the simple form:

$$\mathcal{N} = \begin{pmatrix} M & 0 \\ -K & 1/M \end{pmatrix} \tag{A1.10}$$

Equations (A1.8) and (A1.9) relate δ, δ', and K to the Gaussian constants A,B,C,D. We can also easily derive the inverse equations:

$$A = 1 + K\hat{\delta}', \qquad B = \hat{\delta} - \hat{\delta}' + K\hat{\delta}\hat{\delta}, \qquad C = -K, \qquad D = 1 - K\hat{\delta}$$

Kogelnik (1979) shows several examples of matrices that can be used for computing the parameters of uniform or Gaussian beams.

Appendix 2

Gaussian Beam Ray Tracing

A2.1. BASIC CHARACTERISTICS OF GAUSSIAN BEAMS

Laser beams are referred to as *Gaussian beams* because the distribution of irradiance across the beam (the beam profile) resembles a Gaussian function. The irradiance has a maximum on axis and decreases in a radially symmetric fashion according to

$$E = E_0 e^{-2(r/w)^2} \tag{A2.1}$$

where E_0 is the axial maximum. This function is plotted in Figure A2.1. The parameter w is called the *Gaussian beam radius* and represents the radius at which the irradiance has dropped to $1/e^2$ ($\cong 0.135$) of the maximum. Laser beams may have profiles that are different from the above form, but they still obey similar laws. We restrict our attention to the profile given by Eq. (A2.1), because it is the most common, and normally, the only one desirable.*

While a true Gaussian distribution extends to infinity, in practice the beam will be inevitably truncated due to the aperture of the optical system. A severely truncated beam will behave essentially like a uniform beam; how much can the beam be truncated and still be considered as truly Gaussian? The answer is at $E = 0.01E_0$, which occurs at $r = 1.52w \cong 1.5w$.

* For a diffraction-based treatment of Gaussian beams that includes proof of Eqs. (A2.2)–(A2.6), the reader is referred to A. E. Siegman, *An Introduction to Lasers and Masers*, McGraw-Hill, 1971, or R. Guenther, *Modern Optics*, John Wiley and Sons, Inc., 1990.

Figure A2.1. The irradiance profile of a Gaussian beam. The Gaussian beam radius is taken as the point where the intensity drops to $1/e^2$ of its axial value. This is only a convention; it does not imply that the rest of the beam can be neglected.

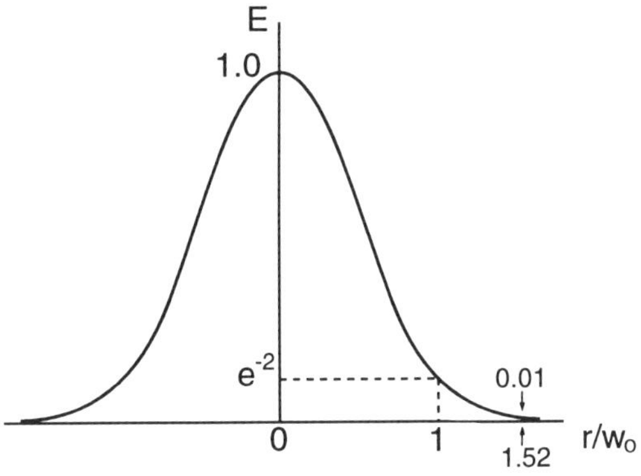

The key characteristic of Gaussian beams is that they remain Gaussian through any axial transformation by an unaberrated optical system, such as focusing. Consider a converging Gaussian beam going through focus, as in Figure A2.2. The *beam waist* has a radius w_0, given by

$$w_0 = \frac{\lambda}{\pi\theta} \tag{A2.2}$$

where θ is the convergence half-angle. The wavefront in the region of the waist is planar; so the wavefront shape changes from spherical to plane to spherical again. The reverse is also true, namely, that a plane wavefront will acquire curvature as it propagates, due to diffraction. The radius of curvature of the wavefront is then given by

$$R(z) = z\left[1 + \left(\frac{\pi w_0^2}{\lambda z}\right)^2\right] \tag{A2.3}$$

and the corresponding beam radius at any distance z is

$$w(z) = w_0\left[1 + \left(\frac{\lambda z}{\pi w_0^2}\right)^2\right]^{1/2} \tag{A2.4}$$

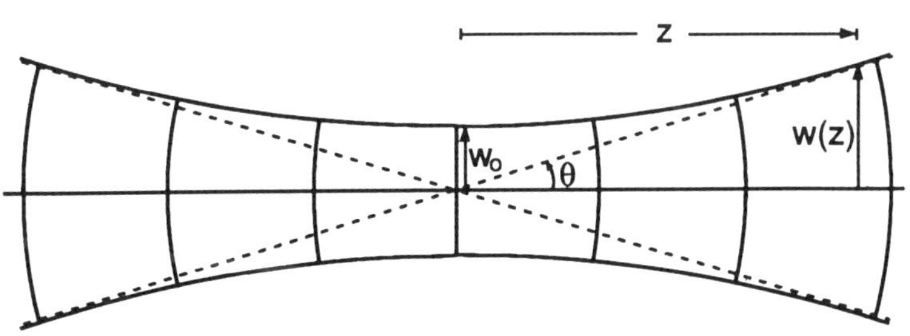

Figure A2.2. Focusing of a Gaussian beam. The narrowest part of the beam is the beam waist. The dashed lines represent rays according to the geometrical optics prediction for uniform beams. The solid hyperbolic lines coincide with the dashed lines far from the waist and represent the locus of the beam radius (irradiance $1/e^2$ of the axial value). The wavefront at the waist is plane.

The region of the beam waist is characterized by the *confocal parameter*, or *Rayleigh range* $\imath$

$$\imath = \frac{\pi w_0^2}{\lambda} \tag{A2.5}$$

which is the parameter that appears in both Eqs. (A2.3) and (A2.4). Alternative expressions for the Rayleigh range are [from Eq. (A2.2)]

$$\imath = \frac{\lambda}{\pi\theta^2} = \frac{w_0}{\theta} \tag{A2.6}$$

Equations (A2.3) and (A2.4) show that at the end of the Rayleigh range ($z = \pm\imath$), the beam radius $w(z)$ has increased by a factor of $\sqrt{2}$, and the absolute value of the wavefront radius $|R(z)|$ has a minimum. These equations may be rewritten as

$$R(z) = z\left[1 + \left(\frac{\imath}{z}\right)^2\right] \tag{A2.7}$$

$$w(z) = w_0\left[1 + \left(\frac{z}{\imath}\right)^2\right]^{1/2} \tag{A2.8}$$

Let us now seek some physical intuition from these equations. Notice first that Eq. (A2.2) shows that the product $w_0\theta$ is a constant (for a given wavelength).* Therefore, for a small beam waist we must have large divergence (or convergence), and for small divergence (i.e., collimated beam) we need a large beam waist. These results are similar to those implied by Eqs. (6.4) and (6.5) for uniform beams: High NA gives small spot size r, and large lens aperture (D) gives small divergence $\Delta\phi$. Notice also that for $\theta = 0$ we have $w_0 = \infty$, which shows that it is impossible to have a perfectly collimated beam with finite extent, as expected from diffraction theory.

A second conclusion is that the behavior of the beam is completely determined if we know only one of the three parameters ($\imath$, w_0, θ) and the wavelength λ. Let us then consider a typical beam from a HeNe laser ($\lambda = 633$ nm), with a radius of 0.4 mm. From Eqs. (A2.2) and (A2.5), we have $\theta = 0.5$ mrad (or 1 mrad full angle) and $\imath = 0.8$ m. At a distance of, say, 10 m, the beam radius will be about 5 mm, or 12.5 times larger than the beam waist.

The 0.5 mrad value for the divergence is roughly one-half of that expected for a uniform beam of $D = 2w_0 = 0.8$ mm: Eq. (6.5) gives $\Delta\phi \cong 1$ mrad. For a converging Gaussian beam of (say) $\theta = 20^\circ$, we obtain a waist radius $w_0 \cong 0.58$ μm, while for a converging uniform beam with $\alpha = 20^\circ$ we have a spot radius $r \cong 1.1$ μm. These comparisons are not in any sense exact, since the

---

* By setting $\lambda = \lambda_0/n$, where λ_0 the wavelength in vacuum, the relation $nw_0\theta =$ const. can be seen as the equivalent of the optical invariant $H = nu\eta$, for $u \to \theta$ and $\eta \to w_0$. Notice, however, that the Gaussian beam equations refer to a single beam (i.e., a pointlike object), whereas our geometrical optical invariant dealt with extended objects.

definition of the divergence or spot size is somewhat arbitrary in both cases, but they do serve to show that there is a difference between the two cases.

While it is tempting to identify the beam waist with the focal spot, it is also quite wrong. For strongly focusing systems (large convergence or small F number), the beam waist is indeed located near the geometrical focus, but in every other case the beam waist location is not predicted by the geometrical optics that we know so far. We therefore will give a set of formulae that can be used to determine the location and size of the beam waist in the general case.

A2.2. PARAXIAL EQUATIONS FOR GAUSSIAN BEAMS

We wish to find the equivalent of the paraxial refraction and transfer equations for Gaussian beams. Recall the paraxial refraction equation

$$\frac{n'}{l'} - \frac{n}{l} = c(n' - n) = K \tag{A2.9}$$

This equation can be thought of as the relation between the curvatures of the incident and refracted wavefronts, $1/l$ and $1/l'$, respectively. The curvature of the wavefronts can be given as $1/R(z)$ from Eq. (A2.3); the action of the refracting surface will be to change the wavefront curvature in accordance with Eq. (A2.9). (In the thin-lens case of Chapter 3, we said that the power of the lens is the algebraic increment in curvature added to the wavefront.)

Figure A2.3 shows the geometry of the problem. The Rayleigh range is always taken as a positive parameter. The distances z and z' that appear in Eq. (A2.3) are measured from the beam waist, at A and A', respectively. In this case, the radius of the wavefront $R(z)$ has the same sign as z; so $R(z)$ would be positive as shown. However, in our Gaussian optics sign convention, we measure all distances from the surface. Using s and s' for the location of the beam waists measured from the surface, and writing

$$R(s) = s\left[1 + \left(\frac{z}{s}\right)^2\right] \quad \text{and} \quad R'(s') = s'\left[1 + \left(\frac{z'}{s'}\right)^2\right] \tag{A2.10}$$

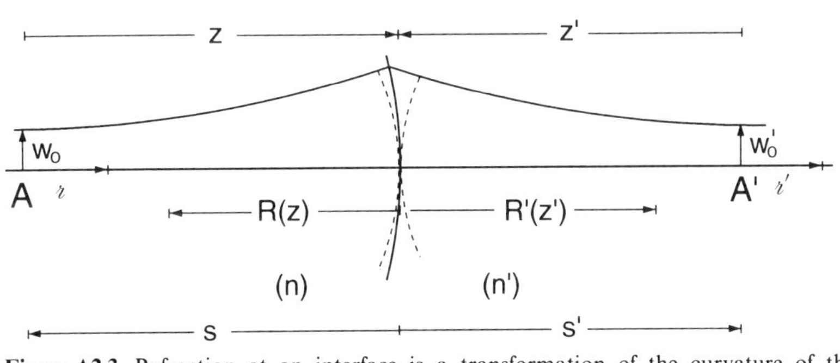

Figure A2.3. Refraction at an interface is a transformation of the curvature of the incident wavefront. We measure the beam waist locations (s, s') from the surface.

we obtain expressions that are consistent with our previous sign convention. From Figure A2.3, s is negative and s' is positive, in which case Eq. (A2.10) gives $R(s)$ as negative and $R(s')$ as positive. We can now apply Eq. (A2.9) to the curvatures of the incident and refracted wavefronts:

$$\frac{n'}{R'(s')} - \frac{n}{R(s)} = K \tag{A2.11}$$

Upon substituting for $R(s)$ and $R'(s')$ from Eqs. (A2.10), we obtain an equation that contains s, z, s', and z', the last two of which can be taken as unknowns. We then need one more equation, which is provided by the fact that the beam radius w will be the same on either side of the surface:

$$w(s) = w_0\left[1 + \left(\frac{s}{z}\right)^2\right]^{1/2} = w'(s') = w'_0\left[1 + \left(\frac{s'}{z'}\right)^2\right]^{1/2} \tag{A2.12}$$

from which we obtain the magnification

$$|m| = \frac{w'_0}{w_0} = \left[1 + \left(\frac{s}{z}\right)^2\right]^{1/2} \Bigg/ \left[1 + \left(\frac{s'}{z'}\right)^2\right]^{1/2} \tag{A2.13}$$

We also have

$$\begin{aligned} z &= \frac{\pi w_0^2}{\lambda} = \frac{n\pi w_0^2}{\lambda_0} \\ z' &= \frac{\pi w_0'^2}{\lambda'} = \frac{n'\pi w_0'^2}{\lambda_0} \end{aligned} \tag{A2.14}$$

where λ_0 is the wavelength in vacuum. By dividing Eq. (A2.14) between them and using Eq. (A2.13), we obtain

$$\boxed{m^2 = \frac{nz'}{n'z}} \tag{A2.15}$$

and through Eq. (A2.12),

$$\left[1 + \left(\frac{s}{z}\right)^2\right] \Bigg/ \left[1 + \left(\frac{s'}{z'}\right)^2\right] = \frac{nz'}{n'z} \tag{A2.16}$$

Equations (A2.11) and (A2.16) can now be solved simultaneously to give s' and z'. The result can be expressed in several different ways, one of which is

$$\frac{n'}{s'\tau} - \frac{n}{s\sigma} = K \tag{A2.17}$$

where

$$\begin{aligned} \sigma &= \sigma(s) = 1 + \left(\frac{z}{s}\right)^2 \\ \tau &= \tau(s) = 1 + \left(\frac{z}{s + (K/n)(s^2 + z^2)}\right)^2 \end{aligned} \tag{A2.18}$$

Notice that for uniform beams in the geometrical optics approximation we have $w_0 = 0 \rightarrow \mathcal{z} = 0$, and therefore, $\sigma = \tau = 1$. We can now solve for s' to obtain (after some tedious algebra)

$$\boxed{s' = n' \frac{s(n + Ks) + K\mathcal{z}^2}{(n + Ks)^2 + K^2\mathcal{z}^2}} \tag{A2.19}$$

We then need to determine $\mathcal{z}'$. From Eqs. (A2.11) and (A2.16), the following relation is derived:

$$s' = \frac{\mathcal{z}'}{\mathcal{z}}\left(s + \frac{K}{n}(s^2 + \mathcal{z}^2)\right) \tag{A2.20}$$

Upon substituting for s' from Eq. (A2.19) and solving for $\mathcal{z}'$, we obtain:

$$\boxed{\mathcal{z}' = \frac{nn'\mathcal{z}}{(n + Ks)^2 + K^2\mathcal{z}^2}} \tag{A2.21}$$

Finally, using Eq. (A2.15), the magnification is given by:

$$\boxed{|m| = \frac{n}{[(n + Ks)^2 + K^2\mathcal{z}^2]^{1/2}}} \tag{A2.22}$$

We can now give iterative relations for tracing a Gaussian beam through any number of surfaces. The numbering of surfaces is as in Chapter 4, with d the distance between successive surfaces. The starting point is the object beam waist location and Rayleigh range, s_1, $\mathcal{z}_1$. To simplify the notation, we use the definition $n + Ks \equiv \zeta$.

$$\begin{aligned}
\zeta_i &= n_i + K_i s_i \\
s'_i &= n'_1 \frac{s_i\zeta_i + K_i\mathcal{z}_i^2}{\zeta_i^2 + K_i^2\mathcal{z}_i^2} \\
\mathcal{z}'_i &= \frac{n_i n'_i \mathcal{z}_i}{\zeta_i^2 + K_i^2\mathcal{z}_i^2} \\
\mathcal{z}_{i+1} &= \mathcal{z}'_i, \qquad n_{i+1} = n'_i \\
s_{i+1} &= s'_i - d_i
\end{aligned} \tag{A2.23}$$

These equations are entirely compatible with the paraxial refraction and transfer equations developed in Chapter 4. We have used s and s' in place of l and l' to emphasize the difference between Gaussian and ordinary beams. Similarly, we use lower case m for the Gaussian beam waist magnification instead of upper case M. Notice that $m \neq ns'/n's$ except in the geometrical optics limit, $\mathcal{z} = 0$. In this last case, Eq. (A2.19) becomes identical with the paraxial conjugate equation, and m becomes the paraxial magnification $ns'/n's$, as can be seen from Eqs. (A2.22) and (A2.19).

In some cases it may be desirable to find the input beam parameters that would result in a given output. Equations (A2.11) and (A2.16) may be solved in terms of s' and z' to give:

$$s = n\,\frac{s'(n' - Ks') - Kz'^2}{(n' - Ks')^2 + K^2 z'^2} \tag{A2.24}$$

$$z = \frac{nn'z'}{(n' - Ks')^2 + K^2 z'^2} \tag{A2.25}$$

We can now examine some special cases:

1. $K = 0$ (plane surface): Equation (A2.19) gives $s' = n's/n$. From Eq. (A2.21) $z' = n'z/n$, but the magnification $m = 1$ [from Eq. (A2.22)] or $w_0' = w_0$, as might have been expected. The beam waist does not change despite the fact that the refraction affects the convergence angle θ; this is because the change in wavelength (by n/n') cancels the change in divergence as far as w_0 is concerned [see also Eqs. (A2.5) and (A2.6)].
2. $K = 0$ and $n' = -n$ (plane mirror) gives $s' = -s$, and $|z'| = |z|$, as expected.
3. $s = -n/K$ (input beam waist at the front focal point of the surface) gives $s' = n'/K$. In words, *if the input beam waist is at the front focal point, the output beam waist is at the back focal point*, rather than infinity. This is a striking difference between the usual geometrical conjugates and the Gaussian beam waists. An additional difference is as follows: In geometrical optics, if the conjugate locations (l, l') are fixed, the magnification (M) is also fixed. However, here we have both object- and image-space beam waist locations fixed at the focal points, yet the beam waist magnification $m = n/Kz$ [from Eq. (A2.22)] is a variable that depends on z (or w_0).
4. $s = 0$ (input beam waist at the surface) gives $s' = (n'Kz^2)/(n^2 + K^2z^2)$, rather than zero; so the output beam waist is not at the surface. This also illustrates the difference between Gaussian beam waists and geometrical conjugates.
5. It is critical to understand that knowledge of the input beam waist location s alone is not sufficient to determine the output beam waist location s'; one additional parameter must be known (z, θ, or w_0). The exception to this is case 3, where the object and image beam waists are at the focal points; this holds independent of the value of z.

A2.3. THIN LENS IN AIR

In principle, we can treat the thin-lens problem as two consecutive surfaces separated by zero distance, and apply the iterative refraction and transfer equations. That approach, however, is rather tedious and unnecessary. Instead, we notice the formal similarity between Eq. (A2.9) and the thin-lens equation, $1/l' - 1/l = K = 1/F$. The two equations are the same for $n = n' = 1$ [but with the power K being that of the thin lens, $K = (n - 1)(c_1 - c_2)$]. Also, Eq. (A2.12) holds for the thin lens as well. Therefore, the solution for the thin lens in air is the same as Eq. (A2.19), but with $n = n' = 1$ and $K = 1/F$. We have:

$$s' = \frac{s(1 + s/F) + z^2/F}{(1 + s/F)^2 + z^2/F^2} \tag{A2.26}$$

which can also take the more often quoted form

$$\boxed{s' = F - \frac{(s + F)F^2}{(s + F)^2 + z^2}} \tag{A2.27}$$

The signs in this equation are consistent with our usual geometrical optics convention, which (unfortunately) is rarely employed in Gaussian beam tracing. Also, from Eq. (A2.22) the magnification can be shown to be

$$\boxed{|m| = \frac{|F|}{[(s + F)^2 + z^2]^{1/2}}} \tag{A2.28}$$

Another interesting difference between beam waists and geometrical conjugates can be seen from Eq. (A2.27). Let us first define the beam waist throw t as

$$t = s' - s \tag{A2.29}$$

By taking the derivative $\partial s'/\partial s = 0$ from Eq. (A2.27), we obtain conditions for minimum and maximum s'. The following conclusions emerge:

1. s' is a *maximum* for $s = -(F + z)$. The value of the maximum is $s' = F + F^2/2z$. The beam waist throw t is equal to $(F + 2z)^2/2z$ in this case.
2. s' is a *minimum* for $s = -(F - z)$. The value of the minimum is $s' = F - F^2/2z$. The beam waist throw t is equal to $(F - 2z)^2/2z$ in this case. Notice that this is an algebraic minimum (s' may be <0), corresponding possibly to a virtual beam waist.
3. The magnification is equal to $F/\sqrt{2z}$ *in both cases above* [see Eq. (A2.28)].
4. The minimum $|s'|$ is zero (output beam waist at the lens). This can be seen directly from Eq. (A2.27); setting $s' = 0$ gives $s = -F/2 \pm (F^2 - 4z^2)^{1/2}/2$.
5. There is no minimum throw condition for t corresponding to $T = 4F$ ($l' = -l = 2F$) as we had for the geometrical conjugates.
6. For $s = -2F$, we have $s' = F + F^3/(F^2 + z^2)$, which becomes equal to $2F$ only at the geometrical optics limit $z = 0$.

A2.4. THE GENERAL SYSTEM: PRINCIPAL PLANES

The entire treatment developed so far can be immediately extended to a general optical system, characterized by its principal planes and power. This is because our starting point, Eqs. (A2.11) and (A2.12), hold for the general system as well as a single surface. Equation (A2.11) is seen as a transformation of wavefront curvatures that is effected by any system described through its principal planes, and Eq. (A2.12) is an immediate consequence of the unit magnification property of the principal planes. Therefore, our final equations for s', z', and $|m|$, Eqs. (A2.19)–(A2.22), as well as the thin-lens versions Eqs. (A2.27)–(A2.28), also hold for a general optical system. We illustrate this with the following example.

Example A1. A thin positive lens of focal length F_1 is followed by a thin negative lens located at $d = F_1$. The input beam waist is at $s = -2F_1$ from the first lens. What is the focal length F_2 if the output beam waist is located at the second lens?

To solve this problem, all we need to do is to apply the refraction and transfer equations for the two lenses. Specifically, Eq. (A2.27) applied for the second lens with $s'_2 = 0$ gives

$$s_2^2 + F_2 s_2 + z_2^2 = 0 \tag{A2.30}$$

The Rayleigh range $z_2 = z'_1$ and is given by Eq. (A2.21) with $n = n' = 1$, $K_1 = 1/F_1$, and $s_1 = -2F_1$:

$$z_2 = \frac{F_1^2 z_1}{F_1^2 + z_1^2} \tag{A2.31}$$

Again, Eq. (A2.27) for $s_1 = -2F_1$ gives

$$s'_1 = F_1 + \frac{F_1^3}{F_1^2 + z_1^2}$$

The transfer equation is

$$s_2 = s'_1 - d = s'_1 - F_1$$

From the last two equations, we immediately find

$$s_2 = \frac{F_1^3}{F_1^2 + z_1^2} \tag{A2.32}$$

Substituting from Eqs. (A2.32) and (A2.31) into Eq. (A2.30) yields (in a few lines indeed):

$$F_2 = -F_1$$

Notice that the problem was solved even though we had no information about z. The explanation for this lies in the fact that the object and image beam waists are at $-F$ and F away from the principal planes of the system, where F is the EFL of the combination; in this case, the solution is independent of z.

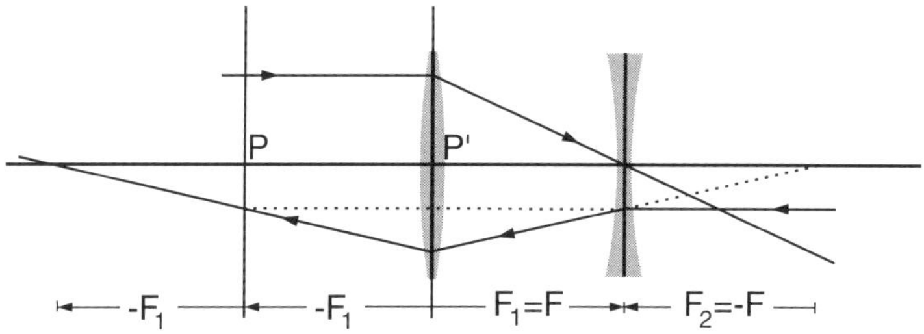

Figure A2.4. For this example, the total power $K = K_1$ and $F = F_1$. The input beam waist is located at a distance $-2F_1$ from the first lens, which is the same as $-F$ from the first principal plane. The output beam waist is then located at F from the second principal plane, that is, on the second lens.

Figure A2.4 shows the EFL $F = F_1$, as well as the location of the principal planes P and P'.

Example A2. Laser beam expander revisited. In Section 6.1 we discussed briefly the laser beam expander as a reversed (Galilean or inverting) telescope. We can assume approximately that the output beam waist will be given by the input beam waist times the magnification of the telescope, and that the beam divergence will be reduced by the same factor, since a larger beam size will give rise to small diffraction spread. That kind of approximation is often adequate, but it will be shown that the error involved may be as large as $\sqrt{2}$.

For a full treatment, one would need to start from the beam waist of the laser beam itself. This is not always known, but a reasonable first assumption is that it is located at the output mirror of the laser cavity (this is often a design goal for the laser cavity). We then need to trace the beam through both lenses of the telescope. For our present purpose, it suffices to consider the second lens only (the "objective"), which is always of positive power.

How are we to obtain a collimated beam? If we knew nothing about Gaussian beams, we would expect to place the input beam waist at the front focal point $\mathscr{F}$ of the objective, expecting that the output waist would then be at infinity. However, we know from the preceding discussion that if the object beam waist is at $\mathscr{F}$, then the image beam waist will be at $\mathscr{F}'$; so the "image" did not end up at infinity. The maximum distance between the image beam waist and the objective is $F + F^2/2\imath$ from above; this is the best we can do, if our definition of collimation is to have the beam waist as far as possible. In this case, the input beam waist will be at $-(F + \imath)$ instead of $-F$. Let us compare the two cases:

Case 1: Input beam waist at $s = -F$, output beam waist at $s' = F$. From Eq. (A2.28), the magnification is $F/\imath$. We also have, for the Rayleigh range, beam waist, and divergence, respectively:

$$z' = m^2 z \qquad \text{or} \qquad z' = F^2/z$$

$$w'_0 = \left(\frac{\lambda z'}{\pi}\right)^{1/2} = F(\lambda/\pi z)^{1/2}$$

$$\theta' = (\lambda/\pi z')^{1/2} = \frac{1}{F}(\lambda z/\pi)^{1/2}$$

Case 2: Input beam waist at $s = -(F + z)$, output beam waist at $s' = F + F^2/z$. Again, from Eq. (A2.28), we have $m = F/\sqrt{2}z$, which is different by a factor of $1/\sqrt{2}$. z is the same in both cases, since it is determined by the laser and first lens; the telescope is focused by moving the objective only. Also,

$$z' = F^2/2z$$

$$w'_0 = F(\lambda/2\pi z)^{1/2}$$

$$\theta' = \frac{1}{F}(2\lambda z/\pi)^{1/2}$$

Thus the beam waist is smaller and farther away in the second case, but the divergence is smaller in the first case. This is of course expected since the smaller beam waist will cause larger diffraction spread. Thus we have two distinct possibilities for defining a collimated beam: Either minimize the beam divergence, or place the beam waist as far as possible. For minimum beam divergence, the beam waist must be as large as possible.*

Which of the two conditions would we choose? This depends on our aim. If we want to maintain minimum beam size very far from the lens, then the minimum divergence condition is preferable. But if we are interested in minimizing the variation of the beam radius over a given range Δz, then the beam waist must be positioned in the middle of Δz.

Let us illustrate the above numerically. Consider a laser beam with 0.4 mm radius, giving 0.5 mrad divergence, and $z = 0.8$ m. We wish to expand and recollimate this beam with a $10\times$ telescope. Suppose that the telescope is focused in such a way as to form a beam waist of 4 mm at the second focal point of the objective: The divergence will be reduced by a factor of 10 to 0.05 mrad [Eq. (A2.2)], and the Rayleigh range will be increased by a factor of 100, to 80 m [Eq. (A2.5)]. Notice that for reasonable values of F we have $F \ll z'$; so we can consider that the output beam waist is effectively at the lens. The beam radius changes from 4 mm near the lens to $4\sqrt{2}$ mm at the end of the Rayleigh range (80 m away).

If we focus the telescope in such a way as to place the end of the Rayleigh range near the output lens, we have

$$w(z) = w_0\sqrt{2} = 4 \text{ mm} \rightarrow w_0 = 2\sqrt{2} \text{ mm}$$

* For a given lens diameter, the maximum waist size is obtained when the waist is at the lens and has a size equal to the lens diameter (or an appropriate fraction thereof, so as to maintain the Gaussian profile). This could be accomplished in a way similar to that of the previous example. However, in this problem we are not concerned with the size of the lens; we only wish to show the effect of focusing the second lens, without regard for the waist size produced by the first lens.

The beam divergence is now larger by a factor $\sqrt{2}$ ($0.05 \times \sqrt{2}$ mrad), and the Rayleigh range is smaller by a factor of 2 (40 m). However, over the distance $2z = 80$ m, the beam varies in radius between $2\sqrt{2}$ mm at the waist and 4 mm at the end of the range: The percent variation can be given as $(4 - 2\sqrt{2})/4 \cong 30\%$. This variation is the same as the one obtained in the previous case, over the same distance of 80 m, where the beam radius changed from 4 mm at the waist to $4\sqrt{2}$ at the end of the range. Therefore, we could say that the two beams are equally well collimated over this distance; but the diameter of the beam is now smaller by a factor of $\sqrt{2}$.

Suppose now that we are interested in maintaining a collimated beam over a range $\Delta z = \pm 20$ m, with the middle of the range located 40 m away from the lens. We can determine the beam radius variation from Eq. (A2.7). In the first case (beam waist near the lens), we have $z = 80$ m and $z = 20$ m, 60 m; so $w(z)$ varies between $1.03w_0$ and $1.25w_0$, or about 22%. In the second case (beam waist at 40 m), we have $z = 40$ m and $z = \pm 20$ m; so $w(z)$ varies between w_0 and $1.12w_0$, or 12%. Therefore, the second case is now preferable, despite the fact that the beam divergence is larger.

A2.5. TWO-RAY FORMULATION OF GAUSSIAN BEAMS

The paraxial ray-tracing equations given earlier solve the problem of determining the beam waist size and location. Often, however, as in Example A2, the beam size is of interest not only at the waist, but at other locations as well. In order to determine that, we must employ Eq. (A2.4) [or (A2.8)], after having determined the last beam waist. An alternative method,* shown in the following, allows us to determine the Gaussian beam parameters by performing only ordinary paraxial ray tracing, as in Chapter 5. We show this method for two reasons: 1. Its elegance provides a beautiful demonstration of the power of the optical invariant, and 2. It exploits the paraxial ray-tracing calculations performed by any optical design program, however simple.

The method rests on the following observation: We notice the formal similarity, as pointed out earlier, between Eq. (A2.2) and the optical invariant, $nu\eta$. We may represent the product $nw_0\theta$ $(=\lambda_0/\pi)$ as the optical invariant associated with two rays: Ray 1 crosses the axis at the waist and forms an angle θ with the axis, and ray 2 is parallel to the axis at a height w_0, as in Figure A2.5. These two rays are equivalent to the paraxial pupil and marginal rays of Chapter 5. We can then take advantage of all the different forms that the optical invariant may take.

In order to return to a notation that recalls the paraxial optics of Chapter 5, we will call ray 1 the divergence ray, and ray 2 the waist ray. The corresponding paraxial angle variables and heights will have subscripts θ and w, respectively, and the Gaussian optical invariant will be called $\mathscr{H}$. Before proceeding further, we should point out clearly that these two rays are fictitious,

* See also R. Herloski, S. Marshall, and R. Antos, "Gaussian beam ray-equivalent modeling and optical design," *Applied Optics* **22**, 1168–74 (1983), and J. A. Arnaud, "Hamiltonian theory of beam mode propagation," in *Progress in Optics*, Vol. XI, E. Wolf, Ed., North Holland, Amsterdam, 1973.

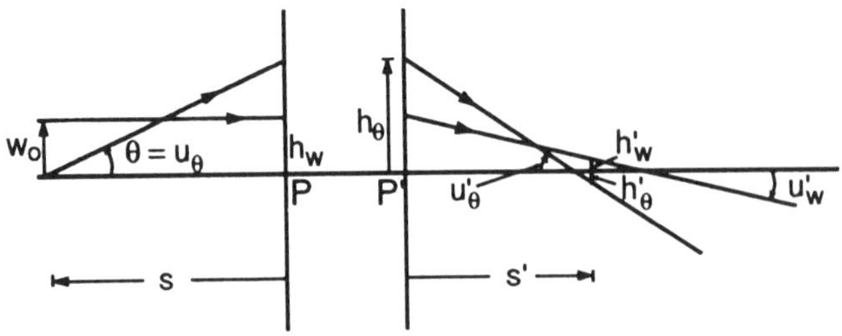

Figure A2.5. Illustration of the two-ray formulation of Gaussian beams.

since they are not normal to the wavefronts. Therefore, no physical meaning should be sought here; this is simply a method of convenience.

Figure A2.5 shows an optical system characterized by its principal planes, and a Gaussian beam with the object-space and image-space waists, located at s and s' away from the corresponding principal planes. The optical invariant gives:

$$\begin{aligned}\mathscr{H} &\equiv n w_0 \theta \equiv n h_w u_\theta = n'(h_w u'_\theta - h_\theta u'_w) \\ &= n'(h'_w u'_\theta - h'_\theta u'_w) = n' w'_0 \theta' \end{aligned} \tag{A2.33}$$

where w'_0 and θ' are the image-space beam waist and divergence as usual. We now seek to determine all the image-space beam parameters in terms of the quantities associated with the divergence and waist rays. We start first with θ'. Use Eq. (A2.21), and make the following substitutions

$$\imath = \frac{\lambda_0}{n\pi\theta^2}, \qquad \imath' = \frac{\lambda_0}{n'\pi\theta'^2}, \qquad \frac{\lambda_0}{\pi} = n w_0 \theta$$

to obtain

$$\theta'^2 = \frac{1}{n'^2}[(n + Ks)^2\theta^2 + K^2 w_0^2]$$

Setting $\theta \equiv u_\theta$, $s = -h_\theta/u_\theta$, we have

$$\theta'^2 = \frac{1}{n'^2}(n'^2 u'^2_\theta + K^2 w_0^2) \tag{A.2.34}$$

The definition of the power of a system gives

$$K = -\frac{n'u'_\infty}{h_1} \equiv -\frac{n'u'_w}{w_0}, \qquad \text{for } u_w = 0 \tag{A2.35}$$

Upon substitution of K into Eq. (A2.34), we obtain

$$\boxed{\theta' = \sqrt{u'^2_\theta + u'^2_w}} \tag{A2.36}$$

Using Eqs. (A2.33) and (A2.36), we may now obtain easily the image-space beam waist, as

$$\boxed{w_0' = \frac{\mathscr{H}}{n'\theta'} = \frac{\mathscr{H}}{n'\sqrt{u_\theta'^2 + u_\mathrm{w}'^2}}} \tag{A2.37}$$

We have already proved something quite remarkable: The image-space waist size and divergence can be calculated by performing ordinary paraxial ray tracing, using the divergence and waist rays. But more is possible. Let us now determine s' in a similar fashion. Use Eq. (A2.19), with the following substitutions

$$\imath = \frac{w_0}{\theta}, \qquad \theta \equiv u_\theta, \qquad s = -h_\theta/u_\theta$$

to obtain

$$s' = n' \frac{-h_\theta(nu_\theta - h_\theta K) + Kw_0^2}{(nu_\theta - h_\theta K)^2 + K^2 w_0^2}$$

From simple paraxial ray tracing, the bracketed term in the above equation equals $n'u_\theta'$. Using this substitution and the definition of power, Eq. (A2.35), we finally have

$$\boxed{s' = -\frac{h_\theta u_\theta' + h_\mathrm{w} u_\mathrm{w}'}{u_\theta'^2 + u_\mathrm{w}'^2}} \tag{A2.38}$$

where $h_\mathrm{w} \equiv w_0$, the object-space beam waist size. Therefore, by tracing the divergence and waist rays, we can also determine the location of the waist.

Finally, we prove the following theorem: The beam size $w(z)$ at any point z is given by the square root of the sum of the squares of the paraxial height variables of the divergence and waist rays at the same point; that is

$$\boxed{w(z) = \sqrt{h_\theta^2(z) + h_\mathrm{w}^2(z)}} \tag{A2.39}$$

where we have dropped the primes from h_θ and h_w for simplicity. This theorem allows us to determine the beam size at any point along the axis without first having to resort to determining the beam waist. We must first prove the following lemma: Referring to Figure A2.5, the relation

$$\frac{h_\mathrm{w}'}{h_\theta'} = -\frac{u_\theta'}{u_\mathrm{w}'} \tag{A2.40}$$

holds, where the heights are measured at the image-space waist. To prove this, form the ratio

$$\frac{h_\mathrm{w}'}{h_\theta'} = \frac{h_\mathrm{w} + s'u_\mathrm{w}'}{h_\theta + s'u_\theta'}$$

by the paraxial transfer formula. Substitution of s' from Eq. (A2.38) into this equation and cancellation of terms and common factors lead to Eq. (A2.40). Using the above lemma, we can now prove that

$$\boxed{w_0'^2 = h_\theta'^2 + h_w'^2} \tag{A2.41}$$

at the image-space waist. The reader may easily verify this equation, by using Eq. (A2.37) written as follows

$$w_0'^2 = \frac{(h_w' u_\theta' - h_\theta' u_w')^2}{u_\theta'^2 + u_w'^2}$$

and then substituting this value into Eq. (A2.41). Finally, we are ready to consider the beam size away from the waist. Notice that Eqs. (A2.8) and (A2.6) can be combined to give

$$w^2(z) = w_0^2 + \theta^2 z^2 \tag{A2.42}$$

A similar equation can of course be written in the image space with primed quantities. The paraxial transfer equation gives

$$h_\theta(z) = h_\theta' + z u_\theta', \qquad h_w(z) = h_w' + z u_w'$$

where $h_\theta(z)$ and $h_w(z)$ are the paraxial height variables for the divergence and waist rays at a distance z from the image-space beam waist. Upon squaring and adding the last two equations, we have

$$h_\theta^2(z) + h_w^2(z) = h_\theta'^2 + h_w'^2 + z^2(u_\theta'^2 + u_w'^2) + 2z(h_\theta' u_\theta' + h_w' u_w')$$

The bracketed terms are equal to θ'^2 and zero, respectively, because of Eqs. (A2.36) and (A2.40). Therefore,

$$h_\theta^2(z) + h_w^2(z) = h_\theta'^2 + h_w'^2 + \theta'^2 z^2 = w_0'^2 + \theta'^2 z^2 = w'^2(z)$$

that is, we obtain Eq. (A2.39) (upon dropping the prime from the right-hand side). It should be noted that although the distance z was measured from the waist, it is not necessary to determine thc beam waist location first in order to apply Eq. (A2.39). The location of the beam waist was used simply as a reference point for measuring z.

An interesting corollary of this theorem is that the beam size is given by the paraxial ray height of only one of the two rays, at the points where the other ray crosses the axis.

There are alternative treatments of Gaussian beams; the most common one makes use of a complex form of the confocal parameter and the *ABCD* matrix formalism of Appendix 1. The treatment we have given enables one to trace a Gaussian beam through an arbitrary system; also it rests on (and is consistent with) the paraxial ray-tracing concepts developed earlier, requiring in addition only a knowledge of the basic characteristics of Gaussian beams in free space.

PROBLEMS

1. For a positive lens of focal length F, show that the image-space beam waist cannot be virtual if the input beam Rayleigh range satisfies $z = F/2$. (Hint: Use the condition for minimum s'.)

2. For a thin lens in air, under what conditions can the output beam waist be located on the lens? (Example positive and negative lenses separately.)

3. For a thin lens, let the axial points B and B' be arbitrary geometrical conjugates, at distances l and l' from the lens, respectively. Let s and s' be the object- and image-space beam waist locations, respectively. Prove that the beam size at point B', $w'(l' - s')$, is given by the beam size at B, $w(l - s)$, and the ordinary paraxial magnification, $M = l'/l$, through

$$\frac{w'(l' - s')}{w(l - s)} = |M|$$

Prove this algebraically first. Then generalize the conclusion for any system by using the two-ray formulation. [Note: $w(l - s)$ is not a product. It is $w(z)$ for $z = l - s$. The absolute value of M is used because the beam size is always taken as a positive number.)

4. A $0.23p$ GRIN lens has the following characteristics: $n_0 = 1.5$, $\alpha = 0.365\ \text{mm}^{-1}$. We wish to use this lens for focusing the light from a HeNe laser ($\lambda \cong 0.633\ \mu\text{m}$) onto an optical fiber that has a modal field (input beam) diameter of 6 μm and an NA = 0.1. The laser beam has a divergence $\theta = 0.5$ mrad. (a) Determine the position of the beam waist in the image space of the GRIN lens. You may assume that the input beam waist is located 1 cm in front of the lens. Compare the output beam waist position with the back focal length of the GRIN, given by $\cot(\alpha d)/n_0\alpha$. Conclusions? (b) Determine the waist size w'_0 and divergence θ' of the output beam. Compare with the fiber parameters, and estimate the amount of light coupled into the fiber. (c) Comment on the effect of the position of the input beam waist s, on the output beam waist size w'_0. (d) What is the axial tolerance in positioning the fiber face, given as $\pm\delta z$ from the beam waist location in (a)?

5. We wish to relay a beam waist ($m = 1$) using a lens of $F = 50$ cm. The wavelength is 600 nm, and the object-space divergence is 1.5 mrad. Determine the distance between the two beam waists. Compare with geometrical optics.

6. Two thin lenses of focal length F are separated by a distance of $2F$. The input beam waist is w_0 (divergence θ_0) and is located at the front focal

plane of the first lens. Where is the output beam waist, and what is its size? Now add a third identical lens after the second, also located at $2F$ from the second. Determine again the beam waist location and size after the third lens. Comment on the symmetry of the system about its middle in each case, and generalize for any even or odd number of lenses. (Note: This problem becomes obvious if you use the two-ray method!)

Appendix

3

Finite Ray Tracing

Finite ray tracing is the basic tool through which we decide whether a given system has the desired image quality. The algorithm to be shown here has been described in essentially the same form by Spencer and Murty (1962) and Welford (1986). In order to perform finite ray tracing, it is necessary to have a detailed design prescription of the system, in terms of curvatures, separations, and refractive indices at the appropriate wavelength. Our treatment is restricted to rotationally symmetric systems. In addition to spherical surfaces, the algorithm also includes conicoids of revolution. The extension to general aspheric surfaces, and even cylindrical surfaces, toric surfaces, or diffractive elements, do not involve any fundamentally new concepts. These cases are described by Welford (1986).

Finite ray tracing comprises the same two steps as paraxial ray tracing, namely, refraction and transfer. One additional complication is that the image plane may not be well defined, since not all rays from the axial object point may cross the axis at the same point. Thus prior to performing finite ray tracing, a paraxial trace is normally performed, in order to determine the paraxial focal plane. This may not be the optimum focal plane in terms of image quality, but it provides a reference point.

Finite ray tracing used to be very time consuming; the ability and speed of human operators to perform these calculations was the limiting factor in lens design, before the proliferation of computers. Many techniques evolved for tracing rays using log tables, but they are now obsolete. Meridional rays were easier to trace manually. Skew rays were very difficult. Nowadays, there is little reason for distinguishing between the two types, but the necessity of including skew rays means that the ray-tracing operation must be performed in three dimensions.

A3.1. VECTOR FORM OF SNELL'S LAW

If $\mathbf{r}$ and $\mathbf{r}'$ are unit vectors along the incident and refracted rays, and $\mathbf{v}$ is the unit vector along the normal to the surface at the point of incidence, then Snell's law is written as

$$n'(\mathbf{r}' \times \mathbf{v}) = n(\mathbf{r} \times \mathbf{v}) \tag{A3.1}$$

For computation, we need to reduce this equation into component form. Multiplying both sides vectorially by $\mathbf{v}$, we have:

$$n'\mathbf{v} \times (\mathbf{r}' \times \mathbf{v}) = n\mathbf{v} \times (\mathbf{r} \times \mathbf{v})$$

or

$$n'(\mathbf{v}\cdot\mathbf{v})\mathbf{r} - n'(\mathbf{v}\cdot\mathbf{r})\mathbf{r} = n(\mathbf{v}\cdot\mathbf{v})\mathbf{r} - n(\mathbf{v}\cdot\mathbf{r})\mathbf{v}$$

by the vector triple product theorem. But $\mathbf{v}\cdot\mathbf{r} = \cos i$ and $\mathbf{v}\cdot\mathbf{r}' = \cos i'$, where i, i' are angles of incidence and refraction (for which $n \sin i = n' \sin i'$). Therefore,

$$n'\mathbf{r}' - n\mathbf{r} = \mathbf{v}(n' \cos i' - n \cos i) \tag{A3.2}$$

or, in component form,

$$\begin{aligned} n'L' - nL &= \alpha(n' \cos i' - n \cos i) \\ n'M' - nM &= \beta(n' \cos i' - n \cos i) \\ n'N' - nN &= \gamma(n' \cos i' - n \cos i) \end{aligned} \tag{A3.3}$$

where (L, M, N), (L', M', N'), and (α, β, γ) are the (x, y, z) components (direction cosines) of $\mathbf{r}$, $\mathbf{r}'$, and $\mathbf{v}$, respectively. Equations (A3.3) are the required form of Snell's law, suitable for tracing a ray through any surface (not necessarily spherical). To find L', M', and N', we need to know L, M, N for the incident ray, and α, β, γ for the surface normal.

A3.2. THE SURFACE EQUATION AND THE SURFACE NORMAL

A conicoid of revolution around the z axis is given by the equation

$$f(x, y, z) = c[x^2 + y^2 + (1 + Q)z^2] - 2z = 0 \tag{A3.4}$$

where c is the vertex curvature and Q the eccentricity parameter (Figure A3.1). For $Q = 0$ we have a sphere, for $Q = -1$ a paraboloid, and for $Q < -1$ a

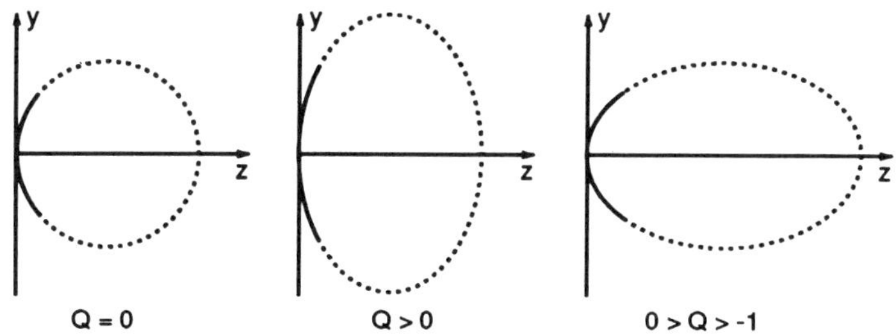

Figure A3.1. Various surfaces of revolution about the z axis.

hyperboloid. The other two cases (oblate and prolate ellipsoid) are shown in the figure. The equations of the line normal to this surface are

$$\frac{x_0 - x}{A} = \frac{y_0 - y}{B} = \frac{z_0 - z}{C}$$

where A, B, C, are the direction numbers of the line, which are proportional to the direction cosines α, β, γ. Analytic geometry gives

$$A = -\frac{\partial f}{\partial x} = -2cx, \qquad B = -\frac{\partial f}{\partial y} = -2cy, \qquad C = -\frac{\partial f}{\partial z} = 2 - 2c(1 + Q)z$$

from Eq. (A3.4), and for the direction cosines,

$$\alpha = -\frac{cx}{V}, \qquad \beta = -\frac{cy}{V}, \qquad \gamma = \frac{1 - c(1 + Q)z}{V}$$

where (A3.5)

$$V = [1 - 2cQz + c^2z^2Q(1 + Q)]^{1/2}$$

a normalizing factor, such that $\alpha^2 + \beta^2 + \gamma^2 = 1$.

A3.3. SURFACE TRANSFER

Referring to Figure A3.2, assume the starting point to be on the ray, with coordinates (X, Y, Z), representing typically the point of intersection with the previous surface. We have

$$x = X + LD, \qquad y = Y + MD, \qquad z = Z + ND \tag{A3.6}$$

The new point (x, y, z) must satisfy the surface equation (A3.4). Substituting from Eq. (A3.6) into Eq. (A3.4) and after a bit of algebra, we obtain a quadratic in D:

$$D^2[c(1 + QN^2)] - 2GD + F = 0$$

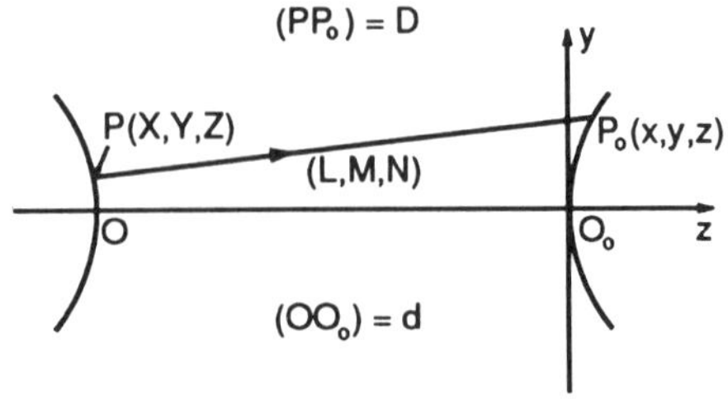

Figure A3.2. Surface-to-surface transfer for finite ray tracing.

where

$$F = c[X^2 + Y^2 + Z^2(1 + Q)] - 2Z \tag{A3.7}$$

$$G = N - c[LX + MY + NZ(1 + Q)] \tag{A3.8}$$

The solution of the quadratic is

$$D = \frac{G \pm [G^2 - (1 + QN^2)Fc]^{1/2}}{c(1 + QN^2)} \tag{A3.9}$$

To choose between the two possible signs, notice that we want $D = 0$ when $F = 0$ (i.e., when point P_0 is on the surface); therefore, we must take the negative sign. For numerical computation, the form of Eq. (A3.9) is not ideal, as it involves the subtraction of two large numbers to obtain a small result. Therefore, we multiply the numerator and denominator by the conjugate surd to obtain

$$D = \frac{F}{G + [G^2 - (1 + QN^2)Fc]^{1/2}} \tag{A3.10}$$

Thus the transfer equations are Eq. (A3.6), with D defined through Eqs. (A3.10), (A3.8), and (A3.7). During transfer, the origin of coordinates must be shifted to the apex of the next surface, since Eq. (A3.4) only holds if the apex is at the origin. This is accomplished simply by defining $z_{i+1} = z_i - d_i$, where d_i is the axial separation between surfaces i and $i + 1$.

A3.4. REFRACTION

In the refraction equation (A3.3), we need to determine $\cos i$ and $\cos i'$. We have

$$\cos i = \mathbf{v} \cdot \mathbf{r} = \alpha L + \beta M + \gamma N \tag{A3.11}$$

All quantities on the right-hand side are known. Also, from Snell's law,

$$n' \sin i' = n \sin i \rightarrow n'^2(1 - \cos^2 i') = n^2(1 - \cos^2 i)$$

or

$$n' \cos i' = [n'^2 - n^2(1 - \cos^2 i)]^{1/2} \tag{A3.12}$$

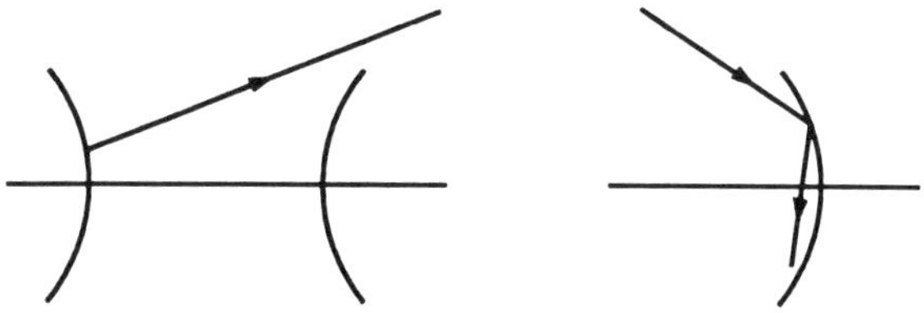

Figure A3.3. Two forms of ray failure: The ray misses the next surface, or it is totally internally reflected.

The positive sign in Eq. (A3.12) is chosen because we want i and i' to be acute angles. Thus we have determined the equations necessary for tracing a ray through a system of centered surfaces. Two checks for ray failure are incorporated in a computer program:

1. If the argument of the square root in Eq. (A3.12) becomes negative, we have total internal reflection.
2. If the argument of the square root in Eq. (A3.10) becomes negative, the ray misses the surface.

These two failure tests must be made at every surface. They are shown in Figure A3.3.

How do we begin and end a ray trace? This will depend on the requirements, that is, what one wants to learn through ray tracing. A ray may originate at the object, or, if the object is at or near infinity, at the entrance pupil or the first surface. If the ray aberration is sought, the ray must be traced to the image plane; if the wave aberration is sought, a ray may be traced up to the exit pupil (reference sphere). If the image is at or near infinity, ray tracing is normally done by reversing the system and tracing rays toward the finite conjugate.

Once we have selected an object point, we then have to choose the direction cosines of the rays emanating from that point and, of course, the number of rays from that point. Typically, one might want to choose rays that connect the object point with a regular square grid at the entrance pupil, corresponding to a uniform illumination of the pupil by the object point.

Appendix

4

Shift of Focus

We examine here the effect of a shift of the center of the reference sphere. Depending on the type of aberration present, judicious choice of the center of the reference sphere can provide a better fit between the reference sphere and the wavefront, thus giving a better idea of the true aberrations present. In other words, the Gaussian image plane may not be the best image plane; similarly, the Gaussian extra-axial image point may not coincide with the point of maximum irradiance. These two cases correspond to longitudinal and transverse focal shift, respectively.

A4.1. LONGITUDINAL FOCAL SHIFT

Referring to Figure A4.1, consider that the focal plane shifts from O_0 to O. For convenience, we will omit the primes from all quantities, even though we refer to the image space. Let the wavefront aberration referred to the sphere centered at O_0 be W_0, and that referred to the sphere at O be W. We wish to calculate the difference $\delta W_L = W_0 - W$ (L for longitudinal), in terms of quantities more directly related to the magnitude of the shift of focal plane.

For an axial object point producing an image free of spherical aberration at O_0, a shift in focal plane will defocus the image, and thence degrade its quality. If some spherical aberration is present, then a judicious shift in focal plane can considerably improve the image quality by introducing an appropriate amount of δW_L to balance (in some measure) the W_0 due to spherical aberration. Hence

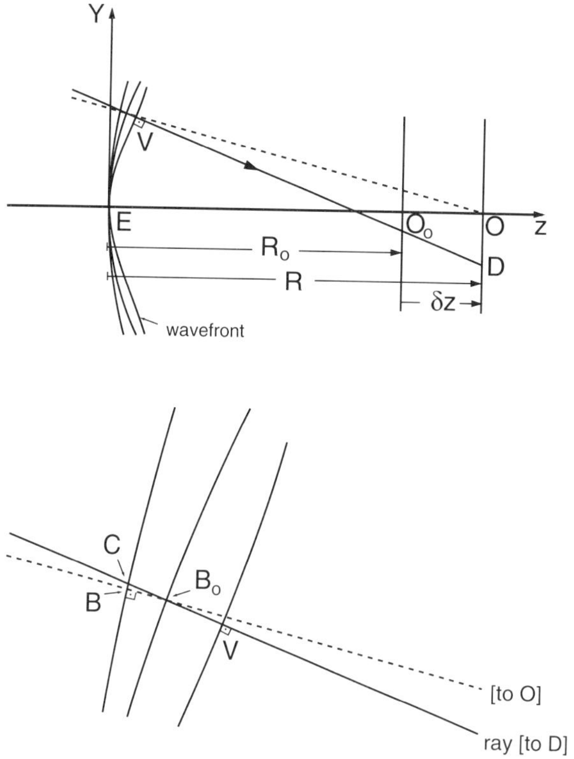

Figure A4.1. Longitudinal focal shift.

the importance of the focal shift formulae, in terms of both calculating how much shift of focal plane might be tolerable and calculating the best position of the focal plane.

From Figure A4.1,

$$\delta W_L = n(BB_0) \cong n(CB_0)$$

on the assumption that the focal shift is small. Let $CB_0 = p$, and let the coordinates of B_0 be (X, Y, Z), with respect to the origin at E. The coordinates of O_0 are $(0, 0, R_0)$, and those of O are $(0, 0, R)$. Therefore:

$$(B_0O)^2 = (R - p)^2 = X^2 + Y^2 + (Z - R)^2$$

or

$$p^2 - 2Rp = X^2 + Y^2 + Z^2 - 2ZR \tag{A4.1}$$

The equation of the reference sphere at O_0 is

$$X^2 + Y^2 + (Z - R_0)^2 = R_0^2$$

or

$$\frac{X^2 + Y^2 + Z^2}{2R_0} = Z \tag{A4.2}$$

Substitution of Z into Eq. (A4.1) gives

$$p = \frac{X^2 + Y^2 + Z^2}{2R - p} \frac{R - R_0}{R_0}$$

Setting $p \ll 2R$ and $\delta W_L = np$, we finally have

$$\delta W_L = \frac{n}{2}(X^2 + Y^2 + Z^2)\left(\frac{1}{R_0} - \frac{1}{R}\right) \tag{A4.3}$$

Carefully note the two approximations made in this derivation. If fractional pupil coordinates are used, $x = X/h_e$, $y = Y/h_e$, $z = Z/h_e$, then

$$\delta W_L = \delta W_{20}(x^2 + y^2 + z^2)$$

where

$$\delta W_{20} = \frac{n}{2} h_e^2 \left(\frac{1}{R_0} - \frac{1}{R}\right) \tag{A4.4}$$

Since z is of the second degree in x and y, we can write

$$\delta W_L = \delta W_{20}(x^2 + y^2) + O(4) \tag{A4.5}$$

or, in polar coordinates,

$$\delta W_L = \delta W_{20} r^2 + O(4) \tag{A4.6}$$

and for small focal shifts,

$$\delta W_{20} = \frac{n}{2} \frac{h^2}{R_0^2} \delta z \tag{A4.7}$$

where

$$\delta z = R - R_0$$

so that

$$\delta W_{20} = \frac{n}{2} \sin^2 \alpha \, \delta z \tag{A4.8}$$

Finally, we note that the fourth-order term in Eqs. (A4.5) and (A4.6) is negligible except for extreme aperture systems.

A4.2. TRANSVERSE FOCAL SHIFT

In Figure A4.2 is shown a system with some comatic aberration at a given extra-axial image point. Clearly, if the reference sphere were tilted downward, corresponding to a shift in the center of the reference sphere from Q to Q_0, then a better fit to the wavefront would be obtained. This reduction in phase error corresponds to the increased irradiance observed at Q_0, compared with Q.

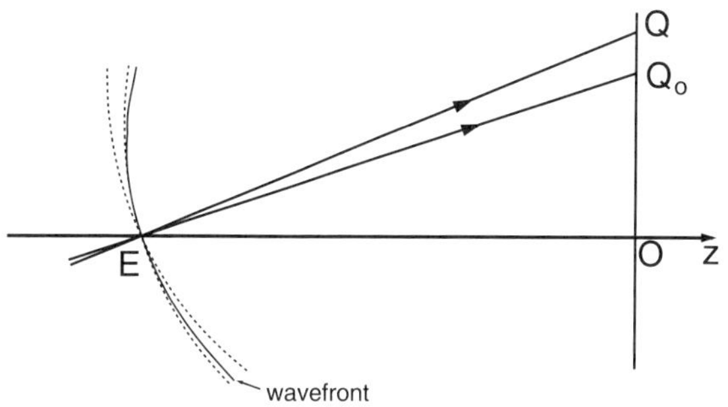

Figure A4.2. Better fit of the reference sphere by moving its center.

This is normally the purpose of a transverse focal shift: To locate the point of maximum irradiance in the image plane for an extra-axial point object.

In the calculation to follow the technique adopted is very similar to that for the case of longitudinal focal shift. Because of meridional plane symmetry, we can consider the points Q_0 and Q to be on the (η,Z) plane (Figure A4.3). The coordinates of the various points are:

$$Q_0\colon (0,\bar{\eta},D), \qquad Q\colon (0,\eta,D), \qquad B_0\colon (X,Y,Z)$$

with the origin at E. The focal shift is

$$\delta\eta = \eta - \bar{\eta}$$

We also have: $W_0 = n(B_0V)$, $W = n(BV)$; so

$$\delta W_{\mathrm{T}} = W - W_0 = n(BB_0) = np$$

Notice that B_0Q is only a construction line, not a ray. The sphere equation gives

$$(B_0Q)^2 = (\bar{R} - p\cos\delta U)^2 = X^2 + (Y - \eta)^2 + (Z - D)^2$$
$$(B_0Q_0)^2 = \bar{R}_0^2 = X^2 + (Y - \bar{\eta})^2 + (Z - D)^2$$

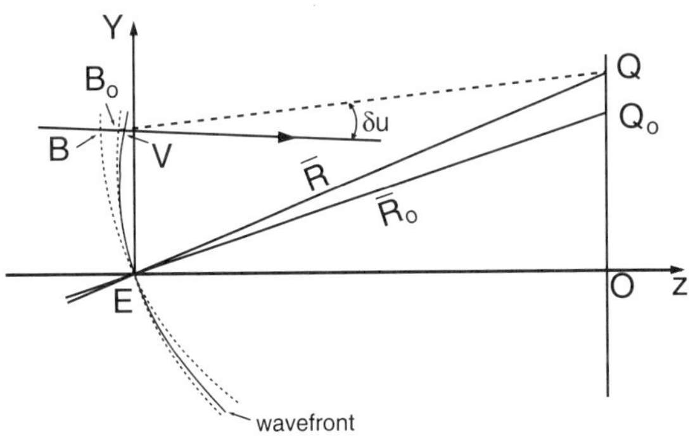

Figure A4.3. Transverse focal shift.

We assume δU to be small, so that $\cos \delta U \cong 1$. We also note that

$$\bar{R}^2 = \eta^2 + D^2, \qquad \bar{R}_0^2 = \bar{\eta}^2 + D^2$$

Therefore,

$$(p^2 - 2\bar{R}p) = X^2 + Y^2 + Z^2 - 2Y\eta - 2ZD$$

$$0 = X^2 + Y^2 + Z^2 - 2Y\bar{\eta} - 2ZD$$

By subtracting the last two equations,

$$\bar{R}p\left(1 - \frac{p}{2\bar{R}}\right) = (\eta - \bar{\eta})Y$$

For small focal shifts, $p/2\bar{R} \ll 1$, and

$$\frac{\bar{R}_0}{\bar{R}} = \left(1 + \frac{\bar{R} - \bar{R}_0}{\bar{R}_0}\right)^{-1} \cong 1$$

Therefore,

$$\delta W_{\mathrm{T}} = \frac{n}{\bar{R}_0} Y \,\delta\eta \tag{A4.9}$$

Using the fractional pupil coordinates, $y = Y/h_e$. Also, for an axial image point, we can write $h_e/\bar{R}_0 = -u$, the PMR angle variable. Thus

$$\delta W_{\mathrm{T}} = -(nu\,\delta\eta)y \tag{A4.10}$$

or

$$\delta W_{\mathrm{T}} = \delta W_{11} y \tag{A4.11}$$

where

$$\delta W_{11} = -nu\,\delta\eta \tag{A4.12}$$

A different normalization scheme must be used off-axis, but Eq. (A4.12) still holds to a first approximation. The extra-axial normalization is described in Section 10.5.3.

Appendix

5

Two Computer Programs

The following computer programs are written in BASIC. No advanced features are used; so they should run on any computer. The programming style is intentionally simplistic, to facilitate comprehension and future modification.

A5.1. THIN APLANATIC DOUBLET DESIGN (STOP AT THE LENS)

This program gives the surface curvatures for a thin aplanatic doublet (zero primary spherical aberration and coma) with the stop at the lens. There are two possibilities: Either the lens should be also achromatic or not. If the lens is not achromatic, the program asks for the powers of the individual elements K1 and K2. For an achromatic lens, the powers are computed automatically. It is assumed that the lens is in air.

The following information must be supplied in both cases:

total doublet power K

refractive index of first and second elements, $N1$, $N2$

dispersive indices $V1$ and $V2$

object distance from lens $L1$ (l)

Care must be taken to use a consistent unit of length. The object distance will normally be negative (in front of the lens). If the object is at infinity, the distance should be entered as a large value (e.g., −1E20). Two solutions are printed out. In general, four solutions are possible, because the order of the glasses may be inverted. To determine the second solution, the order of the glasses must be inverted and the program run again.

Table A5.1.
Additional variable list

Program	Text	Meaning
L1P	l'_1	image distance for first element
L2P	l'_2	image distance for second element
Y (Y1, Y2)	Y	magnification parameter
X (X1A, etc.)	X	shape factor
C (C11A, etc.)	c	surface curvature

Several other intermediate variables used in the calculation are unimportant (lines 310–560 of Program Listing).

Program Listing

```
10 REM      APLANATIC DOUBLET-PRIMARY ABRN DESIGN

100 INPUT "ACHROMATIC LENS";A$

110 IF (A$="Y") OR (A$="y") THEN 140

120 IF (A$="N") OR (A$="n") THEN 150

130 PRINT "TYPE Y OR N": GOTO 100

140 INPUT "TOTAL POWER =";K

150 INPUT "REFRACTIVE INDEX N1=";N1

160 INPUT "N2=";N2

170 IF (A$="N") OR (A$="n") THEN 200

180 INPUT "INDEX OF DISPERSION V1=";V1

190 INPUT "V2=";V2

200 INPUT "OBJECT DISTANCE";L1

210 IF (A$="N") OR (A$="n") THEN 250

220 K1 = V1*K/(V1-V2)

230 K2 = -V2*K/(V1-V2)

240 GOTO 270
```

```
250 INPUT "K1 = ";K1
260 INPUT "K2 = ";K2
270 L1P = L1/(1+K1*L1)
280 L2P = L1P/(1+K2*L1P)
290 Y1 = -(L1+L1P)/(L1-L1P)
300 Y2 = -(L2P+L1P)/(L1P-L2P)
310 A1 = (N1+1)/(N1*(N1-1))
320 A2 = (N2+1)/(N2*(N2-1))
330 B1 = (2*N1+1)/N1
340 B2 = (2*N2+1)/N2
350 F1 = (N1+2)/(N1*(N1-1)^2)
360 F2 = (N2+2)/(N2*(N2-1)^2)
370 C1 = 2*(N1*N1-1)/(N1+2)
380 C2 = 2*(N2*N2-1)/(N2+2)
390 E1 = N1/(N1-1)
400 E2 = N2/(N2-1)
410 G1 = N1/(N1+2)
420 G2 = N2/(N2+2)
430 D1 = F1*C1*C1-G1
440 D2 = F2*C2*C2-G2
450 D = B2*K2*K2*Y2+B1*K1*K1*Y1
460 F = (D1*Y1*Y1+E1*E1)*K1^3 + (D2*Y2*Y2+E2*E2)*K2^3
470 G = F1*D*D/(K1*A1*A1) - 2*K1*F1*C1*Y1*D/A1 + F
480 HH = F2*C2*Y2*K2^3
490 P = F1*D*A2*K2*K2/(K1*A1*A1) - K1*F1*C1*Y1*A2*K2*K2/A1 + HH
500 B = (F1*(A2*K2*K2/A1)^2)/K1 + F2*K2^3
510 X2A = (-P + SQR(P*P-B*G))/B
520 X2B = (-P - SQR(P*P-B*G))/B
530 X1A = -(D+A2*K2*K2*X2A)/(K1*K1*A1)
540 X1B = -(D+A2*K2*K2*X2B)/(K1*K1*A1)
550 NN1 = 2*(N1-1)
560 NN2 = 2*(N2-1)
```

```
570 C11A = K1*(X1A+1)/NN1
580 C12A = K1*(X1A-1)/NN1
590 C21A = K2*(X2A+1)/NN2
600 C22A = K2*(X2A-1)/NN2
610 C11B = K1*(X1B+1)/NN1
620 C12B = K1*(X1B-1)/NN1
630 C21B = K2*(X2B+1)/NN2
640 C22B = K2*(X2B-1)/NN2
650 PRINT TAB(10);"SOLUTION 1"
660 PRINT
670 PRINT "LENS 1"
680 PRINT "C1 = ";C11A; TAB(25);"C2 = ";C12A
690 PRINT "LENS 2"
700 PRINT "C1 = ";C21A; TAB(25);"C2 = ";C22A
710 PRINT
720 PRINT
730 PRINT TAB(10);"SOLUTION 2"
740 PRINT
750 PRINT "LENS 1"
760 PRINT "C1 = ";C11B;TAB(25);"C2 = ";C12B
770 PRINT "LENS 2"
780 PRINT "C1 = ";C21B;TAB(25);"C2 = ";C22B
790 END
```

Sample Run

(operator responses in boldface)

```
ACHROMATIC LENS? Y
TOTAL POWER =? .2
REFRACTIVE INDEX N1=? 1.48749
N2=? 1.78472
INDEX OF DISPERSION V1=? 70.41001
V2=? 25.76
OBJECT DISTANCE? -275
         SOLUTION 1

LENS 1
C1 =  .7906222          C2 =  .1436626
LENS 2
C1 =  .9906532          C2 =  1.137695


         SOLUTION 2

LENS 1
C1 =  .3320787          C2 = -.3148809
LENS 2
C1 = -.3526215          C2 = -.2055801
```

A5.2. PARAXIAL RAY TRACING AND SEIDEL ABERRATION COMPUTATION

This program traces the PMR and the PPR through an optical system, specified in terms of surface curvatures, separations, refractive indices, and dispersions. The opening ray-trace data consist of the paraxial ray height and angle variable at the first surface for both the PPR and the PMR. The numbering of surfaces is consistent with the convention adopted throughout this book (e.g., the first refractive index is n_1, the first surface is c_1, etc.). If desired, the stop may be entered as a dummy surface of zero curvature and the same refractive index on both sides. Clearly, for k surfaces where will be $k + 1$ refractive indices and dispersions and $k - 1$ separations.

The following options are offered: printout of ray-trace data, printout of surface-by-surface aberration contribution, total primary wavefront aberration, and ray aberration. Once the system data have been entered, the system may be retraced with different rays if required. Also, before exiting, the user is offered the option of altering individual curvatures, separations, etc.

Ray tracing is scale independent: The unit of the output quantities is the same as that of the input, whatever that may be. Both wavefront and ray aberrations have units of length. A consistent unit of length must be used at the input.

Table A5.2.
Variable list

Program	Text	Meaning
H(I)	$\underline{h}_i$	PMR intersection height
HBAR(I)	h_i	PPR intersection height
U(I)	u_i	PMR angle variable
UBAR(I)	$\bar{u}_i$	PPR angle variable
C(I)	c_i	surface curvature
N(I)	n_i	refractive index
D(I)	d_i	surface-to-surface separation
DN(I)	δn_i	dispersion
BIGH	H	optical invariant
S1,S2,...	$S_{\mathrm{I}},S_{\mathrm{II}},\ldots$	Seidel sums
CL,CT	$C_{\mathrm{L}},C_{\mathrm{T}}$	chromatic aberration coefficients
S1C(I),...,	—	contribution of surface to Seidel sum
A	A	PMR refraction invariant
ABAR	$\bar{A}$	PPR refraction invariant
EFL	F (EFL)	Equivalent focal length
RELAP	$F\#$ ($\mathcal{N}$)	F number (image space only)
W400	${}_0W_{40}$	spherical aberration coefficient (other coefficients similarly defined)
WAV	λ	design wavelength
Ray aberrations		
TRSPH	—	transverse spherical aberration
SACOM	—	sagittal coma
TACOM	—	tangential coma
ASLIN	—	length of astigmatic line
FRDIS	—	fractional distortion
LOCHR	—	longitudinal chromatic aberration
TRCHR	—	transverse chromatic aberration

Program Listing

```
10 REM       PARAXIAL RAYTRACING AND SEIDEL ABERRATIONS

20 PRINT

30 PRINT "PARAXIAL RAYTRACING AND PRIMARY ABERRATION CALCULATION"

40 PRINT

50 GOTO 100

60 INPUT "DO YOU WANT TO MODIFY THE SYSTEM"; A$

70 IF (A$="N") OR (A$="n") THEN 1860

80 IF (A$="Y") OR (A$="y") THEN 1730

90 PRINT "TYPE Y OR N": GOTO 60

100 INPUT "NO OF SURFACES K="; K

110 L = K+1

120 DIM H(L),U(L),HBAR(L),UBAR(L),N(L),C(L),D(L),DN(L)

130 DIM S1(L),S2(L),S3(L),S4(L),S5(L),CL(L),CT(L)

140 PRINT "SYSTEM SPECIFICATION"

150 PRINT

160 PRINT "INPUT SURFACE CURVATURES"

170 FOR I=1 TO K

180 PRINT "C"; I; "=";

190 INPUT C(I)

200 NEXT

210 PRINT

220 PRINT "INPUT REFRACTIVE INDICES"

230 FOR I=1 TO L

240 PRINT "N"; I; "=";

250 INPUT N(I)

260 NEXT

270 PRINT

280 PRINT "INPUT SURFACE SEPARATIONS"

290 FOR I=1 TO K-1

300 PRINT "D"; I; "=";
```

```
310 INPUT D(I)
320 NEXT
330 PRINT
340 PRINT "INPUT DISPERSIONS"
350 FOR I=1 TO L
360 PRINT "DN"; I; "=";
370 INPUT DN(I)
380 NEXT
390 PRINT
400 INPUT "PMR DATA  H1="; H(1)
410 INPUT "U1="; U(1)
420 INPUT "PPR DATA  HBAR1="; HBAR(1)
430 INPUT "UBAR1="; UBAR(1)
440 INPUT "WAVELENGTH="; WAV
450 READ S1,S2,S3,S4,S5,CL,CT
460 DATA 0,0,0,0,0,0,0
470 RESTORE
480 FOR I=1 TO K
490 U(I+1) = (-H(I)*C(I)*(N(I+1)-N(I))+N(I)*U(I))/N(I+1)
500 H(I+1) = H(I)+D(I)*U(I+1)
510 UBAR(I+1) = (-HBAR(I)*C(I)*(N(I+1)-N(I))+N(I)*UBAR(I))/N(I+1)
520 HBAR(I+1) = HBAR(I)+D(I)*UBAR(I+1)
530 NEXT
540 INPUT "DO YOU WANT H,U OUTPUT"; A$
550 IF (A$="Y") OR (A$="y") THEN 1170
560 IF (A$="N") OR (A$="n") THEN 580
570 PRINT "TYPE Y OR N": GOTO 540
580 BIGH = N(1)*(HBAR(1)*U(1)-H(1)*UBAR(1))
590 EFL = BIGH/(N(1)*N(L)*(U(L)*UBAR(1)-U(1)*UBAR(L)))
600 RELAP = 1/(2*N(L)*U(L))
610 PRINT
620 PRINT "BIGH="; BIGH, "EFL="; EFL, "F-NUMBER="; RELAP
```

```
630 FOR I=1 TO K

640 A = N(I)*(H(I)*C(I)+U(I))

650 ABAR = N(I)*(HBAR(I)*C(I)+UBAR(I))

660 T = U(I+1)/N(I+1)-U(I)/N(I)

670 F = 1/N(I+1)-1/N(I)

680 S1(I) = -A*A*H(I)*T

690 S2(I) = -A*ABAR*H(I)*T

700 S3(I) = -ABAR*ABAR*H(I)*T

710 S4(I) = -BIGH*BIGH*C(I)*F

720 G = H(I)*UBAR(I)^2 + C(I)*U(I)*HBAR(I)^2

730 S5(I) = -ABAR*((H(I)*UBAR(I+1)^2 + (HBAR(I)^2)*C(I)*U(I+1))-G)

740 P = DN(I+1)/N(I+1)-DN(I)/N(I)
750 CL(I) = A*H(I)*P

760 CT(I) = ABAR*H(I)*P

770 NEXT

780 PRINT

790 INPUT "DO YOU WANT SURFACE-BY-SURFACE ABERRATION OUTPUT"; A$

800 IF (A$="Y") OR (A$="y") THEN 1360

810 IF (A$="N") OR (A$="n") THEN 830

820 PRINT "TYPE Y OR N": GOTO 790

830 FOR I=1 TO K

840 S1 = S1+S1(I)

850 S2 = S2+S2(I)

860 S3 = S3+S3(I)

870 S4 = S4+S4(I)

880 S5 = S5+S5(I)

890 CL = CL+CL(I)

900 CT = CT+CT(I)

910 NEXT

920 W040 = S1/8

930 W131 = S2/2

940 W222 = S3/2
```

```
950 W220 = S3/4 + S4/4
960 W311 = S5/2
970 DW20 = CL/2
980 DW11 = CT
990 AP = 1/RELAP
1000 GOTO 1530
1010 INPUT "DO YOU WANT TRANSVERSE RAY ABERRATION"; A$
1020 IF (A$="Y") OR (A$="y") THEN 1050
1030 IF (A$="N") OR (A$="n") THEN 1130
1040 PRINT "TYPE Y OR N": GOTO 1010
1050 TRSPH = S1/AP
1060 SACOM = 2*S2/AP
1070 TACOM = 3*SACOM/2
1080 ASLIN = ABS(4*S3/AP)
1090 FRDIS = .5*S5/BIGH
1100 AXCHR = 2*CL/AP
1110 TRCHR = 2*CT/AP
1120 GOTO 1630
1130 INPUT "RE-TRACE SYSTEM"; A$
1140 IF (A$="Y") OR (A$="y") THEN 60
1150 IF (A$="N") OR (A$="n") THEN END
1160 PRINT "TYPE Y OR N": GOTO 1130
1170 PRINT
1180 PRINT
1190 PRINT TAB(15); "PMR DATA"
1200 PRINT "SURFACE"; TAB(15); "H"; TAB(30); "U"
1210 PRINT
1220 PRINT TAB(25); U(1)
1230 FOR I=1 TO K
1240 PRINT TAB(3); I; TAB(10); H(I); TAB(25); U(I+1)
1250 NEXT
```

```
1260 INPUT "HIT ANY KEY TO CONTINUE"; A$
1270 PRINT
1280 PRINT TAB(15); "PPR DATA"
1290 PRINT "SURFACE"; TAB(15); "HBAR"; TAB(30); "UBAR"
1300 PRINT
1310 PRINT TAB(25); UBAR(1)
1320 FOR I=1 TO K
1330 PRINT TAB(3); I; TAB(10); HBAR(I); TAB(25); UBAR(I+1)
1340 NEXT
1350 GOTO 580
1360 PRINT
1370 PRINT
1380 PRINT "ABERRATION CONTRIBUTION BY SURFACE"
1390 PRINT
1400 FOR I=1 TO K
1410 PRINT "SURFACE "; I
1420 PRINT "S1="; S1(I)
1430 PRINT "S2="; S2(I)
1440 PRINT "S3="; S3(I)
1450 PRINT "S4="; S4(I)
1460 PRINT "S5="; S5(I)
1470 PRINT "CL="; CL(I)
1480 PRINT "CT="; CT(I)
1490 INPUT "HIT ANY KEY TO CONTINUE"; A$
1500 NEXT
1510 GOTO 830
1520 PRINT
1530 PRINT "TOTAL PRIMARY WAVEFRONT ABERRATION"
1540 PRINT
1550 PRINT "W040="; W040, "("; W040/WAV; " WAVES)"
1560 PRINT "W131="; W131, "("; W131/WAV; " WAVES)"
1570 PRINT "W222="; W222, "("; W222/WAV; " WAVES)"
```

```
1580 PRINT "W220="; W220, "("; W220/WAV; " WAVES)"

1590 PRINT "W311="; W311, "("; 100*W311/BIGH; " PERCENT)"

1600 PRINT "DW20="; DW20, "("; DW20/WAV; " WAVES)"

1610 PRINT "DW11="; DW11, "("; DW11 / WAV; " WAVES)"

1620 GOTO 1010

1630 PRINT

1640 PRINT "PRIMARY TRANSVERSE RAY ABERRATION"

1650 PRINT "TRANSV. SPHERICAL:"; TRSPH

1660 PRINT "SAGITTAL COMA:"; SACOM

1670 PRINT "TANGENTIAL COMA:"; TACOM

1680 PRINT "LENGTH OF ASTIG. LINE:"; ASLIN

1690 PRINT "FRACTIONAL DISTORTION:"; FRDIS

1700 PRINT "AXIAL COLOR:"; AXCHR

1710 PRINT "TRANSVERSE COLOR:"; TRCHR

1720 GOTO 1130

1730 PRINT "INPUT PROMPT FOR SYSTEM DATA"

1740 PRINT "ALLOWED PROMPTS ARE: C, D, N, DN (CAPITAL LETTERS)"

1750 INPUT Q$

1760 IF (Q$="C") OR (Q$="D") OR (Q$="N") OR (Q$="DN") THEN 1780

1770 GOTO 1740

1780 INPUT "INDEX (FOLLOW NUMBERING OF INPUT DATA)"; KK

1790 IF ABS(INT(KK))<>KK THEN PRINT "INPUT POSITIVE INTEGER": GOTO 1780

1800 IF KK>L THEN PRINT "NUMBER TOO LARGE": GOTO 1780

1810 IF Q$="C" THEN PRINT "CURRENT VALUE IS:"; C(KK): INPUT "NEW VALUE"; C(KK)

1820 IF Q$="D" THEN PRINT "CURRENT VALUE IS:"; D(KK): INPUT "NEW VALUE"; D(KK)

1830 IF Q$="N" THEN PRINT "CURRENT VALUE IS:"; N(KK): INPUT "NEW VALUE"; N(KK)

1840 IF Q$="DN" THEN PRINT "CURRENT VALUE IS:"; DN(KK): INPUT "NEW VALUE"; DN(KK)

1850 GOTO 60

1860 INPUT "CHANGE RAY DATA"; A$

1870 IF (A$="N") OR (A$="n") THEN 450

1880 IF (A$="Y") OR (A$="y") THEN 400

1890 PRINT "TYPE Y OR N": GOTO 1860
```

Sample Run

(operator responses in boldface)

```
PARAXIAL RAYTRACING AND PRIMARY ABERRATION CALCULATION


NO OF SURFACES K=?4

SYSTEM SPECIFICATION


INPUT SURFACE CURVATURES

C 1 =? .1666

C 2 =? -.2258

C 3 =? -.2258

C 4 =? -.07475


INPUT REFRACTIVE INDICES

N 1 =? 1

N 2 =? 1.5168

N 3 =? 1

N 4 =? 1.6727

N 5 =? 1


INPUT SURFACE SEPARATIONS

D 1 =? .3

D 2 =? 0

D 3 =? .2


INPUT DISPERSIONS

DN 1 =? 0
```

```
DN 2 =? .008054

DN 3 =? 0

DN 4 =?.020884

DN 5 =? 0


PMR DATA  H1=? .75

U1=?  0

PPR DATA  HBAR1=? 0

UBAR1=? .05

WAVELENGTH? .56E-4

DO YOU WANT H,U OUTPUT? N


BIGH = -0.0375        EFL = 9.938921        F-NUMBER = -6.625947

DO YOU WANT SURFACE-BY-SURFACE ABERRATION OUTPUT? N

TOTAL PRIMARY WAVEFRONT ABERRATION


W040 = 4.68779E-06                  (8.371054E-02 WAVES)

W131 = 1.233813E-06                 (2.203238E-02 WAVES)

W222 = 6.770695E-05                 (1.209053 WAVES)

W220 = 5.950016E-05                 (1.062503 WAVES)

W311 = 2.150889E-06                 (-5.735705E-03 PERCENT)

DW20 = 5.735375E-06                 (.1024174 WAVES)

DW11 = 3.36743E-06                  (6.013267E-02 WAVES)


DO YOU WANT TRANSVERSE RAY ABERRATION? N

RE-TRACE SYSTEM? N
```

Appendix

6

Thin-Lens Bending Program

This is a BASIC program which can be used to compute the aberrations of a thin lens in air as a function of the shape factor X. The prime function of the program is to produce a printout of the Seidel coefficients S_{I}, S_{II}, S_{III}, and S_{V} for different values of X, spanning the range $X = -8$ to 8 in steps of 0.05 (the user may, of course, modify the program to change these limits and/or the step). The "PRINT" command produces a printout on a monitor, which may not be very useful: The user may change "PRINT" to "LPRINT" in lines 750, 770 (and/or elsewhere) to produce hard copy output. All bendings and aberration values are stored in arrays. As a secondary function, the program can save the user some time by identifying the shape that fits a particular design target best. For example, we show as a sample run below the camera objective that was designed in Section 8.1.6. If we specify zero targets for coma and astigmatism, as was the case there, the program automatically finds the shape that was described in Section 8.1.6 (or the appropriate pupil location). Also, the program allows the user to shift the stop and then calculates the new aberration values.

Determining the shape that fits an aberration target best is a form of optimization, hence line 1140 defines (somewhat pretentiously) a merit function to be the weighted sum of the absolute values of the aberrations minus the targets. If S_{1t}, S_{2t}, etc., are the targets for S_{I}, S_{II}, etc., respectively, and w_1, w_2, etc., the corresponding weight factors, the "merit" function is

$$w_1|S_{\mathrm{I}} - S_{1t}| + w_2|S_{\mathrm{II}} - S_{2t}| + w_3|S_{\mathrm{III}} - S_{3t}| + w_5|S_{\mathrm{V}} - S_{5t}|$$

It should be clear that only one aberration can be controlled by a single parameter such as the shape. Therefore, the results will be most meaningful if at least two of the four weight factors are set to zero. A second aberration can be controlled through the pupil location, as described by E, but the program does not offer an automatic way of doing this.

Although this program was used to produce the operating spectacles design example of Section 10.3, it may be more useful as an instructional tool: The student can experiment with values of the parameters and discover what happens to the four aberration parabolas under different circumstances.

The input to the program are the lens power and refractive index, the object and entrance pupil distances, and one each of the following two pairs of parameters: u or h_e, and $\bar{u}$ or η, where all symbols have their usual meaning. The variable names are self-expanatory and in many cases identical with those used in the text, with the following caution: H stands for the paraxial height h, while the optical invariant is called BIGH.

Program Listing

```
10 REM THIN LENS ABERRATION COMPUTATION AND OPTIMUM SHAPE DETERMINATION
20 DIM S1(321), S2(321), S3(321), S5(321), X(321), MERIT(321)
30 PRINT "THIN LENS ABERRATIONS AND BENDING"
40 PRINT
50 INPUT "DO YOU WANT TO SPECIFY THE LENS"; A$
60 IF (A$="Y") OR (A$="y") THEN 100
70 IF ((A$="N") OR (A$="n")) AND (Q=0) THEN END
80 IF ((A$="N") OR (A$="n")) AND (Q=1) THEN PRINT "HBAR="; E*H, "UBAR="; E*U-
BIGH/H: END
90 PRINT "TYPE Y OR N": GOTO 50
100 INPUT "POWER"; K
110 INPUT "REFRACTIVE INDEX N="; N
120 INPUT "OBJECT DISTANCE"; L
130 INPUT "ENTRANCE PUPIL DISTANCE"; LBAR
140 PRINT "INPUT EITHER THE MARGINAL RAY ANGLE U OR THE ENTRANCE PUPIL RADIUS"
150 PRINT "TYPE 'U' OR 'HE' TO SELECT ONE"
160 INPUT A$
170 IF (A$="U") OR (A$="u") THEN 200
180 IF (A$="HE") OR (A$="he") THEN 230
190 GOTO 150
```

```
200 INPUT "U="; U
210 IF U=0 THEN INPUT "ENTRANCE PUPIL RADIUS MUST BE GIVEN. HE="; HE: GOTO 340
220 HE = -U*(L-LBAR): GOTO 240
230 INPUT "HE="; HE: U = -HE/(L-LBAR)
240 IF U = 0 THEN 340
250 PRINT "INPUT EITHER THE IMAGE HEIGHT OR THE PUPIL RAY ANGLE"
260 PRINT "TYPE 'ETA' OR 'UBAR' TO SELECT ONE"
270 INPUT A$
280 IF (A$="ETA") OR (A$="eta") THEN 310
290 IF (A$="UBAR") OR (A$="ubar") THEN 340
300 GOTO 260
310 INPUT "ETA="; ETA
320 BIGH = U*ETA
330 UBAR = -U*ETA/HE: GOTO 360
340 INPUT "UBAR="; UBAR: BIGH = -HE*UBAR
350 IF U=0 THEN H=HE: GOTO 380
360 H = -U*L
370 IF UBAR = 0 THEN INPUT "IMAGE HEIGHT MUST BE GIVEN. ETA="; ETA: HBAR = ETA:
BIGH = U*ETA: GOTO 390
380 HBAR = -UBAR*LBAR
390 PRINT "MINIMUM LENS APERTURE = "; 2*(ABS(H)+ABS(HBAR))
400 UP = U-H*K
410 E = HBAR/H
420 Y = (U+UP)/(U-UP)
430 PRINT "BIGH="; BIGH: PRINT "E="; E
440 INPUT "WAVELENGTH"; WAVL: PRINT
450 S3C = BIGH*BIGH*K: PRINT "S3C="; S3C, "("; S3C/WAVL; " WAVES)"
460 S4 = S3C/N: PRINT "S4="; S4, "("; S4/WAVL; " WAVES)"
470 N1 = (N+2)/(N*(N-1)^2)
480 N2 = 2*(N*N-1)*Y/(N+2)
490 N3 = (N/(N-1))^2
500 N4 = N3-N*Y*Y/(N+2)
```

```
510 N5 = (N+1)/(N*(N-1))
520 N6 = (2*N+1)*Y/N
530 A1 = (H^4)*(K^3)/4
540 A2 = -.5*BIGH*(H*K)^2
550 X(1) = -8
560 FOR I = 1 TO 321
570 S1(I) = A1*(N1*(X(I)+N2)^2+N4)
580 S2C = A2*(N5*X(I)+N6)
590 S2(I) = S2C+E*S1(I)
600 S3(I) = S3C+2*E*S2C+E*E*S1(I)
610 S5(I) = E*(3*S3C+S4)+3*E*E*S2C+S1(I)*E^3
620 IF I = 321 THEN 640
630 X(I+1) = X(I)+.05
640 NEXT
650 PRINT
660 IF Q = 1 THEN 1110
670 INPUT "DO YOU WANT ABERRATION PRINTOUT"; A$
680 IF (A$="Y") OR (A$="y") THEN 710
690 IF (A$="N") OR (A$="n") THEN 840
700 PRINT "TYPE Y OR N": GOTO 670
710 PRINT "SPECIFY STEP IN X. INPUT A VALUE (G) BETWEEN 0.05 AND 1 IN STEPS OF
0.05."
720 PRINT "NUMBER OF PRINTED VALUES WILL BE 320/(20*G)"
730 INPUT G
740 G = INT(20*G)
750 PRINT : PRINT TAB(2); "I"; TAB(8); "X"; TAB(18); "S1"; TAB(30); "S2";
TAB(42); "S3"; TAB(54); "S5"
760 FOR I = 1 TO 321 STEP G
770 PRINT I; TAB(7); INT(1E4*X(I))/1E4; TAB(17); INT(1E6*S1(I))/1E6; TAB(29);
INT(1E6*S2(I))/1E6; TAB(41); INT(1E6*S3(I))/1E6; TAB(53); INT(1E6*S5(I))/1E6
780 NEXT
790 INPUT "IS PRINTOUT ADEQUATE"; A$
```

```
800 IF (A$="Y") OR (A$="y") THEN 830
810 IF (A$="N") OR (A$="n") THEN 710
820 PRINT "TYPE Y OR N": GOTO 790
830 PRINT
840 INPUT "DO YOU WANT TO SPECIFY TARGETS"; A$
850 IF (A$="Y") OR (A$="y") THEN 880
860 IF (A$="N") OR (A$="n") THEN 50
870 PRINT "TYPE Y OR N": GOTO 840
880 INPUT "TARGET FOR S1 (WAVES)"; S1T
890 INPUT "TARGET FOR S2 (WAVES)"; S2T
900 IF E = 0 THEN 930
910 INPUT "TARGET FOR S3 (WAVES)"; S3T
920 INPUT "TARGET FOR S5 (PERCENT)"; S5T
930 S1T = S1T*WAVL: S2T = S2T*WAVL: S3T = S3T*WAVL: S5T = S5T*(2*BIGH)
940 IF FLAG = 1 THEN 1300
950 INPUT "DO YOU WANT TO SPECIFY WEIGHTS"; A$
960 IF (A$="Y") OR (A$="y") THEN 990
970 IF (A$="N") OR (A$="n") THEN 1100
980 PRINT "INPUT Y OR N": GOTO 950
990 PRINT "WEIGHTS MUST BE POSITIVE"
1000 INPUT "WEIGHT FOR S1"; W1
1010 IF W1 < 0 THEN 990
1020 INPUT "WEIGHT FOR S2"; W2
1030 IF W2 < 0 THEN 990
1040 IF E = 0 THEN W3 = 0: GOTO 1120
1050 INPUT "WEIGHT FOR S3"; W3
1060 IF W3 < 0 THEN 990
1070 INPUT "WEIGHT FOR S5"; W5
1080 IF W5 < 0 THEN 990
1090 GOTO 1110
1100 W1 = 1: W2 = 1: W3 = 1: W5 = 1
1110 IF E = 0 THEN W3 = 0
```

```
1120 MER = 1E+20: J = 0

1130 FOR I = 1 TO 321

1140 MERIT(I) = W1*ABS(S1(I)-S1T)+W2*ABS(S2(I)-S2T)+W3*ABS(S3(I)-S3T)+
W5*(S5(I)-S5T)

1150 IF MERIT(I) > MER THEN 1180

1160 J = I

1170 MER = MERIT(I)

1180 NEXT

1190 IF J = O THEN PRINT "INCORRECT RESULT. CHECK INPUT.": GOTO 840

1200 PRINT "OPTIMUM SHAPE IS X="; INT(1E3*X(J))/1E3

1210 PRINT "S1="; INT(1E6*S1(J))/1E6, "S2="; INT(1E6*S2(J))/1E6, "S3=";
INT(1E6*S3(J))/1E6, ="; "S5="; INT(1E6*S5(J))/1E6

1220 PRINT

1230 PRINT "C1="; INT(1E5*K*(X(J)+1)/(2*(N-1)))/1E5

1240 PRINT "C2="; INT(1E5*K*(X(J)-1)/(2*(N-1)))/1E5

1250 INPUT "DO YOU WANT TO RE-DEFINE TARGETS"; A$

1260 FLAG = 0

1270 IF (A$="Y") OR (A$="y") THEN FLAG = 1: GOTO 880

1280 IF (A$="N") OR (A$="n") THEN 1300

1290 PRINT "TYPE Y OR N": GOTO 1250

1300 INPUT "DO YOU WANT TO RE-DEFINE WEIGHTS"; A$

1310 IF (A$="Y") OR (A$="y") THEN 990

1320 IF ((A$="N") OR (A$="n")) AND (FLAG=1) THEN 1110

1330 IF ((A$="N") OR (A$="n")) AND (FLAG=0) THEN 1350

1340 PRINT "TYPE Y OR N": GOTO 1300

1350 INPUT "DO YOU WANT STOP SHIFT"; A$

1360 IF (A$="N") OR (A$="n") THEN 50

1370 IF (A$="Y") OR (A$="y") THEN 1390

1380 PRINT "TYPE Y OR N": GOTO 1350

1390 PRINT "CURRENT E="; E

1400 INPUT "NEW E="; E: Q = 1

1410 GOTO 560
```

Sample Run

(operator responses in boldface)

Note: Numbers taken from Section 8.1.6. In this example, we avoid printing the aberration values to save space.

```
THIN LENS ABERRATIONS AND BENDING


DO YOU WANT TO SPECIFY THE LENS? Y

POWER? 0.01

REFRACTIVE INDEX N=? 1.5

OBJECT DISTANCE? -1E20

ENTRANCE PUPIL DISTANCE? 22.22

INPUT EITHER THE MARGINAL RAY ANGLE U OR THE ENTRANCE PUPIL RADIUS

TYPE 'U' OR 'HE' TO SELECT ONE

? HE

HE=? 2.94

INPUT EITHER THE IMAGE HEIGHT OR THE PUPIL RAY ANGLE

TYPE 'ETA' OR 'UBAR' TO SELECT ONE

? UBAR

UBAR=? .5

MINIMUM LENS APERTURE = 28.1

BIGH = -1.47

E = -3.778911

WAVELENGTH? .56E-3


S3C = .021609    (38.5875 WAVES)
```

```
S4 = .0144        (25.725 WAVES)


DO YOU WANT ABERRATION PRINTOUT? N

DO YOU WANT TO SPECIFY TARGETS? Y

TARGET FOR S1 (WAVES)? 0

TARGET FOR S2 (WAVES)? 0

TARGET FOR S3 (WAVES)? 0

TARGET FOR S5 (PERCENT)? 0

DO YOU WANT TO SPECIFY WEIGHTS? Y

WEIGHTS MUST BE POSITIVE

WEIGHT FOR S1? 0

WEIGHT FOR S2? 1

WEIGHT FOR S3? 1

WEIGHT FOR S5? 0


OPTIMUM SHAPE IS X = 3.5

S1 = .001512     S2 = 0    S3 = -.000001      S5 = -.136106


C1 = .045

C2 = .025


DO YOU WANT TO RE-DEFINE TARGETS? N

DO YOU WANT TO RE-DEFINE WEIGHTS? N

DO YOU WANT STOP SHIFT? N

DO YOU WANT TO SPECIFY THE LENS? N
```

(Notice that the curvatures are identical with those found in Section 8.1.6.)

BIBLIOGRAPHY

1. REFERENCES CITED

(for which bibliographic information is not given in the text)

B. N. Begunov, N. P. Zakaznov, S. I. Kiryushin, and V. I. Kuzichev (1988), *Optical Instrumentation: Theory and Design*, MIR Publishers (Moscow).

W. Brouwer (1964), *Matrix Methods in Optical Instrument Design*, Benjamin (New York).

M. Born and E. Wolf (1975), *Principles of Optics*, Pergamon Press.

A. Cox (1964), *A System of Optical Design*, Focal Press.

H. A. Haus (1984), *Waves and Fields in Optoelectronics*, Prentice Hall.

M. P. Keating (1988), *Geometric, Physical, and Visual Optics*, Butterworths.

R. Kingslake (1962), "Automatic predesign of the Cooke triplet lens," *Proc. 1961 ICO Conference on Optical Instruments and Techniques*, K. J. Habell, Ed., pp. 107–20, Chapman and Hall Ltd. (London).

R. Kingslake (1978), *Lens Design Fundamentals*, Academic Press.

H. Kogelnik (1979), "Propagation of laser beams," in *Applied Optics and Optical Engineering*, Vol. VII, R. Shannon and J. Wyant, Eds., Academic Press.

R. S. Longhurst (1973), *Geometrical and Physical Optics*, Longman.

L. Levi (1968), *Applied Optics*, John Wiley and Sons, Inc.

Military Handbook 141, *Optical Design*, U.S. Government Printing Office (1963).

M. Minnaert (1954), *The Nature of Light and Colour in the Open Air*, Dover Publications.

K. D. Möller (1988), *Optics*, Oxford University Press.

W. J. Smith (1990), *Modern Optical Engineering*, McGraw-Hill.

W. J. Smith (1992), *Lens Design: A Resource Manual*, McGraw-Hill.

G. H. Spencer and M. V. R. K. Murty (1962), "General ray-tracing procedure," *J. Opt. Soc. Am.* **52**, 672–78.

W. T. Welford (1986), *Aberrations of Optical Systems*, Adam Hilger (London).

2. ADDITIONAL REFERENCES (GENERAL OPTICS)

G. R. Fowles (1975), *Introduction to Modern Optics*, Dover Publications.

R. Guenther (1990), *Modern Optics*, John Wiley and Sons, Inc.

E. Hecht (1987), *Optics*, Addison Wesley.

K. Iizuka (1987), *Engineering Optics*, Springer Verlag.

J. R. Meyer-Arendt (1984), *Introduction to Classical and Modern Optics*, Prentice Hall.
F. L. Pedrotti and L. S. Pedrotti (1993), *Introduction to Optics*, 2nd ed., Prentice Hall.
F. G. Smith and J. H. Thomson (1988), *Optics*, John Wiley and Sons, Inc.
M. Young (1986), *Optics and Lasers*, Springer Verlag.

3. ADDITIONAL REFERENCES (OPTICAL, LENS, AND SYSTEM DESIGN)

(* denotes out of print, but worth hunting for)

H. A. Buchdahl (1968), *Optical Aberration Coefficients*, Dover.
*H. H. Hopkins (1950), *Wave Theory of Aberrations*, Oxford University Press.
*R. E. Hopkins (1963), *Optical Design*, Military Handbook 141.
R. Kingslake (1983), *Optical System Design*, Academic Press.
R. Kingslake (1989), *A History of the Photographic Lens*, Academic Press.
M. Laikin (1991), *Lens Design*, Marcel Dekker.
V. N. Mahajan (1991), *Aberration Theory Made Simple*, SPIE Optical Engineering Press (Bellingham, WA).
D. Malacara (1988), *Geometrical and Instrumental Optics*, Academic Press.
D. Malacara and Z. Malacara (1994), *Handbook of Lens Design*, Marcel Dekker.
E. W. Marchand (1978), *Gradient Index Optics*, Academic Press.
*L. C. Martin and W. T. Welford (1961), *Technical Optics* (2 vols.), Pitman (London).
J. M. Palmer (1971), *Lens Aberration Data*, American Elsevier.
D. C. O'Shea (1985), *Elements of Modern Optical Design*, John Wiley and Sons, Inc.
G. G. Slyusarev (1984), *Aberration and Optical Design Theory*, Adam Hilger (London).
W. J. Smith (Ed.) (1992), *Lens Design*, SPIE Vol. CR41, Optical Engineering Press (Bellingham, WA).
O. N. Stavroudis (1982), *Modular Optical Design*, Springer Verlag.
B. H. Walker (1995), *Optical Engineering Fundamentals*, McGraw-Hill.
W. T. Welford (1962), *Geometrical Optics*, North Holland.
C. L. Wyatt (1992), *Electro-Optical System Design*, McGraw-Hill.
P. R. Yoder, Jr. (1986), *Opto-Mechanical Systems Design*, Marcel Dekker.
H. G. Zimmer (1970), *Geometrical Optics*, Springer Verlag.

Special mention must be made of the series *Applied Optics and Optical Engineering*, published by Academic Press, which contains several volumes with relevant articles; also the *Handbook of Optics*, published by McGraw-Hill.

Index